Sustainable Development Goals Series

World leaders adopted Sustainable Development Goals (SDGs) as part of the 2030 Agenda for Sustainable Development. Providing in-depth knowledge, this series fosters comprehensive research on these global targets to end poverty, fight inequality and injustice, and tackle climate change.

The sustainability of our planet is currently a major concern for the global community and has been a central theme for a number of major global initiatives in recent years. Perceiving a dire need for concrete benchmarks toward sustainable development, the United Nations and world leaders formulated the targets that make up the seventeen goals. The SDGs call for action by all countries to promote prosperity while protecting Earth and its life support systems. This series on the Sustainable Development Goals aims to provide a comprehensive platform for scientific, teaching and research communities working on various global issues in the field of geography, earth sciences, environmental science, social sciences, engineering, policy, planning, and human geosciences in order to contribute knowledge towards achieving the current 17 Sustainable Development Goals.

This Series is organized into eighteen subseries: one based around each of the seventeen Sustainable Development Goals, and an eighteenth subseries, "Connecting the Goals," which serves as a home for volumes addressing multiple goals or studying the SDGs as a whole. Each subseries is guided by an expert Subseries Advisor.

Contributions are welcome from scientists, policy makers and researchers working in fields related to any of the SDGs. If you are interested in contributing to the series, please contact the Publisher: Zachary Romano [Zachary.Romano@springer.com].

More information about this series at http://www.springer.com/series/15486

Godwell Nhamo
Muchaiteyi Togo • Kaitano Dube
Editors

Sustainable Development Goals for Society Vol. 1

Selected topics of global relevance

Springer

Editors
Godwell Nhamo
Exxaro Chair in Business and Climate Change
Institute for Corporate Citizenship
University of South Africa
Pretoria, South Africa

Muchaiteyi Togo
Department of Environmental Sciences
University of South Africa
Florida, South Africa

Kaitano Dube
Department of Hospitality
Tourism and PR
Vaal University of Technology
Vanderbijlpark, South Africa

ISSN 2523-3084 ISSN 2523-3092 (electronic)
Sustainable Development Goals Series
ISBN 978-3-030-70947-1 ISBN 978-3-030-70948-8 (eBook)
https://doi.org/10.1007/978-3-030-70948-8

This Springer imprint is published by the registered company Springer Nature Switzerland AG
The registered company address is: Gewerbestrasse 11, 6330 Cham, Switzerland

Preface

This book is part of the Sustainable Development Goals for Society book series, which comes in two volumes. This particular volume (Volume I) presents selected topics of global relevance. The volume comes in 5 major parts and 19 chapters. The five parts are, namely, (1) Introduction and Background, (2) Addressing Poverty in the Context of SDGs, (3) Leadership in Implementing SDGs, (4) SDGs in Service Delivery and Local Government and (5) Conclusion and Recommendations. The history of the Sustainable Development Goals (SDGs) dates back to the United Nations Conference on Sustainable Development (Rio+20) held in 2012 in Rio de Janeiro. The need for a post-2015 Agenda was highlighted in the Rio+20 outcome document 'The Future We Want'. From the Rio+20 Conference, the United Nations General Assembly reaffirmed the desire to continue mainstreaming sustainable development at all levels. Such an approach had to consider the integration of the three main pillars of sustainable development including economic, social and environmental aspects. However, people were to be at the centre of such development. Therefore, the commitment to accelerate the achievement of the Millennium Development Goals (MDGs) was made. The MDGs implementation process, therefore, remains the foundation for implementing the SDGs. Designed along with a similar architecture to the MDGs, the SDGs seek to complete what the MDGs could not attain based on a balance of the three dimensions of sustainable development at a much more enhanced and ambitious level. The 17 SDGs as embedded into the 2030 Agenda for Sustainable Development are universal, more inclusive and they include a well-defined means of implementation. The SDGs are interlinked and supported by 169 targets. Through the SDGs, the world should involve every citizen and organisation, hence the motto 'Let no one be left behind'. This volume presents selected case studies that are of global importance, with certain implementation lessons that can be drawn from Africa, Europe and Latin America.

Pretoria, South Africa — Godwell Nhamo
Florida, South Africa — Muchaiteyi Togo
Vanderbijlpark, South Africa — Kaitano Dube

Acknowledgements

The editors, Prof. Godwell Nhamo, Dr. Muchaiteyi Togo and Dr. Kaitano Dube, wish to thank all the blind peer reviewers for their invaluable inputs during the writing and publishing process of the book. We thank Springer for taking this book project on board and for a product of high standard and quality. We also wish to extend gratitude to our families for their ongoing support of our work. The book project was coordinated through the Exxaro Chair in Business and Climate Change, a research chair that is funded by the Exxaro Resources (Pty) Ltd Chairman's Fund and hosted by the Institute for Corporate Citizenship at the University of South Africa. The Exxaro Chair was established in 2008 and is currently in its fourth term ending in 2022.

This book is double-blind peer reviewed. Apart from this being the international best practice norm, this double-blind peer review process is mandatory for South Africa-based authors to fulfil the requirements of the Department of Higher Education and Training's (DHET) policy for recognised research outputs for subsidy purposes. The authors invested their time to incorporate observations from the blind peer review process, an aspect that enhanced the quality of the product.

Contents

About the Editors

Godwell Nhamo, PhD (Editor-in-Chief and Accounting Officer), is a Full Professor and Exxaro Chair in Business and Climate Change at the University of South Africa (UNISA). He is a National Research Foundation (NRF) C-rated researcher undertaking research in the fields of climate change and governance, sustainable tourism, green economy and sustainable development. Prof. Nhamo has conceptualised and completed 11 book projects, the most recent being Counting the Cost of COVID-19 on the Global Tourism Industry by Springer (2020); Scaling up SDGs Implementation: Emerging Cases from the State, Development and Private Sectors published by Springer (2020); SDGs and Institutions of Higher Education published by Springer (2020); and SDG 7—Ensure Access to Affordable, Reliable, Sustainable and Modern Energy by Emerald (2020). Prof. Nhamo has also published over 90 journal articles. Since 2013, Prof. Nhamo has graduated 11 PhDs and hosted 10 postdoctoral fellows. Currently, Prof. Nhamo is leading a mega research project on Cyclones, Tornados and Floods in the era of SDGs in Southern Africa. Professor Nhamo also sits in a number of both international and national boards and has received several awards and recognitions for his outstanding work both locally and internationally. Finally, Prof. Nhamo was one of the four-member African Union High-Level Panel drafting the Green Innovation Framework for the continent. Email: nhamog@unisa.ac.za

Muchaiteyi Togo, PhD (Co-editor), holds a PhD from Rhodes University (South Africa). She works for UNISA in the Environmental Science Department. Dr. Togo's current research focuses on SDGs implementation in universities and low-carbon/greening strategies at household level (with a special focus on water-energy-food nexus) and in other economic sectors. Dr. Togo's publications include a resource book, journal articles, peer-reviewed book chapters and conference proceedings on the same topics. Her work has influenced practices in universities in Africa and beyond when a resource book she published as part of her PhD became part of resource material for the UNEP Mainstreaming Environment and Sustainability Universities (MESA) Partnership. She currently lectures a green economy module at honours level and supervises Honours, Masters and PhD students in Environmental Sciences, with a few Masters and PhD students having been supervised to completion. Dr. Togo has done research for international organisations like the UNESCO, the UNEP, the AAU, SARUA and the SADC-REEP and several South Africa-based organisations as a consultant. She has also received a few research awards at local and international levels. Since 2010 to early 2020, Dr. Togo has been part of the Southern African Journal for Environmental Education editorial collective. She was a council member of the Environmental Education Association of Southern Africa holding the publications portfolio from 2016 to 2019. Email: togom@unisa.ac.za

Kaitano Dube, PhD (Co-editor), is an Ecotourism Management Lecturer at the Vaal University of Technology, South Africa. He is one of the African leading tourism geographers researching in the area of tourism, climate change, sustainability and green aviation. He has published in high-impact international peer-reviewed journals with his work receiving global attention. His work has received extensive media coverage including in National Geographic, Wunderground, Atlas Obscura and Agence France-Presse (AFP) to mention but a few. He has granted several international television and radio interviews. Dr. Dube holds a PhD and MSc from the University of South Africa. He graduated with a BSc Hons from Midlands State University in Gweru, Zimbabwe. He holds several other qualifications from the UNISA, the Vaal University of Technology and the University of the Witwatersrand Business School in Johannesburg, South Africa. Dr. Dube is an executive member of the Tourism Educators Association of South Africa and Tourism Sector Human Resource Development Governance and Institutional Coordination Forum hosted by South Africa's National Department of Tourism. Email: kaitanod@vut.ac.za

Aldo Palacios Courret is currently the Executive Chair of the NGO SINERGIAS and former Chair of Global Water Partnership-Chile and the Global Water Partnership-South America (2017–2018). Email: ahpc@sinergias.cl

Alejandro Valencia-Arias, PhD is a professor who holds a doctorate in engineering (Industry and Organizations) and a master's degree in systems engineering. Prof. Valencia-Arias is editor-in-chief for Revista CEA. He is a published author in the fields of entrepreneurship, technology adoption models and mobile learning. Prof. Valencia-Arias has over 8 years of work experience in the academic field. Among his offerings are 3 books and over 40 journal articles. Email: javalenciar@gmail.com

Alejandra Mujica is the Global Water Partnership (GWP) Regional Coordinator for South America. She is a biologist (University of the Republic, Uruguay) with postgraduate studies in science and technology management. She has more than 20 years of experience in science and innovation field developed in academic and governmental areas. She was a programme manager and contributed to the creation of national programmes aimed at fostering innovation in companies and public-private alliances. Her experience includes being an advisor to the Direction of Science and Technology of the Ministry of Education and Culture as well as a researcher of the Commission of Scientific Investigation of the University of the Republic. She has teaching experience and was a professor in several courses related to science, society and development. She authored several research and review articles, book chapters and essays on the subject. Email: alejandra.mujica@gwpsudamerica.org

Ana Claudia de Almeida (master's degree) has a bachelor's degree in international relations and a master's degree in sustainable development from the University of Brasília (UnB). Her research interests are around vulnerability and adaptation to climate change, environmental migration and the link between social protection policies and climate change. Email: anaclaud.al@gmail.com

Bjanka Korb, Pr. Eng is a senior environmental engineer at SRK Consulting, who works primarily in mine water management and water use optimisation. Bjanka's experience also includes manufacturing process management and optimisation, due diligence reporting and feasibility studies. A chemical engineer by training, Bjanka is passionate about sustainable development and completed her Master's in Development Practice in 2019. Bjanka's fields of interest include sustainable resource use, circular economy and gender equality. Bjanka strives to make a positive impact in the fields of water and sanitation and to demonstrate that women can thrive in STEM disciplines. She is registered as a professional engineer with the Engineering Council of South Africa.

Carlos Hiroo Saito, PhD is a biologist working in interdisciplinary research involving environmental education, spatial analysis and water security. He works with conceptual modelling for science literacy in a systemic approach and studies how the understanding of social and environmental processes can strengthen social participation. He is a research fellow of CNPq and executive coordinator of the National Institute of Science and Technology (INCT-Observatory of Socio-Environmental Dynamics). He is a full professor at the University of Brasília, Brazil, linked to the Department of Ecology/Institute of Biological Sciences and the Centre for Sustainable Development. He is a researcher associated with UMR Espace-DEV, Montpellier, France. Email: carlos.h.saito@hotmail.com

Corina Piaggio is the communication officer of Global Water Partnership (GWP), South America. She is a field communication professional with almost 30 years of experience, implementing strategic planning to meet the objectives of various local and international companies and organisations. She is a Communication Technician (Catholic University of Uruguay) and Marketing Analyst (ORT Uruguay University), with Diploma in Crisis Management (Pontifical Catholic University of Lima), MBA (University of Cádiz) and Master's in Management Systems (University of Cádiz). Email: corina.piaggio@gmail.com

Carola Calabuig, PhD has a doctorate in industrial engineering from the Polytechnic University of Valencia (UPV). She is a lecturer in the Department of Engineering Projects at the UPV and currently the Academic Director of the Master's Degree in Development Cooperation at the UPV, specialising in project management and development processes. Her main areas of work and interests are in the field of development studies that include development planning, international cooperation, sustainable development and critical approaches, democratic governance, education for development and global citizenship. Email: cacator@dpi.pv.es

Cowen Dziva, PhD has a doctorate in development studies from the UNISA. He teaches in the Julius Nyerere School of Social Sciences at the Great Zimbabwe University. He considers himself a researcher and development practitioner with an in-depth background in the application of human rights-based approach in development work. His human rights experience spans from working with the Zimbabwe Human Rights Commission and NGOs to researching, promoting and protecting human rights of disadvantaged groups in Zimbabwe. He has co-authored and published several book chapters and articles on democratisation and development and human rights in Zimbabwe. Email: cowendziva@gmail.com

Dawieş (D.A.J.) Bornman, PhD is a senior lecturer, researcher and postgraduate supervisor in the Department of Business Management at the University of Pretoria, South Africa. Dr. Bornman acts as a guest lecturer at various international universities, including Vorarlberg University of Applied Sciences in Austria and Karel de Grote University College in Belgium, and he has presented his research at numerous national and international conferences, doctoral seminars and personal development workshops focusing on leadership communication, business management and entrepreneurial innovation. Dr. Bornman is passionate about small, medium and micro-sized enterprises (SMME) development projects and has provided consulting at the Mamelodi Business Clinic and the South African Creative Industries Council (SACII), which is structured to empower, develop and build a framework that covers all topics critical to establish, manage, grow and sustain a small business. Email: dawie.bornman@up.ac.za

Daniel Cardona Valencia has a master's degree in finance and is currently studying for a PhD in Local Development and International Cooperation at the Polytechnic University of Valencia. He is a professor at the Instituto Tecnológico Metropolitano de Medellín, Colombia. Prof. Cardona has a special interest in issues related to financial inclusion, financial education and its relationship with ODS and has more than 6 years of work experience in the academic field. Email: danielcardona@itm.edu.co

Daniela Morales-Zapata has a double major in industrial and management engineering from the Colombia National University and an MBA from the University of Essex in the UK. She has worked in the financial industry for over 6 years across different markets in Latin America. Daniela is a volunteer researcher that has published around higher education topics and is keen on connecting the academic and the industry world in the light of the 2030 agenda. Email: dm19409@essex.ac.uk

Dominique E. Uwizeyimana, PhD is an Associate Professor of Public Policy Management and Monitoring and Evaluation (M&E) in the School of Public Management, Governance and Public Policy, in the College of

Business and Economics (CBE) at the University of Johannesburg. He holds a Bachelor's (cum laude) in Public Administration from the University of the Western Cape (UWC), a Honours (cum laude) in Public Administration from the University of Stellenbosch, a Master's of Social Sciences (MSc) in Public Affairs and Administration from the University of Cape Town (UCT) and a PhD in Public Management and Governance from the University of Johannesburg (UJ). His lecturing and research interests include public administration theories, public policy monitoring and evaluation (M&E), sustainable development and food security.

Donald Chikoto is a Graduate Teaching Assistant with the School of Social Sciences at Great Zimbabwe University. Currently, he is pursuing a master's degree in gender and policy studies. His areas of interest are research on displaced persons and sustainable livelihoods. He has vast experience in conducting training workshops, gained through working with organisations such as Population Services International in Zimbabwe and Real Agenda for Youth Transformation Trust. Email: chikotodonald@gmail.com

Emma Terämä, DSc, Tech is a civil servant at the Ministry of Finance, Finland, and a director and researcher on sabbatical from the Finnish Environment Institute (SYKE). In her current role, she is senior advisor working as a chief specialist in the fields of urban strategy and urban policy. She has over 10 years of international research experience on population, climate, resources and urbanisation as well as policy from institutions such as the International Institute for Applied Systems Analysis (IIASA), University of Amsterdam and University College London, in addition to SYKE. She is keen to advance science-based decision-making, open policymaking and the impact of science for the public good. E-mail: emma.terama@vm.fi

Eliana Villa is a Professor at Instituto Tecnológico Metropolitano, in Medellín, Colombia. She is currently researching in the field of transformative innovation policies, a framework for the construction of public policies in science, technology and innovation to support the changes needed to achieve the SDGs. Specifically, the field in which she researches is in the role of universities in these processes: how can they generate dynamics that allow, through the CTI, to solve problems that have a specific direction and intention, based not only on economic growth but also on the reduction of existing gaps, and accentuated more and more in the world. Email: elianavilla@itm.edu.co

Ernest Dube, PhD is a senior lecturer and researcher for Development Sciences at Marondera University of Agricultural Sciences and Technology (MUAST), Zimbabwe. Dr. Dube has an interest in the fields of disaster risk reduction and climate change, poverty and livelihoods, as well as food security and sustainable development. Dr. Dube published extensively in his areas of interest and has done review work for more than 15 journal publishers. He has been in the academic field for more than 8 years. Among Dr. Dube's research output are over 15 publications in internationally accredited journals. Dr. Dube is also a consultant in disaster risk reduction and development-related issues. He has also carried out funded research projects to completion. Email: edube@muast.ac.zw

Esaú Toro-Vanegas is a forestry engineer and has an MSc in forests and environmental conservation. He has also been a teacher and leading researcher for more than 10 years in the field of protected areas, agroforestry systems, ethnobiology, conservation of species of fauna and flora (*Juglans neotropica* and *Ceroxylon quindiuense*) and in alternatives for the sustainable use of forested, rural and urban areas. He has published about 16 articles and has advised more than 50 specialised jobs in environmental management with an emphasis on biodiversity. He has obtained two honourable mentions as a best high school student and teacher of excellence at the Fundación Universitaria Católica del Norte. Email: etorov@ucn.edu.co

Fray Betancur is a graduate in financial and business engineering from the Metropolitan Technological Institute of Medellín, Colombia. He has more than 10 years of experience in the academic field as a teacher, five of them with vulnerable populations. He has written a scientific article and participated in different presentations. Fray Betancur is interested that his work has the purpose of positively impacting the communities; he currently works in a non-governmental organisation. Email: fraybetancur@gmail.com

Gabriela Zamignan has a PhD in sustainable development and is a substitute professor of the Centre of Excellence in Tourism at the University of Brasília (UnB). In research, she has an interest in the fields of sustainable development and its interactions with tourism, environmental education, IWRM from a systemic approach by conceptual maps and network analysis. Email: gzamignan@gmail.com

Itai Kabonga holds academic qualifications in development studies, sociology, project management and monitoring and evaluation (M and E). He has also taught several courses in development studies that include non-governmental organisations (NGOs) and civil society in Africa, gender studies and International Development Corporations at Zimbabwe Ezekiel Guti University (ZEGU) and ARRUPE Jesuit University in Zimbabwe. Itai's research interests include NGOs and development, civil society-state relations, volunteerism in Africa and Zimbabwe's development crisis since 1980. Focusing on these areas, Itai has published ten journal articles in reputable journals. Email: vakabonga@gmail.com

Jani Päivänen is a Team Leader in the Human Environment at Finnish Consulting Group. He is a social impact and sustainability specialist with over 20 years' experience in planning and research. Päivänen has worked as a consultant and project manager/director since 1999 in large consulting companies, in charge of SIA, citizen involvement practices, social research and co-design practices. Jani has led several projects dealing with public space, city walkability and liveability, social inclusion, civic participation face-to-face and online, surveys, sterification and gamification. Email: jani.paivanen@fcg.fi

Jefferson Quiroz-Fabra is a chemical engineer with knowledge on industrial processes, interested in renewable energies and process safety management, a field where he won an international essay contest addressing the importance of process safety. His experience and research interests are around management technology, education and innovation with some articles and book chapters published. Email: jsquirozf@unal.edu.co

Joe de Beer is the Deputy-Director General for Economic Statistics at Statistics South Africa. He has more than 25 years of experience in the field of official statistics, specialising in the measurement of the dynamics of the economy and prices. His interest in sustainable development has seen him serve as a past member of the Bureau of the United Nations Committee of Experts of Environmental Economic Accounting as well as the Chair of the African Group of National Accountants. He has qualifications in economics as well as development practice from the University of Pretoria and completed various leadership and management courses at the Wits Business School and Harvard Business School. Email: joedb@statssa.gov.za

Joseph Tinarwo is an experienced social scientist and researcher with more than 10 years of demonstrated experience in public policy analysis, food and nutrition security, and governance. He is currently the coordinator and lecturer of politics and public management programme at Great Zimbabwe University and is about to complete his PhD in public management and governance with the University of Johannesburg in South Africa researching on food policy systems in Zimbabwe. Joseph has presented research and policy-relevant papers in both national and international conferences and has published in peer-reviewed journals. Over the years, Joseph has been actively involved in various assignments as a consultant to the Government of Zimbabwe and international development partners in the field agriculture and food security, health and development sector with key result areas including research, knowledge management, institutional capacity assessments, training, designing and managing special studies as well as data management. Email: josetinarwo@gmail.com

Juan Bussi, PhD is a chemical engineer, graduated in 1980 from the University of the Republic of Uruguay, Public University (Udelar). He later earned his PhD degree in chemical engineering in the Louis Pasteur University, Strasbourg, France (1984). Since 1977, he works in the area of physical chemistry in the Faculty of Chemistry (Udelar), doing teaching and research in several topics including those related to remediation of air and water using advanced oxidation techniques (AOTs). He has held the presidency of Global Water Partnership (GWP), Uruguay, from 1917 to date. Email: bussijuan85@gmail.com

Kaisa Schmidt-Thomé, PhD is a senior expert at Demos Helsinki, working with the questions of fair and sustainable urban futures. She has extensive experience in interdisciplinary urban studies (with LicSc in planning geography from the University of Helsinki, and PhD on urban regeneration from

Aalto University) and strong expertise in societal interaction of research. Her current research interests include urban sustainability, collaborative problem-solving and scenario methods. Email: kaisa.schmidt-thome@demosheisinki.fi

Karembe F. Ahimbisibwe is currently a doctoral student in the Department of Social Sciences and Philosophy at the University of Jyväskylä, Finland. He is an Assistant Lecturer and former Head, Department of Adult and Community Education, Makerere University, Uganda. He is also a member of the Sustainable Futures in Africa, a multidisciplinary research network coordinated by the University of Glasgow, Scotland. His main research interests are in areas of citizenship, participatory community learning, sustainable agrarian livelihoods and development. He has vast experience in conducting research using community empowering participatory methodologies. Email: hkarembe@gmail.com

Larisa Ho Bech Gaivizzo, PhD has a degree in agronomic engineering, a master's and a doctorate in soil science in the area of soil bioremediation and recovery of degraded areas. She has worked as coordinator and researcher of the Watershed Project with the State Foundation for Agricultural Research (FEPAGRO); as coordinator and professor in the area of environmental management at the State University of Rio Grande do Sul (UERGS); and as officer of the environment unit of the United Nations Development Programme (UNDP). She is currently a postdoctoral student in environmental sciences at the Observatory of Social and Environmental Dynamics (INCT-Odisseia) at the Center for Sustainable Development (CDS) of the University of Brasília (UnB) in the area of adaptation to climate change. Email: larisabech@gmail.com

Leandro Raúl Díaz is a civil engineer (awarded by the UNT of Argentina) and an international specialist in the territory and environmental planning (granted by the UNT of Argentina and the Polytechnic University of Valencia). He is full professor of hydrology at the National University of Tucumán (UNT) and the Director of the Laboratory of Hydraulic Constructions of the UNT, with experience in the Regulatory Body of Public Water and Sewer Services and the Tucumán Television. His main research interest is Water Resource Management with special emphasis on sanitation, regulation of public water and sewer services, watershed and demand management, and flood and drought management. He is presently the Chair of the Global Water Partnership-South America (2019–2020). Email: ldiaz@herrera.unt.edu.ar

Luis Fernando Garcés-Giraldo, PhD has a doctorate from the Pontifical Bolivarian University in Medellín and a master's degree in environmental engineering from the University of Antioquia; he graduated as Sanitary Engineer from the University of Antioquia. He is a senior researcher at the Ministry of Science, Technology and Innovation (MinCiencias) and a professor in the Directorate of Education and Humanities of the Remington University Corporation. He has published 7 books, 40 book chapters resulting from research and about 120 national and international articles in indexed journals in environmental sciences, philosophy and bioethics. Email: luis.garces@uniremington.edu.co

Mavis Thokozile Macheka, PhD is a lecturer for development studies in the Department of Archaeology and history and development studies at Great Zimbabwe University. She is also a part-time lecturer at Zimbabwe Open University. Dr. Macheka's teaching and research straddle the fields of rights of vulnerable groups of the society, political ecology, community development, livelihoods and sustainability, agriculture and development, and human rights and development. To date, she has published with the Journal of Cultural Heritage Management and Sustainable Development and South African Geographical Journal, among others. Email: mavythoko@gmail.com

Melanie Malan, B.Bibl. Hons. is the personal librarian for the School of Management Sciences, the Department of Public Administration and the Institute for Corporate Citizenship. Mrs. Malan holds an honours degree from the University of Pretoria and has worked at Unisa for the College of Human Sciences and the College of Science, Engineering and Technology before joining the College of Economic and Management Sciences. Mrs. Malan supports researchers in the College of Economic and Management Sciences doing research on sustainable development. Email: malanmm@unisa.ac.za

Nyasha Chaminuka is a lecturer and researcher at Zimbabwe Open University (ZOU). His research interests are in urban agriculture, food security, disaster risk reduction and gender-based violence. Mr. Chaminuka has published in the field of urban agriculture and has more than three publications in internationally accredited journals. He has been in the academic field for over 7 years. Email: chaminukan2015@gmail.com

Romero Gomes Pereira da Silva, PhD is a forest engineer and researcher at the Centre for Sustainable Development at the University of Brasília, Brazil. He is interested in geographic information systems and remote sensing applied to the planning and management of protected areas (conservation units and indigenous lands) and territorial planning and ordering with a focus on urban green spaces. Email: romerogomes1@hotmail.com

Samantha Castle is the senior manager for Alumni Relations at the University of Pretoria and is a postgraduate supervisor. She is also currently completing her PhD in business management. She holds a BA Hons from the University of the Western Cape and a master's in development practice from the University of Pretoria. She attained her master's degree in development practice cum laude and has an in interest in the field of leadership, partnerships, adaptability and the implementation of the Sustainable Development Goals by corporations. She has over 10 years' working experience in corporate tertiary environments, focused on university development, serving in management portfolios related to institutional relations and advancement. She prides herself in managing complex projects and teams, conceptualising new ideas, developing business processes, analysing data, and collaborating and connecting with people. She has served on non-profit boards and has received awards for both academic and outstanding work in the institutional relations and advancement field. Email: samantha.castle@up.ac.za

Satu Lähteenoja is a senior expert at the Nordic think tank Demos Helsinki. She has 15 years of experience in research, consultancy and facilitation on topics such as climate change, sustainable lifestyles, circular economy and the SDGs. Satu is also a doctoral researcher at Aalto University School of Arts, Design and Architecture. Her research interests include sustainable development at different policy levels and the role of design in accelerating sustainability transformations. E-mail: satu.lahteenoja@demoshelsinki.fi

Tiina Kontinen, PhD is an Associate Professor in International Development Studies and a Coordinator of the International Master's Degree Programme in Development, Education and International Cooperation at the University of Jyväskylä, Finland. She has published on NGOs and development, civil society, impact assessment and North-South partnership and currently leads a Research Group on Civil Society and Citizenship in Development. She has been a principal investigator of several research projects conducted in collaboration with universities in Tanzania, Uganda and Zambia. Currently, she leads a 4-year consortium on Theory and Practice of Learning to Be a Citizen. Among her latest publications is an edited volume Practices of Citizenship in East Africa Perspectives from Philosophical Pragmatism (2020) published by Routledge.

Wendy van Eyck holds a master's in development practice from the University of Pretoria, South Africa. She was a recipient of the Baden-Württemberg STIPENDIUM and spent 3 months at Zeppelin University in Germany. She is interested in understanding why some development projects succeed when others fail and how leadership plays a role in this. Her background as a communications professional for several non-governmental organisations across Africa has given her unique access to study leadership in the context of a pioneering multisectoral partnership. Email: wendy@believagency.com

Willem Fourie, PhD established the South African SDG Hub at the University of Pretoria as well as the SDG-focused master's in development practice. He is an Alexander von Humboldt fellow and has published numerous peer-reviewed articles and chapters. Email: willemfourie@gmail.com

Yolande Steenkamp, PhD is a senior researcher at the Albert Luthuli Leadership Institute at the University of Pretoria, South Africa. Her research approach is largely interdisciplinary and addresses contemporary challenges. She contributes from a theological/philosophical perspective to questions of ethics, meaning and value in leadership and explores the ethical dimensions of the imagination, narrative and hospitality as response to diversity in transcultural engagement. She is interested in how narrative and storytelling impact on encounters with racial, economic, sexual, cultural and religious otherness and investigates how this may address inequality, justice and development in postcolonial countries. She collaborates with Zeppelin University in Germany on two projects. The first studies leadership in multi-sector partnerships and the SDGs through the project Relational Leadership for the SDGs: Implications for a developed and a developing economy, while the second studies transcultural leadership and transcultural competence. Email: yolande.steenkamp@up.ac.za

Part I

Introduction and Background

Making Sustainable Development Goals Relevant for, in and with Societies

1

Godwell Nhamo, Muchaiteyi Togo, and Kaitano Dube

Abstract

This introductory chapter sets the platform for how the global leadership that includes governments, territories, business, civil society, development partners and individuals can make the Sustainable Development Goals (SDGs) work and be more relevant for, in and with societies. This way, the world can move quickly towards the attainment of the 2030 Agenda for Sustainable Development, for which by the time of publishing this book, there will be 10 years left. After presenting the history leading to the ratification of the SDGs in New York on 25 September 2015, the chapter profiles the need to address poverty in the context of SDGs, leadership in implementing the SDGs and a focus on service delivery and the attainment of SDGs at the local government level. The chapter concludes with the book and chapter outline.

Keywords

SDGs · Sustainable development · Poverty · Leadership · Partnerships · Stakeholders

1 Introduction

Societies do not stop developing and growing. Hence, with the current advancement in technology and the phenomenon of Industry 4.0 (Fourth Industrial Revolution or 4IR), this is happening at an increasingly fast rate (Hofmann and Rüsch 2017; Kamble et al. 2018). Growth and development do not happen in a vacuum; advancements in society are associated with negative externalities. The society and the economy are at stake from impacts of development, while the environment bears the brunt of such externalities. In 2012, the global community, therefore, gathered for the United Nations Conference on Sustainable Development (Rio+20) held in Rio de Janeiro. From the conference, global leaders made a decisive attempt to reconcile socio-economic development and environmental goals. Rio+20 is the third of the three major world conferences on sustainable development that shaped the SDGs. The first one was the 1992 United Nations

G. Nhamo (✉)
Exxaro Chair in Business and Climate Change, Institute for Corporate Citizenship, University of South Africa, Pretoria, South Africa
e-mail: nhamog@unisa.ac.za

M. Togo
Department of Environmental Sciences, University of South Africa, Florida, South Africa
e-mail: togom@unisa.ac.za

K. Dube
Department of Hospitality, Tourism and PR, Vaal University of Technology, Vanderbijlpark, South Africa
e-mail: Kaitanod@vut.ac.za

G. Nhamo et al. (eds.), *Sustainable Development Goals for Society Vol. 1*, Sustainable Development Goals Series, https://doi.org/10.1007/978-3-030-70948-8_1

Conference on Environment and Development, also held in Rio de Janeiro, followed by the 2002 World Summit on Sustainable Development in Johannesburg, South Africa. The foundation for both the 2030 Agenda for Sustainable Development (AfSD) and the 17 SDGs was developed as part of the outcomes of the Rio+20 conference.

The SDGs intend to guide the global community in addressing sustainable development challenges. These cover current global challenges that society faces, including poverty, inequality, climate change, environmental degradation, peace and justice (United Nations 2015). Society has a pivotal role in changing its circumstances through the implementation of SDGs as these goals are the ultimate guide for achieving a better and more sustainable future for all. This book documents some of the key efforts to implement the 2030 AfSD by, in, with and for society (SDGs for society) in pushing towards achieving sustainable development.

2 Historical Development of the SDGs: From Rio to Rio+20

This section tracks the historical background of the development of SDGs. It focuses on three major United Nations conferences—the Rio Earth Summit, the World Conference on Sustainable Development and Rio+20. The section particularly focuses on how these major gatherings contributed to the SDGs but more specifically how society is included or reflected in the outcomes, agendas, declarations and so forth.

The Rio Earth Summit, also known as the United Nations Conference on Environment and Development (UNCED), was held in Rio de Janeiro, Brazil, from 3 to 14 June 1992. At that time, it was historically the largest gathering of national leaders to put their heads together in resolving the disparity between economic development and environmental well-being. As many as 700 voluntary commitments were made at the Rio Earth Summit and many partnerships were formed to drive the goal of sustainable development (Parson et al. 1992). Some of the outcome documents of the conference are the Rio Declaration on Environment and Development and Agenda 21. The Rio Declaration on Environment and Development is an embodiment of 27 principles that recognise the importance of environmental well-being and set out guiding principles for national governments on issues pertaining to environment and development.

The Rio Declaration on Environment and Development is seen as an essential part of the evolution of thinking where environmental issues are concerned as well as the envisaged solutions. Unlike the 1972 Stockholm Conference on Human Environment, which attributed environmental challenges to industrialisation and technology and emphasised technological solutions to development challenges (Hens 2005), the Rio Declaration establishes the centrality of humanity in sustainable development concerns. This is highlighted in Principle 1, which articulates that "Human beings are at the centre of concerns for sustainable development. They are entitled to a healthy and productive life in harmony with nature" (United Nations 1992a, p. 1). Citizen participation is recognised as crucial in policy development and implementation in the declaration, and the youth, women, children and indigenous people all have attributed roles in sustainable development. There is a strong anthropocentric vision of the environment, where the environment is viewed as a "means to an end"; in other words: "human well-being depends upon the quality of the environment, and therefore it is in the interest of humans to preserve their environment" (Hens 2005, p. 6). Secondly, Hens (2005, p. 6) argues that the principle asserts that "care of people is the main aim of the measures taken to provide a stable environment", which is an attempt to recognise the right of humanity to a safe environment by making it a foundation, though a weak one, for the development of environmental standards.

Agenda 21 (simply interpreted as the 21st-Century Agenda) is the other outcome document of the 1992 Rio Earth Summit. The heads of states adopted it at the Rio Earth Summit. It embraces both the environment and development agenda and states goals related to specific major resources as well as social and environmental issues, among others (Parson et al. 1992). Agenda 21 is an all-inclusive global plan of action for governments, the United Nations and other major groups in all areas where human beings have an impact on the environment. This is where the centrality of people is reflected in the document. Overall, the document has 40 chapters, which are subdivided into three sections focusing on (1) social and economic dimensions, (2) conservation and management of resources for development and (3) strengthening the role of major groups and means of implementation (United Nations 1992b). The first section addresses social issues, including poverty, health and consumption patterns, and is biased towards developing countries where issues of poverty are concerned. The second section focuses on specific environmental challenges and the third section emphasises the participation of different sectors of society. The youth, women, children and indigenous people, non-governmental organisations (NGOs), local authorities, workers and so forth are specifically mentioned. The means of implementation, like the 1972 Stockholm Declaration on the Human Environment, still focuses on science and technology but includes other social structures like education, international institutions and financial mechanisms (United Nations 1992b). Agenda 21 recognises the importance of genuine involvement of different social groups. Broad public participation in decision-making, especially on issues that potentially affect their communities, is argued in Agenda 21 as a prerequisite for achieving sustainable development. While Agenda 21 was not legally binding, it was a crucial document as signatories were politically obliged to pursue the set goals and targets (Parson et al. 1992) with the beginning of the twenty-first century as the ambitious target date, hence the name "Agenda 21".

Overall, the Rio Earth Summit is said to have been "a beneficial start and a necessary step" in discussing important environmental protection issues that have meaning beyond the meeting and hence with great potential for future implementation in real-world contexts (Affan 2017, p. 2). It brought about a paradigm shift in the way people think about sustainable development. The conference managed to raise public awareness on the need to integrate the environment and development leading up to the World Summit on Sustainable Development (WSSD).

The WSSD, which is unofficially called Rio+10, was held in 2002 in Johannesburg, South Africa, 10 years after the 1992 Rio Earth Summit. The WSSD gathered more or less the same audience as the Rio Earth Summit to establish the means of implementation of the same goals established 10 years earlier as well as Millennium Development Goals (MDGs). The focus was more on progress review since the Rio Earth Summit, rather than on establishing new commitments (Von Schirnding 2005). The gathering surpassed the Rio Earth Summit in terms of attendance.

Prior to the WSSD in 2000, a Millennium Summit was held to establish the MDGs that would be used to benchmark progress in the implementation of sustainable development. The eight MDGs focus on social issues and are integrative of poverty-related challenges, save for two. The odd two MDGs address environmental sustainability and partnerships. The MDGs had targets and indicators, which were meant to help in monitoring progress in the implementation of the goals (Nhamo et al. 2019) and were carried over in the 2030 AfSD.

Since the WSSD was not intended to develop new commitments or conventions, but to implement existing commitments, it follows that the goals and targets of poverty, environment and sustainable development in the Millennium Declaration were the same goals that were recalled in one of the WSSD outcomes documents, the Johannesburg Plan of Implementation (JPI). The other outcome of the WSSD was the Johannesburg Declaration on Sustainable

Development—JPI (La Viña et al. 2003). The JPI is a political document, which is not legally binding, and was meant to guide development and government decisions, among other stakeholders. The JPI put more emphasis on local issues and social issues. This was a departure from the Rio Earth Summit where climate change and other related global environmental challenges took precedence. The JPI called for action by various stakeholders and there was less emphasis on science and technology and more emphasis on human action. Hence, sustainable development was positioned as both a public and political agenda and local sustainability challenges like health energy, water and sanitation were receiving considerable attention (Von Schirnding 2005). The WSSD was momentous in achieving a significant transition in the environmental movement where ecological problems were embedded in social realities. The importance of society was brought to the fore. The fact that priority was on local versus global challenges embeds sustainability challenges in local socio-economic contexts. This was one of the major contributions of the WSSD where society can be said to have been fronted in terms of both causal factors and resolutions to sustainable development challenges.

In 2012, 20 years after the Rio Earth Summit, the international community gathered in Rio De Janeiro again at what became known as the Rio+20 conference. The need for a post-2015 agenda was highlighted in the Rio+20 outcome document "The Future We Want". From the Rio+20 Conference, the United Nations General Assembly reaffirmed "the need to further mainstream sustainable development at all levels, integrating economic, social and environmental aspects and recognizing their interlinkages" and noted the fact that people are at the centre of sustainable development (United Nations 2012, p. 2). The commitment to accelerate the achievement of the MDGs was also made. The MDGs' implementation process, therefore, remains the foundation for implementing the SDGs. Designed along with a similar architecture to the MDGs, the SDGs seek to complete, top-up and broaden what the MDGs could not attain based on a balance of the three dimensions of sustainable development at a much more enhanced and ambitious level. The SDGs are universal and more inclusive and they include a well-defined means of implementation. The SDGs are interlinked, and through them, the world looks to involve every citizen and organisation, hence the motto "Let no one be left behind" (Nhamo et al. 2020).

The 2030 AfSD is broad and enshrines the 17 interwoven SDGs and their 169 targets and many more indicators. It covers a wide range of goals (United Nations 2015) including poverty eradication, economic growth, social inclusion, environmental sustainability and peace for all by 2030. Post-2015, countries have been working towards meeting the SDGs. Across the globe, the challenge of domesticating and localising the SDGs in terms of national and local development priorities requires a combination of technical, scientific as well as administrative and political input. A collaborative research approach is needed to stay true to the SDGs' inclusive and bottom-up approach. Of interest is the notion that the SDGs represent a development agenda that should be realised by both the developed and developing countries. This provides researchers across and within disciplines with endless novel opportunities to engage with the SDGs. Given the foregone discussion, the SDGs remain an agenda for society, hence the need to document their implementation across societies in this book *Sustainable Development Goals for Society* (*SDGs4S*).

3 Addressing Poverty in the Context of SDGs

The need to address poverty in the context of SDGs comes out clearly under SDG 1 that looks at the desire to "End poverty in all its forms everywhere" (United Nations 2015, p. 14). Aligned with this goal are seven targets with many indicators that assist in the attainment of the goal by 2030. See Box 1.1 for more details about the targets.

Box 1.1: Targets for SDG 1: Ending poverty in All Its Manifestations Everywhere by 2030

- 1.1 By 2030, eradicate extreme poverty for all people everywhere, measured as people living on less than $1.25 a day.
- 1.2 By 2030, reduce at least by half the proportion of men, women and children living in poverty in all its dimensions according to national definitions.
- 1.3 Implement nationally appropriate social protection systems and measures for all, including floors, and by 2030 achieve substantial coverage of the poor and the vulnerable.
- 1.4 By 2030, ensure that all men and women, in particular the poor and the vulnerable, have equal rights to economic resources, as well as access to basic services, ownership and control over land and other forms of property, inheritance, natural resources, appropriate new technology and financial services, including microfinance.
- 1.5 By 2030, build the resilience of the poor and those in vulnerable situations and reduce their exposure and vulnerability to climate-related extreme events and other economic, social and environmental shocks and disasters.
- 1.a Ensure significant mobilisation of resources from a variety of sources, including through enhanced development cooperation, to provide adequate and predictable means for developing countries, in particular least developed countries, to implement programmes and policies to end poverty in all its dimensions.
- 1.b Create sound policy frameworks at the national, regional and international levels, based on pro-poor and gender-sensitive development strategies to support accelerated investment in poverty eradication actions.

Source: United Nations (2015, p. 15)

There are several matters worth highlighting from SDG 1 targets outlined in Box 1.1. The call is to have all global citizens out of extreme poverty by 2030. In addition, although extreme poverty was defined at the time of approving the SDGs in September 2015 as people living below US$1.25 per day, Target 1.2 brings up another perspective on the need to reduce global citizens living in poverty, a measure that would be determined nationally. There is also the dimension of protecting citizens from extreme weather events induced by climate change, as well as any other recognised economic, social and environmental disasters and shocks. This last category brings us right into the 2020 Covid-19 pandemic that left the foundations of the world shaken and threatening to derail the attainment of SDG 3 and all other SDGs.

Durizzo et al. (2021) put across the matter of managing Covid-19 in poor urban neighbourhoods, focusing on Accra (Ghana) and Johannesburg (South Africa). From a survey of 1400 poor households, the findings confirmed that the lack of compliance with social distancing and protective hygiene had more to do with lack of adequate infrastructure and poverty than the unwillingness to change behaviours. In the USA, Finch and Finch (2020) found that in the earlier stages of Covid-19, higher infections and death rates were associated with disadvantaged (poorer) counties. Drawing from work done in the United Kingdom, Patel et al. (2020, p. 110) maintain that the commonly heard statement that "Covid-19 does not discriminate has been repeated. This, however, is a dangerous myth, side-lining the increased vulnerability of those most socially and economically deprived". Economically disadvantaged populations were likely to live in overcrowded accommodation, compounding the chances of getting infected. Linked to the poor are matters pertaining to inequality (SDG 10) (Ahmed et al. 2020). It is a situation where the Covid-19-poverty-inequality vicious cycle is created, seriously retarding progress towards the 2030 AfSD. As people stayed at home, with companies closing or wages reduced, more households were pushed into poverty putting governments under pressure to put up social security measures. Subsequently, other governments had to borrow money to finance relief and recovery measures (Republic of South Africa 2020). Hence, poverty remains the main issue that demands resolute leadership in the lead to the 2030 AfSD.

4 Leadership in Implementing SDGs

Speaking during the World Government Summit's SDGs in Action event on 9 February 2019 in Dubai, the United Nations Deputy Secretary-General, Amina Mohammed, raised critical pillars on leadership. It emerged that the SDGs remained the global blueprint for a sustainable, inclusive and just future (Mohammed 2019). To this end, leadership was required at all spatial levels and from all stakeholders, including development partners and philanthropists. Hence, at times, unpopular decisions had to be taken to move away from the business-as-usual pathways, and stakeholders must "move out of comfort zones and embrace innovative ways of working, thinking and leading" (Mohammed 2019). Good leadership meant that the governments and other stakeholders should commit to sharing experiences and tools, be they good or bad for the greater public good of advancing the SDGs. Sharing is part of the United Nations High-Level Political Forum on SDGs that receives Voluntary National Reviews annually, which stocktake implementation progress on SDGs from members. Further, there is a need for effective and accountable public institutions built on transformational leadership (Manzoor et al. 2019).

SDGs' leadership demands that governments at various spatial levels ensure commitment and ownership of programmes and project implementation by taking such to all corners of society (Gornitzka and Pipa 2018). This approach supports Meuleman and Niestroy's (2015) early work on the principle of Common But Differentiated Governance (CBDG) and how CBDG could be applied in implementing the SDGs. Drawing from Nigeria, Akinloye (2018) introduces the need to maximise the influence of religious leaders in implementing the SDGs. Religious leadership is viewed as an enabler that will bring more actors into the SDG space, making sure that those from the grassroots are not left behind. There is a growing body of literature demanding good leadership for SDG implementation amid the Covid-19 pandemic (Filho et al. 2020). The authors highlight that governments and other stakeholders were being pressurised to divert resources (financial, human and other) towards the pandemic, an aspect that could lead to many other SDGs not being practised for a while.

Evidence of good leadership in SDG implementation was coming up from emerging economies such as South Africa. In his address to the joint sitting of Parliament in October 2020, President Cyril Ramaphosa highlighted what the government had done and was going to urgently do to mitigate Covid-19 impacts through the proposed Economic Reconstruction and Recovery Plan (ERRP) (Republic of South Africa 2020). The ERRP identified four main areas, namely (1) the infrastructure build programme; (2) rapid expansion of energy generation capacity (with a substantial increase in renewables, battery storage and gas); (3) employment stimulus package to create jobs and support livelihoods, with 800,000 job opportunities envisaged in a few months; and (4) propelling of industrial growth. The identified main areas both directly and indirectly link to several SDGs, particularly SDGs 1–3 (ending poverty, ending hunger and health and well-being), SDGs 7 and 8 (sustainable energy and sustainable and decent work), SDGs 9–12 (industry and infrastructure, reducing inequality, sustainable cities and sustainable consumption and production), SDG 13 (climate action) and SDG 17 (partnerships). The ERRP presented a policy commitment to SDG 7 (gender), noting the desire to end gender-based violence and the commitment to work "with women-empowered companies to progressively reach our target of directing at least 40% of procurement spend to such enterprises" (Republic of South Africa 2020, p. 11). Lastly, the ERRP is linked to the government's National Development Plan: Vision 2030 and was to be driven by an Economic Recovery Leadership Team.

Leadership in implementing SDGs for, in and with societies is critical in the business sector. To this end, the United Nations Global Compact (UNGC) came up with a plan for business leadership in the SDGs. The plan highlights that business leadership on the SDGs is informed and evolves through three cyclical steps that harness the desire to prioritise, act and learn (UNGC 2017). Companies are supposed to prioritise their engagements drawing from their multiplier and effective contribution towards the SDGs, the idea being to maximise the positive impacts. To act, five leadership qualities should be embodied, namely intentionality, being ambitious, consistent, collaborative and accountable. The leading companies should learn about their actions' impact on the SDGs. This takes companies back to the need to report and share lessons like the state actors, revealing both the good and bad testimonies. The involvement and collaboration with stakeholders become part of responsible leadership (Muff et al. 2020). Further elaborations of the five SDG leadership qualities are reflected in Fig. 1.1.

Drawing from a sample of 25 multinational companies (MNCs) from Brazil, Russia, India, China and South Africa (BRICS), Ali et al. (2018) determined the extent to which the MNCs had adopted the SDGs in their vision and mission statements. They found that while progress was being made, there was a gap in that crucial SDGs were missing. Such important SDGs included quality education (SDG 4), climate action (SDG 13) and life below water (SDG 14). Yu et al. (2020) made similar findings from a study involving 100 Chinese companies that were on the Shanghai Stock Exchange from 2016 to 2018. These companies were found to focus more on infrastructure development, industrial innovation and economic growth and placed some emphasis on affordable and clean energy (SDG 7), dignified and respectable working environment (SDG 8), as well as peace, justice and strong institutions (SDG 16). From Australia, Noh (2020) raises the concept of creating shared value (CSV). Although the companies were managing to frame SDGs around the CSV, the operationalisation left a gap as there was a weak connection to the SDGs.

Pedersen (2018) presents the notion that SDGs are a huge gift to business as these present opportunities and guidance for long-term investment, as well as new business opportunities. The SDGs are said to have presented global priorities for the business, with big and quick wins potentially availing themselves for harvesting. Hence, the private sector is a key stakeholder and should shoulder responsibility in accelerating SDG implementation (Rashed and Shah 2020). The authors present engagements in corporate social responsibility, the circular economy and the environmental initiatives as pathways towards the realisation of the 2030 AfSD. The lack of influential leadership and appropriate indicators to measure implementation are flagged as some of the major drawbacks for business entities. However, the 2030 AfSD and its 17 SDGs may not be attained from a business-as-usual trajectory and leadership approach (ElAlfy et al. 2020). Therefore, to be visible and effective, the business sector should address some of the actions listed in Table 1.1 as proposed by the UNGC.

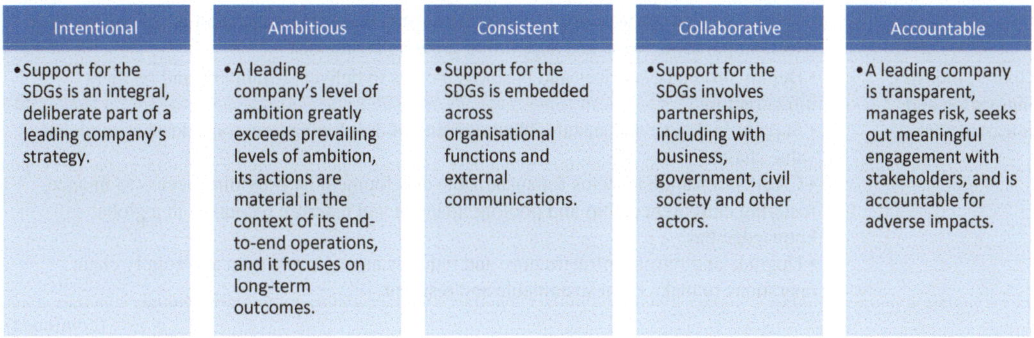

Fig. 1.1 Five qualities of SDG leadership. Source: Authors, based on UNGC (2017, p. 5)

Table 1.1 Business actions in support of SDGs

The SDGs	Selected business actions
Goal 1: No poverty	• Create sustainable and decent jobs, especially in the least developed countries and for vulnerable populations • Implement programmes to economically empower disadvantaged groups • Ensure decent working conditions for all employees • Create and market goods and services that improve the lives of vulnerable groups
Goal 2: Zero hunger	• Support small-scale farmers to increase yields and incomes from sustainable agriculture • Alter food logistics to contribute to ending malnutrition and hunger in all communities surrounding company operations and eliminate food waste and loss
Goal 3: Good health and well-being	• Ensure the best possible health outcomes for employees and surrounding communities across own and supply chain operations • Deploy products, services and business models for improved health outcomes and lead initiatives encouraging healthy behaviours and improve access to healthcare
Goal 4: Quality education	• Ensure that all employees have access to vocational training and lifelong learning opportunities • Ensure that all employees earn a wage that allows them to support the education of dependents and that there is zero child labour • Implement programmes to support higher education and access to free, equitable and inclusive primary and secondary education • Deploy products and services that improve educational access and learning outcomes
Goal 5: Gender equality	• Implement policies and practices that are free from and prevent gender-based discrimination across the workplace, marketplace and community • Support women's employment and strive for gender balance at all levels across the business and supply chain • Develop products and services and implement marketing practices that empower women • Promote gender equality through investment, community initiatives and advocacy
Goal 6: Clean water and sanitation	• Implement holistic water strategies socially equitable, environmentally sustainable and economically beneficial in watersheds around company and supply chain operations • Protect and/or restore water-based ecosystems around its own operations and supply chain • Ensure access to water and sanitation by addressing impacts of company and supply chain operations on local water supplies and supporting stakeholders to deliver clean water and sanitation
Goal 7: Affordable and clean energy	• Increase energy efficiency, source remaining energy needs from renewable sources and promote the same action across the supply chain • Deploy affordable, sustainable energy and energy efficiency products and services • Develop and implement business models to deliver sustainable energy and energy-efficient technologies to new markets and communities
Goal 8: Decent work and economic growth	• Support decent working conditions for all employees across the business and supply chain, with partnerships to build suppliers' capacity to do the same • Educate and train the labour force, focusing on vulnerable and economically disadvantaged groups • Create decent formal sector jobs in labour-intensive sectors • Drive economic growth and productivity by investing in research and development (R&D), upgrading skills and supporting growing businesses
Goal 9: Industry, innovation and infrastructure	• Deploy products, services and business models to deliver sustainable and resilient infrastructure • Support inclusive and sustainable upgrading of developing country industries in global value chains • Create innovation systems for sustainable development by providing access to finance, fostering entrepreneurship and pooling financial and research resources in a global knowledge base • Upgrade and retrofit infrastructure and industry assets across own and supply chain operations to make them sustainable and resilient

(continued)

Table 1.1 (continued)

The SDGs	Selected business actions
Goal 10: Reduced inequalities	• Assess the distribution of economic value across stakeholder groups and implement policies and practices to make it more equal • Support the establishment and expansion of social protection measures at national level • Implement policies and practices to support equality of opportunity, treatment and outcome for all across own and supply chain operations • Design and implement products, services and business models that explicitly target the needs of disadvantaged and marginalised populations
Goal 11: Sustainable cities and communities	• Deploy products and services that improve access to resilient buildings, transport, green spaces and utilities • Protect and invest in cultural and natural heritage and support access to essential services across the workplace, marketplace and community
Goal 12: Responsible consumption and production	• Design and adopt a responsible, circular business model and narrow or close material and energy loops across own and supply chain operations • Shift to a portfolio of goods and services that require and promote negligible use of resources and produce negligible waste • Develop, implement and share solutions for tracking and reporting on the sustainability of production and consumption across end-to-end operations
Goal 13: Climate action	• Ensure climate resilience of company and supply chain operations and the communities surrounding them • Reduce emissions associated with own and supply chain operations, in alignment with climate science, and shift to goods and services promoting negligible emissions • Promote conscious climate behaviour and build capacity for climate action
Goal 14: Life below water	• Implement policies and practices to protect ocean ecosystems that are affected by business and supply chain activities • Deploy products and business models that negate impacts on ocean ecosystems and contribute to their restoration • Galvanise finance for the protection and further development of ocean ecosystems and water system flows • Design and implement solutions to accurately value and respect natural capital and drive wider adoption of these solutions
Goal 15: Life on land	• Implement policies and practices to protect natural ecosystems that are affected by business and supply chain activities • Deploy products, services and business models to help decouple economic activity from the degradation of natural ecosystems • Galvanise finance to create awareness of, protect and further develop natural ecosystems • Design and implement solutions to accurately value and respect natural capital and drive wider adoption of these solutions
Goal 16: Peace, justice and strong institutions	• Identify and take robust action against corruption and violence in own operations and the supply chain • Work with the government to strengthen institutions and increase respect and support for the rule of law • Work with the government and/or international institutions in areas of conflict and humanitarian crises to contribute to peace and institution building
Goal 17: Partnerships for the goals	• Lead on partnerships to improve domestic resource mobilisation through responsible tax practices • Galvanise private-sector finance to support sustainable development initiatives in developing countries • Build regulatory, organisational and staff capacity in developing countries • Lead on partnerships to develop and share new and existing technology, knowledge and business models • Lead on partnerships that address systemic challenges for achieving the SDGs

Source: Authors, based on UNGC (2017, pp. 7–163)

Bissinger et al. (2020) bring another dimension of leadership in SDG implementation by linking it to voluntary standards. The authors found that the voluntary sustainability standards (VSS) that were developed by industry, civil society and other key stakeholders extensively cover three SDGs (2, 8, 12). SDGs 13, 14 and 17 were found to have few or no links to the VSS. Such VSS remain useful in the remaining period of implementation of the SDGs to 2030.

5 Service Delivery and SDGs at the Local Government Level

There is no contestation that most goods and services are consumed at the local government level. To this end, the SDGs will be attained through several programmes and projects implemented at this spatial scale (Dawkins et al. 2019). Linking up SDG 2 (ending hunger) and SDG 11 (sustainable cities), Ilieva (2017) brings the concept of urban food systems strategies (UFSS) as a promising tool for the implementation and attainment of the SDGs. The recommendation comes out of research work that looked at the application of the UFSS in the megacities of New York, Philadelphia, Los Angeles, Chicago (all from the USA) and Toronto in Canada. Kawakubo et al. (2018) bring another dimension when they focus on cities in the context of SDGs and greenhouse gas (GHG) emissions. Interfacing this with the 2015 Paris Agreement on climate change, the authors believe that the involvement of local government was significant in promoting sustainable development and attain 2030 AfSD overall.

Zinkernagel et al. (2018) raise additional advantages that the SDGs have brought in tracking progress in implementing sustainable development. In their view, traditional indicators for measuring progress in implementing sustainable development in cities focused on health and safety, environmental sustainability and economic growth. Such indicators lacked dimensions on gender equality (SDG 5) and the need to reduce inequality (SDG 10). To this list, Dawkins et al. (2019) add sustainable consumption and production (SDG 12) that touches on food procurement and waste, water, waste prevention, clothing and other consumables. Talking of water and sanitation (SDG 6) and the World Health Organization's call for "safe hands" for Covid-19, Zvobgo and Do (2020) pick up a real challenge in many cities and settlements from developing nations. Using Chitungwiza in Zimbabwe as a case study, the authors found that the demand for domestic water and sanitation increased by 90%, which added to an already existing water scarcity. An additional 4.5 L of water was needed per person per day to attain "safe hands". The water issue draws us further into the 2030 AfSD as it is also a human rights matter (Carrard et al. 2020) and in many countries, water provision remains a local government mandate.

Krellenberg et al. (2019) argue that urban sustainability strategies should now be guided by the SDGs. Hence, the SDGs have brought up a more balanced and integrated matrix to monitor urban sustainability (Zinkernagel et al. 2018). However, more work still needs to be done to localise the SDGs so that they become context specific and more relevant. This proposal makes sense given that cities are and will be at different stages of development, as well as have differentiated implementation ability and capabilities. Considering the set-up of cities, it is inevitable that authorities continue to move swiftly to localise all the SDGs. Therefore, there are several chapters in this book addressing the role of local governments in propelling the world forward along a sustainable pathway. The next section briefly presents the methodological orientation applied in this book.

6 Methodological Orientation

This book is a compilation of chapters from diverse authors and contexts from Africa, Europe and Latin America. Therefore, there is no single methodological orientation informing this work. The authors have also used diverse methods, which reflect their different research background contexts. The methods were mostly influenced by the research questions addressed by each of the chapters. Specific theoretical, methodological orientations are discussed as part of each chapter. This section will highlight the broad categories of the research orientations.

The focus of the book is society. Hence, the research documents can be categorised as social research. Social scientists employ social research methods to gain a better understanding of society (and people) and find solutions to social issues (Mogalakwe 2006). Social research mostly depends on the collection of primary rather than secondary data; thus, it relies on primary methods of data collection. Research topics are diverse in social research as any aspect of society can be a topic of research. Social research can be either quantitative or qualitative, depending on the problem being researched. In either case, it follows a well-organised research plan. Qualitative research focuses on establishing people's opinions, attitudes, actions and behaviour through interviews, focus group discussion and documentary reviews and the results are textual and non-quantitative (Kothari 2004). Quantitative research relies on measurement and generates numerical data, which is analysed through quantitative analytic methods: for example, student t-test, correlation and regression (Kothari 2004). Examples of quantitative methods include a questionnaire survey and structured observations. The next and last section of the chapter presents the book and chapter outline.

7 Book and Chapter Outline

This book is divided into five parts. Part I provides the introduction and background and is made up of a single chapter looking at making the SDGs relevant for, in and with societies. Part II addresses poverty in the context of SDGs and comprises five chapters. Chapter 2 presents an overview of SDG policies in Brazil, while Chap. 3 looks at the localisation of SDGs in rural Uganda. Chapter 4 considers ending extreme poverty in the Chegutu District of Zimbabwe, with Chap. 5 dedicated to programmes to support SDGs through social, inclusive and transformative innovation. Chapter 6 articulates financial inclusion as a complementary strategy to address the SDGs for society.

Part III is made up of another five chapters that focus on leadership in the implementation of SDGs. Chapter 7 documents mentoring women in the resources sector, while Chap. 8 explores leadership capabilities in a multi-sector road infrastructure and innovation (SDG 9) and partnership (SDG 17) in South Africa. Chapter 9 discusses the role of leadership capacities in the response of the South African National Statistics System to the SDGs, with Chap. 10 presenting the unique role of libraries in promoting the SDGs. Chapter 11 considers leadership capacities in multi-stakeholder partnerships contributing towards the SDGs using a case study of Project Last Mile in Eswatini.

Part IV is dedicated to service delivery and the attainment of SDGs at the local government level and comes in seven chapters. Chapter 12 brings up the Global Water Partnership, South America, and the transboundary implementation of integrated water resources management (SDG target 6.5). Chapter 13 narrows down to sustainability reporting through UNGC looking at opportunities and challenges in mainstreaming the Global Reporting Initiative Standards and the SDGs. Chapter 14 considers leadership and the implementation of SDGs in Finnish municipalities, while Chap. 15 presents opportunities and challenges for local government institutions in localising SDGs in Zimbabwe. Chapter 16 considers water, energy, health and sanitation challenges in Masvingo's low-income urban communities in the context of SDGs in Zimbabwe. Chapter 17 pitches the trends in research around the SDG objectives from a bibliometric analysis, with Chap. 18 dedicated to harnessing the potential of information and communication technologies (ICTs) in agribusiness for youth employment. The last part of the book is also made up of a single chapter, which presents the summary of findings, conclusions and policy recommendations.

References

Affan, F.B. 2017. Earth Summit: From Rio 1992 via Kyoto and Copenhagen, Back to Rio 2012. J Ecosyst Ecography, 7(229). https://doi.org/10.4172/2157-7625.1000229.

Ahmed, F., Ahmed, N., Pissarides, C., Stiglitz, J. 2020. Why inequality could spread COVID-19. The Lancet, 5, e240. https://doi.org/10.1016/S2468-2667(20)30085-2

Ali, S., Hussain, T., Zhang, G., Nurunnabi, M., Li, B. 2018. The Implementation of Sustainable Development Goals in "BRICS" Countries. Sustainability, 10, 2513; https://doi.org/10.3390/su10072513

Akinloye, I.A. 2018. Towards the implementation of sustainable development goals in Nigeria: Maximizing the influence of religious leaders. Stellenbosch Theological Journal, 4(1), 39–60. https://doi.org/10.17570/stj.2018.v4n1.a02

Bissinger, K., Brandi, C., Cabrera de Leicht, S., Fiorini, M., Schleifer, P., Fernandez de Cordova, S., Ahmed, N. 2020. Linking Voluntary Standards to Sustainable Development Goals. Geneva: International Trade Centre.

Carrard, N., Neumeyer, H., Pati, B.K., Siddique, S., Choden, T., Abraham, T., Gosling, L., Roaf, V., Torreano, J.A.S., Bruhn, S. 2020. Designing Human Rights for Duty Bearers: Making the Human Rights to Water and Sanitation Part of Everyday Practice at the Local Government Level. Water, 12, 378; https://doi.org/10.3390/w12020378

Dawkins, E., Andr, K., Axelsson, K., Benoist, L., Swartling, Å.G., Persson, Å. 2019. Advancing sustainable consumption at the local government level: A literature review. Journal of Cleaner Production, 231, 1450–1462. https://doi.org/10.1016/j.jclepro.2019.05.176

Durizzo, K., Asiedu, E., Van der Merwe, A., Van Niekerk, A., Gunther, I. 2021. Managing the COVID-19 pandemic in poor urban neighbourhoods: The case of Accra and Johannesburg. World Development 137, 105175. https://doi.org/10.1016/j.worlddev.2020.105175

ElAlfy, A., Palaschuk, N., El-Bassiouny, D., Wilson, J., Weber, O. 2020. Scoping the Evolution of Corporate Social Responsibility (CSR) Research in the Sustainable Development Goals (SDGs) Era. Sustainability, 12, 5544; https://doi.org/10.3390/su12145544

Filho, W.L., Brandli, L.L., Salvia, A.L., Rayman-Bacchus, L. Platje, J. 2020. COVID-19 and the UN Sustainable Development Goals: Threat to Solidarity or an Opportunity? Sustainability, 12, 5343; https://doi.org/10.3390/su12135343

Finch, W.H. & Finch, M.E.H. 2020. Poverty and Covid-19: Rates of Incidence and Deaths in the United States During the First 10 Weeks of the Pandemic. Frontier in Sociology, 5:47. https://doi.org/10.3389/fsoc.2020.00047

Gornitzka, C.P. & Pipa, A.F. 2018. A new type of leadership from national governments is essential for success of the SDGs. https://www.brookings.edu/blog/up-front/2018/04/24/a-new-type-of-leadership-from-national-governments-is-essential-for-success-of-the-sdgs/ (Accessed 15 October 2020).

Hens, L. 2005. The Rio declaration on environment and development. Regional sustainable development review: Africa. Oxford, UK, Eolss Publishers. http://www.eolss.net/sample-Chapters/C16/E1-48-43.pdf. Accessed 20 Sept 2020.

Hofmann, E. & Rüsch, M. 2017. Industry 4.0 and the current status as well as future prospects on logistics. Computers in Industry, 89: 23-34. https://doi.org/10.1016/j.compind.2017.04.002

La Viña, A. G. M., Hoff, G. and DeRose, A. M. 2003. The outcomes of Johannesburg: Assessing the World Summit on Sustainable Development. SAIS Review: A Journal of International Affairs, XXIII(1), Winter-Spring 2003 Issue.

Ilieva, R. 2017. Urban Food Systems Strategies: A Promising Tool for Implementing the SDGs in Practice. Sustainability, 9, 1707; https://doi.org/10.3390/su9101707

Kamble, S.S., Gunasekaran, A., Gawankar, S.A. 2018. Sustainable Industry 4.0 framework: A systematic literature review identifying the current trends and future perspectives. Process Safety and Environmental Protection, 117: 408-425. https://doi.org/10.1016/j.psep.2018.05.009

Kawakubo, S., Murakami, S., Ikaga, T., Asami, Y. 2018. Sustainability assessment of cities: SDGs and GHG emissions. Building Research & Information, 46:5, 528-539, https://doi.org/10.1080/09613218.2017.1356120

Kothari, C.R. (2004). Research methodology: Methods and techniques (2nd Edition). New Delhi: New Age International Publishers.

Krellenberg, K., Bergsträßer, H., Bykova, D., Kress, N., Tyndall, K. 2019. Urban Sustainability Strategies Guided by the SDGs—A Tale of Four Cities. Sustainability, 11, 1116; https://doi.org/10.3390/su11041116

Manzoor, F., Wei, L., Nurunnabi, M., Subhan, Q.A., Shah, S.I.A., Fallatah, S. 2019. The Impact of Transformational Leadership on Job Performance and CSR as Mediator in SMEs. Sustainability, 11, 436; https://doi.org/10.3390/su11020436

Meuleman, L. & Niestroy, I. 2015. Common But Differentiated Governance: A Metagovernance Approach to Make the SDGs Work. Sustainability, 7, 12295-12321; https://doi.org/10.3390/su70912295

Mogalakwe, M. 2006. The Use of Documentary Research Methods in Social Research. African Sociological Review, 10, (1), 2006, pp. 221-230.

Mohammed, A. 2019. Delivering on Sustainable Development Goals Requires Transformational, Inclusive Leadership, Deputy Secretary-General Tells World Government Summit. https://www.un.org/press/en/2019/dsgsm1249.doc.htm (Accessed 15 October 2020).

Muff, K., Liechti, A., Dyllick, T. 2020. How to apply responsible leadership theory in practice: A competency tool to collaborate on the sustainable development goals. Corporate Social Responsibility and Environmental Management, 27:2254–2274. https://doi.org/10.1002/csr.1962.

Nhamo, G., Nhemachena, C. and Nhamo, S. 2019. Is 2030 too soon for Africa to acheive the water and sanitation sustainable development goal. Science of the Total Environment, 669: 129–139.

Nhamo, G., Dube, K., Chikodzi, D. 2020. Sustainable Development Goals: Concept and Challenges of Global Development Goal Setting. In: Haring R., Kickbusch I., Ganten D., Moeti M. (eds) Handbook of Global Health. Springer, Cham. https://doi.org/10.1007/978-3-030-05325-3_79-1

Noh, J.E. 2020. Promotion of Shared Value for the SDGs (Sustainable Development Goals): A Case Study of Australia. Journal of Sustainability Research, 2(3):e200025. https://doi.org/10.20900/jsr20200025

Parson, E. A.; Haas, P. M, Levy, M. A. 1992. A Summary of the Major Documents Signed at the Earth Summit and the Global Forum. Environment: Science and Policy for Sustainable Development. 34. 12–36.

Patel, J.A., Nielsen, F.B.H., Badiani, A.A., Assi, S., Unadkat, V.A., Patel, B., Ravindrane, R., Wardle, H. 2020. Poverty, inequality and COVID-19: the forgotten vulnerable. Public Health, 183, 110-111. https://doi.org/10.1016/j.puhe.2020.05.006

Pedersen, C.S. 2018. The UN Sustainable Development Goals (SDGs) are a great gift to business! Procedia CIRP, 69, 21 – 24.

Rashed, A.H. & Shah, A. 2020. The role of private sector in the implementation of sustainable development goals. Environment, Development and Sustainability, https://doi.org/10.1007/s10668-020-00718-w

Republic of South Africa. 2020. Address by President Cyril Ramaphosa to the joint sitting of parliament on South Africa's economic reconstruction and recovery plan, 15 October 2020. Pretoria: The Presidency.

United Nations. 1992a. Rio Declaration on Environment and Development 1992. https://www.jus.uio.no/lm/environmental.development.rio.declaration.1992/portrait.a4.pdf. Accessed 20 Sept 2020

United Nations 1992b. United Nations Conference on Environment & Development. Rio de Janeiro, Brazil, 3 to 14 June 1992 AGENDA 21. https://sustainabledevelopment.un.org/content/documents/Agenda21.pdf. Accessed 20 Sept 2020.

United Nations. 2012. The future we want. Outcome document of the United Nations Conference on Sustainable Development. Rio de Janeiro Brazil. https://sustainabledevelopment.un.org/content/documents/733FutureWeWant.pdf. Accessed 17 Aug 2018.

United Nations. 2015. Transforming our world: the 2030 Agenda for Sustainable Development. Resolution adopted by the General Assembly 25 September 2015. Available at: http://www.un.org/ga/search/view_doc.asp?symbol=A/RES/70/1&lANG=e. Accessed 02 July 2019.

UNGC (United Nations Global Compact). 2017. Blueprint for business leadership on the SDGs: A principles-based approach. New York: United Nations Global Compact.

Von Schirnding, Y. 2005. The World Summit on Sustainable Development: reaffirming the centrality of health. Global Health, 1(8). https://doi.org/10.1186/1744-8603-1-8. Accessed 23 Sept 2020.

Yu, S., Sial, M.S., Tran, D.K., Badulescu, A., Thu, P.A., Sehleanu, M. 2020. Adoption and Implementation of Sustainable Development Goals (SDGs) in China—Agenda 2030. Sustainability, 12, 6288; https://doi.org/10.3390/su12156288

Zinkernagel, R., Evans, J., Neij, L. 2018. Applying the SDGs to Cities: Business as Usual or a New Dawn? Sustainability, 10, 3201; https://doi.org/10.3390/su10093201

Zvobgo, L. & Do, P. 2020. COVID-19 and the call for 'Safe Hands': Challenges facing the under-resourced municipalities that lack potable water access - A case study of Chitungwiza municipality, Zimbabwe. Water Research X. https://doi.org/10.1016/j.wroa.2020.100074

Part II

Addressing Poverty in the Context of SDGs

Brazilian National Policies Related to Sustainable Development Goals: An Overview

2

Carlos Hiroo Saito, Gabriela Zamignan,
Ana Claudia de Almeida, Larisa Ho Bech Gaivizzo,
and Romero Gomes Pereira da Silva

Abstract

The integration of the Sustainable Development Goals (SDGs) with public policies is a central element of the 2030 Agenda for Sustainable Development. Brazilian national policies from 1962 to 2019 were analyzed to determine their interactions and synergies with the SDGs through a matrix of interconnections. All national policies have at least one connection to an SDG. SDG 16 was the most connected to national policies, followed by SDG 10. A downscaling policy hierarchy was also performed to identify programs and projects related to national policies, their interactions with the SDGs, and their complexity. At this step, programs and project-level policies in the 2016–2019 Pluriannual Plan were analyzed, and for those with the highest number of connections, didactic qualitative modeling was performed to understand the chain of interdependencies with SDGs. These analyses supported the production of 18 portraits as an educational material to provide systemic comprehension of the relationship among national policies and the SDGs seeking awareness. The studied initiatives are generally within the framework of human rights, which corresponds to the social pillar of sustainability. This means that Brazilian Government mandates, in different periods of the country's history, were in line with the SDG core guideline, based on a social commitment to "leave no one behind" and reduce global inequalities.

Keywords

SDGs · 2030 Agenda · Public policies · Qualitative modeling · Government actions

1 Introduction

Public policies of national scale connected to commitments of international importance demand more effort, especially in a country as large and diverse as Brazil. To implement the commitment made by the state, the entire framework and the context must be internalized, which means that the commitment must be understood as an important and current matter (UN 2014, 2015). This internalization is an issue that is related to the knowledge sphere, while the ordering of processes and actions for implementation is connected to the management sphere.

C. H. Saito (✉) · A. C. de Almeida
L. H. B. Gaivizzo · R. G. P. da Silva
Center for Sustainable Development, University of Brasilia, Asa Norte, Brasília-DF, Brazil
e-mail: saito@unb.br

G. Zamignan
Center of Excellence in Tourism, University of Brasilia, Asa Norte, Brasília-DF, Brazil

G. Nhamo et al. (eds.), *Sustainable Development Goals for Society Vol. 1*, Sustainable Development Goals Series, https://doi.org/10.1007/978-3-030-70948-8_2

Currently, attention is being given to the 2030 Agenda for Sustainable Development (2030 Agenda), which represents the current trajectory of several international sustainability commitments signed by nations (Nhamo et al. 2020). Its predecessors were Agenda 21 and, more recently, the Millennium Development Goals (MDGs) (Sachs 2012). When the projects and the period to adopt measures to reach the MDGs were concluded, there was a need to continue global efforts to recognize that the eradication of poverty in all dimensions and forms represents a great world challenge and an indispensable step in seeking sustainable development (Dedecca 2014; CEPAL 2018).

Despite the results achieved with the MDGs, there was a need to expand the scope of this global action, especially in relation to the many aspects of the environment and sustainability, moving from the people-centered MDGs to a planet-centered SDGs (Liverman 2018). While expanding the scope of international focus, the specific conditions of each country must also be considered (UN 2014, 2015).

In the MDGs, there was a lack of integration between sectors regarding strategy, policies, and implementation. This insufficient understanding and accountability of trade-offs and synergies between sectors led to incoherent policies with divergent trends. This led to the formulation of general goals of sustainable development (Le Blanc 2015a). The new 2030 Agenda specifies 17 Sustainable Development Goals (SDGs), set in 2015 to be met until 2030, detailed in 169 targets aiming to integrate the three dimensions of sustainable development: economic, social, and environmental (UN 2014).

Integrative, indivisible, global, and universally applicable features are basic elements of the 2030 Agenda (UN 2015). Therefore, there were two concerns when creating the SDGs: intersectoral integration, that is, integration between the SDGs themselves, and integration of the SDGs with public policies (Le Blanc 2015a, 2015b).

In this context, Brazilian national policies from 1962 to 2019 were analyzed to determine their adherence to the SDGs, and after downscaling these policies to program level, their intersectoriality and synergies with SDGs are modeled and represented by concept maps to support the production of didactic material. This chapter presents a small part of the results of the project *Scientific Literacy and Integrative Modeling of Policies Associated with the SDGs*, which received the support of the National School of Public Administration (ENAP, acronym in Portuguese), also called *Cátedra Brasil de Pesquisa 2018–2019*. After this introductory section, a background context about SDGs and their relationship with public policies is presented, followed by materials and methods section which presents the research design. The main results are jointly presented with their discussion before a conclusion section.

2 Background and Literature Review

It is important to note that the SDGs were conceived as an integrated set of priorities and objectives, and as such, they are interdependent (Griggs et al. 2017). This study sought to illustrate the interrelations between some of the SDGs, for example, the relationship between SDGs 2, 3, 7, and 14 with the other goals, and not all goals in their totality. Several studies have tried to reaffirm this interdependence, either among all the SDGs or among some of them. For example, a study carried out by UN-Water in 2016 showed the importance of Goal 6, highlighting its interdependence with other goals (UN-Water 2016). The recognition of synergies among the SDGs provides more balanced development trajectories (Nilsson et al. 2016; Weitz et al. 2018). Another type of work seeking to establish the relations among the 17 SDGs divided them into three major categories (Waage et al. 2015): well-being (SDGs 1, 3, 4, 5, 10, and 17), infrastructure (SDGs 2, 6, 7, 8, 9, 11, and 12), and natural environment (SDGs 13, 14, and 15). Norström et al. (2014) proposed an interaction structure for SDGs that goes beyond sectors, focusing on socioecological systems and considering the interdependence of nature and society. In the same way, interdependence is present regarding planetary boundaries, role of the life support environment, and safe limits for a resilient planet (Rockström et al. 2009).

In Brazil, an interesting effort to determine the interactions between the SDGs was done in a study published by the National Confederation of Municipalities (CNM, in Portuguese) (CNM 2017). In the study, in addition to promoting the implementation of SDGs in municipal administrations, the CNM introduces each SDG and its relations to the other SDGs. This is a very isolated initiative considering that SDG indicators were developed at a national scale, and thus they cannot be directly transferred to municipal/city level. In Brazil, most municipalities are not organized to collect the type of data required to fit SDG indicators. Academic articles are thus focused on assessing how far national indicators are from the targets. Local implementation has been neglected.

Despite being classified under an integrative view, in alignment with other international movements and discussions within this perspective, such as the nexus of water, food, and energy (Weitz et al. 2014; Rasul and Sharma 2016), or perhaps because of these previous movements, this integrative view is neither homogeneous nor balanced. This lack of integration occurs because the goals were derived from a political negotiation process in which the connections among targets were in reference to several other targets. According to Le Blanc (2015a), the goals and targets within the SDGs can be seen as a network reflecting the results of negotiations in an intergovernmental context. In this setting, the authors distinguished the possibility of mapping the interaction network between the SDGs at first as a "political mapping" of the sustainable development universe and then as a "scientific mapping" based on insights of the natural and social sciences.

In addition to the concern about seeing the SDGs as an interdependent goal network, the need for the goals to be fulfilled through public policies has been emphasized since their conception (Stafford-Smith et al. 2017; Weymouth and Hartz-Karp 2018). In the World Public Sector Report (UN 2018) one of the studies points out that in a sample of 60 countries examined around the world, at different stages of development, 32 have institutional arrangements in place for the implementation of the SDGs that cover all sectors. This may be an indication of countries' interest in addressing the integrated nature of the SDGs and a determination to bring all parts of governments together around these objectives.

Brazil took an important step towards ensuring this interoperation. The National Committee of Sustainable Development Goals (CNODS 2017) was created to internalize, promote, and clarify the implementation process of the United Nations 2030 Agenda for Sustainable Development, with a plural and multiscale structure. The committee presented an Action Plan for 2017–2019 that noted six strategies based on internalization and assimilation of the 2030 Agenda at country level: territorialization, governance, mobilization, communication, monitoring and integration, and/or transversality. These strategies are meant to support intersectoral practices while respecting regional diversity and to promote social participation, especially to enable the integration of cultural diversity, the environment, and the economy.

Understanding how to adapt institutional structures to deliver integrated policies that effectively address the existing interconnections between the SDGs is essential to carry out actions consistent with sustainable development, with important implications for administrations and public services (UN 2018). It implies finding ways to foster cooperation and common approaches between institutions at different national scales, as well as implementing appropriate institutional arrangements, practices, mechanisms, capacities, budgetary arrangements, and public administration resources. It also encompasses various modalities of engaging non-state stakeholders in decision-making (Stafford-Smith et al. 2017, Tosun and Leininger 2017, UN 2018).

Nhamo and Mjimba (2020) argue that since the SDGs were created they have been scantly internalized by the nations, although they recognize an increase in reports of multiscale agents on amplifying partnerships (public and/or private) and local awareness. In Brazil, a study carried out in its semiarid region showed that the understanding of the SDGs' connections with public policies should also incorporate bottom-up community-based contributions (Lopes et al. 2020). These empirical debates guide this study, in which we aim to develop an integrative methodology to analyze the Brazilian policies associated with the SDGs in a context of scientific literacy, within a systemic view. We concluded from these debates that the intentions of internalizing, territorializing, and interoperating the goals indicate that they can only be carried out once the fundamentals of the SDGs are understood and the interactions between the goals and their connections to other goals and national policies are equally understood (Weitz et al. 2018).

This means that the different stakeholders should understand why a given policy is connected to all the mentioned SDGs and be capable of visualizing the chains of interdependencies among national policies and SDGs and explaining them. At government level, sectorization as a fragmented management should be combated (Saito 2016). Fragmented knowledge can be transformed into a comprehensive understanding of all SDGs together through a systemic view that leads to an integrated management. Folledo (2000) defends the importance of a systemic view and argues that the ignorance or disregard of systemic concepts when a new policy is elaborated with the goal of solving one problem may cause reactions in other parts of the system (a sort of counterstrike) and lead to prolonged debates on unimportant themes.

Fretwell (2009) goes further and states that the lack of clarity regarding connections and interdependences generates shallow arguments, and the educational process (referring to environmental education) is based on fear. That is, according to the author, instead of enlightenment, the adherence to a cause is sought through fear that the lack of support might worsen the current situation, even if the process chain is not clearly defined.

Systemic thinking as a paradigm presents a new point of view that demonstrates that each listed cause appears in a circular process, instead of a linear and unidirectional flow of relations among the consequences (effects) and the other listed causes, called feedback loops (Folledo 2000; Patten et al. 2002). According to Richmond (1993), this change to a systemic view represents a change in perspective, from the view of a set of static relations of stimulus response to the view of an ongoing and self-sustainable dynamic process. In the second view, variables no longer have a fixed weight but present dynamic feedbacks that can lead to distinct temporal dynamics. Systemic thinking is a perspective of the whole, the parts, and their relations, that is, an iteration among structure, function, and process (Ing 2013).

By opposing the juxtaposition of parts and an analysis based on the parts and their correlation, the systemic view challenges the analytical view that regression models present an operational model of how a system works (Gallagher and Appenzeller 1999; Gattie et al. 2007). The systemic view is capable of understanding how the system can adapt to changes in the environment and shows that the system itself can be a functional part of an even larger system (Hieronymi 2013). This idea of an operational integrated model is presented here as a learning process (scientific literacy, in the context of a systemic view), and the operational models chosen are based on concept maps (Novak and Canas 2008).

Scientific literacy refers to a scientific approach to acquiring knowledge and developing a creative way of thinking about the world around us (Miller 1983; Hurd 1998; Laugksch 2000). For the American Association for the Advancement of Science, scientific literacy, while promoting attitudes, values, and skills, corresponds to the establishment of a mental habitude that helps people deal with matters involving evidence, quantitative considerations, logical arguments, and uncertainties in their daily life (AAAS 1990). Scientific literacy is considered a mental habitude because, in addition to providing ways to solve specific problems, it corresponds to the development of a metacognition of conservation and revolution, an attitude of constant openness to what is new, and permanent questioning of what is believed to be crystallized, from the subject competence to meta-competence (Gräber 2000).

Scientific literacy seeks to justify itself through social value while recognizing a number of forces that mobilize and transform life in a society. It concerns not just knowing science but also developing a scientific attitude based on rational and collective choices (Holbrook and Rannikmae 2009). Maurice Bazin, inspired by *Science for the People* movement of early 1970s, featured a reconceptualization of scientific literacy: he called it technical and he aligned the (scientific) literacy attribute with Paulo Freire's view in *Pedagogy of the Oppressed* to develop the power of intellectual analysis in the oppressed in such a way as to awaken political awareness (Bazin 1977). This new approach ascribes scientific literacy to a political plan as an emancipatory quality (Saito 2013; Saito and Bastos 2018).

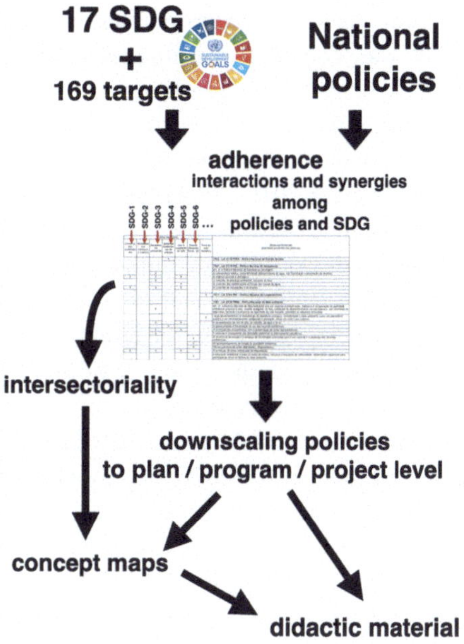

Fig. 2.1 Synthesis of the methodology processes. Source: Authors

3 Materials and Methods

The development of the methodology for the integrative analysis of policies and SDGs was structured into two steps: (i) analyzing the level of adherence of national policies to the SDGs and (ii) developing the educational potential of the analysis results by elaborating communication materials for mobilizing public and collective actions for the 2030 Agenda (Fig. 2.1). The analysis initially consisted of extensive and qualitative data and national policies and laws from 1962 to 2019 were considered accordingly.

Once the policies were identified, their interactions and synergies with the SDGs were analyzed. This analysis was supported by an interaction matrix, arranging the national policies in the rows and the SDGs in the columns. Interaction matrices, which originated in the work of Leopold et al. (1971), have been increasingly used in socioecological studies and literacy, such as Saito (2013), Thalmeinerova et al. (2017), and Zamignan (2018), among others. Basically, a cell with its position (row A, column C) will indicate a relationship between the two variables (expressed in row and column) and its content, the type of interaction.

The national policy-SDG interactions were analyzed based on the principles and objectives of each national policy and the objectives of each SDG. The focus on principles and objectives of each national policy and objectives of each SDG was to identify the "core" concepts on which interactions could be built.

This is especially important in regard to socio-ecological systems in contexts of institutional, economic, social, and environmental changes (Ebbesson 2010) such as the current Brazilian situation, which grants a safety margin for the diagnosis of the policies' adherence levels to the SDGs.

The second step of this study was based on the diagnosis of the interactions between SDGs and the adherence of policies to the goals. National policies may be implemented following a decreasing hierarchical order through plans and/or programs and/or projects. Thus, for the inventory of plans, programs, policies, and projects, the Pluriannual Plans (PPA, acronym in Portuguese) in Brazil were considered the main data source. PPAs are instruments established in the country's Federal Constitution to organize and plan policy implementation (Brazil 1988). The implementation is operated through initiatives, and each initiative corresponds to a specific action (CNODS 2017). After a comprehensive evaluation, a list of current PPA national policies ranging from 2016 to 2019 was compiled encompassing different themes and fields of action by the state.

A downscaling policy hierarchy was performed, and the interactions among programs and projects in the 2016–2019 PPA (concluded in 2015 for planning the following years) and SDGs were analyzed. At this step, the educational potential of the integrative analysis was also discussed to develop materials for communication and mobilization regarding the role of the SDGs and the associated national policies.

The objective was to demonstrate the Brazilian situation in light of the SDG requirements and the consequent challenges to be faced, and also present the recommendations regarding filling the gaps. The educational material was based mainly on the interaction matrix of national policies and the SDGs, and it was produced to demonstrate these interactions, describing the actions and their conceptual foundation through concept maps. A separate concept map was built for each program and project.

CmapTools, a free software specific to building concept maps, was used for this research. The software was developed at the Institute for Human & Machine Cognition (IHMC) in Florida, USA. It provides an environment to draw concept maps and allows the user to standardize connectors, edit the spatial and graphic distribution of the concepts, and add submaps within concepts/elements of a larger concept map; that is, it allows the user to work on different scalar levels (Novak and Canas 2008). The educational material was founded on scientific literacy mediated by the analysis of local scale situations, considering the problem-solution (or conflict-action) concept as a guiding axis for public and collective actions (Freire 1970; Santos et al. 2005; Saito 2013). The idea of scientific literacy based on a systemic vision facilitated by the conceptual maps was adopted as a didactic approach to the understanding of different levels of interdependence between the SDGs and national policies. This approach also informed choices regarding public and collective actions, which in turn supported the development of empirical and scientific knowledge (Miller 1983; Gräber 2000; Laugksch 2000).

It is important to highlight that the problem-solution (conflict-action) concept is anchored on the recommendations of the Intergovernmental Conference on Environmental Education, organized by UNESCO in collaboration with UNEP (UNESCO 1978).

4 Presentation and Discussion of Results

4.1 Analyzing the Adherence Level of National Policies to the SDGs

All national policies have at least one connection to an SDG. There are also national policies connected to up to eight SDGs. An alternative view showed that SDG 16 was the most connected to national policies (21 connections), followed by SDG 10 (17 connections), offering an overview of the main historical agenda (Fig. 2.2).

In addition to the analysis at the macro level of national policies, the interactions between the operationalization of these national policies through programs, projects, and initiatives in relation to SDGs were also analyzed by us. The results show that all national policies have some adherence to some SDGs (Table 2.1).

The national policies that most showed adherence to the SDGs were the National Environmental Education Policy (Brazil 1999), presenting interaction with seven SDGs; the National Policy for Sustainable Development of Traditional Peoples and Communities (Brazil 2007), which has interactions with six SDGs; the National Policy for Homeless People (Brazil 2009); and the National Policy on Agroecology and Organic Production (Brazil 2012), interacting with five SDGs.

The motivation to analyze these interactions in the sphere of public policies is in line with the public action policies in Brazil.

Nevertheless at this national policy level, it is necessary to know how they work at the downscale level of programs and projects where budget and action are associated to make the policy happen in a country's life and how they are connected to SDGs.

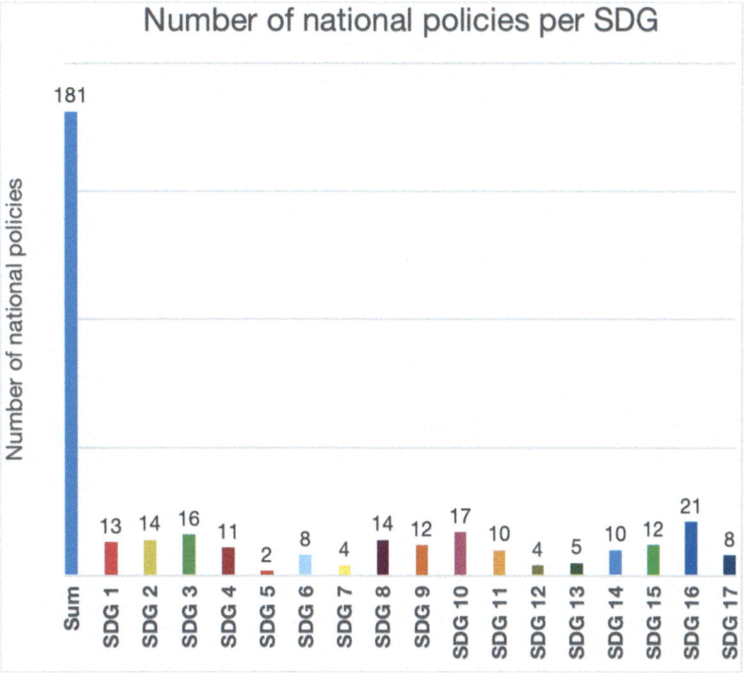

Fig. 2.2 Number of national policies related to each SDG. Source: Authors

A previous work was done at the government level, trying to identify alignment between some of the features in the PPA and the SDG attributes. The results of this effort are described in the Annual Monitoring Report of the SDG Agendas in the 2016–2019 PPA (SIOP 2018).

According to this government report (SIOP 2018), after analyzing 54 PPA programs, 303 objectives in the programs, 1132 targets, and 3094 PPA initiatives, all 2016–2019 PPA programs have attributes related to some SDGs, which matches our findings in this study showing that all national policies adhere to some SDGs (Table 2.1). This convergence of results is fully justified, considering that PPA programs derive from and are inspired by national policy guidelines.

When analyzing the alignment in the opposite direction of SDG targets for PPA programs, and considering all 169 SDG targets, the SIOP (2018) report highlights that 95% of SDG targets have some PPA attributes related to their implementation. This is also in line with this study, which identified that all SDGs are present to some level in national policies (Table 2.1).

We updated SIOP (2018) analysis until 2019 because that one was based on 2017 data and there have been several changes in government policy at the federal level that have directly affected adherence to the SDGs since that. The updates revealed the initiatives with the highest number of simultaneous interactions with different SDGs (Table 2.2). These listed initiatives, far from exhausting existing initiatives in Brazilian politics, play a prominent role in the Brazilian scenario and represent the best evidence of the country's capacity and effort to advance the SDGs. Furthermore, the fact that many of these initiatives started long before the SDGs were defined shows the country's concern with the themes and the early adjustment of its national policy. In brief, the data in Table 2.2 show the alignment of Brazilian politics with the global trends and priorities that are the foundations of the SDGs.

The data in Table 2.2 show that, like national policies, the programs and projects are actions to implement policies (thus, a downscaled level) capable of working in favor of the SDGs and, taken as a whole, are comprehensive where the 17 SDGs are concerned. However, it should be noted that these policies involve downscaled actions related to a limited number of SDGs.

Table 2.1 Interaction of national policies with SDGs. Source: Authors

Year	Policy name	SDG																	Total
		1	2	3	4	5	6	7	8	9	10	11	12	13	14	15	16	17	
1962	Política Nacional de Energia Nuclear							X											1
1967	Política Nacional de Saneamento						X					X							2
1971	Política Nacional de Cooperativismo		X						X		X								3
1981	Política Nacional do Meio Ambiente						X							X	X	X			4
1984	Política Nacional de Informática									X									1
1989	Política Nacional para a Integração da Pessoa Portadora de Deficiência				X						X								2
1991	Política Nacional de Arquivos Públicos e Privados																X		1
1992	Política Nacional de Salários	X							X		X								3
1994	Política Marítima Nacional														X				1
1994	Política National du Idoso			X							X								2
1997	Política Nacional de Recursos Hídricos						X								X	X			3
1997	Política Energética Nacional							X						X					2
1999	Política Nacional de Educação Ambiental				X		X					X	X	X	X	X			7
2001	Estatudo da Cidade											X							1
2001	Política National de Conservação e Uso Racional de Energia							X					X	X					3
2002	Diretrizes e Princípios da Política Nacional da Biodiversidade														X	X			2
2003	Política Nacional de Promoção da Igualdade Racial	X									X						X		3
2003	Política Nacional do Livro				X														1
2005	Política Nacional de Biossegurança		X													X			2
2005	Política Nacional para os Recursos do Mar		X												X				2
2006	Política Nacional da Agricultra Familiar e Empreendimentos Familiares Rurais	X	X								X								3
2006	Política Nacional de Enfrentamento ao Tráfico de Pessoas					X					X						X	X	4
2006	Política Nacional de Plantas Medicinais e Fitoterápicos			X	X						X								3
2007	Política Nacional de Proteção aos Defensores dos Direitos Humanos																X	X	2
2007	Política Nacional sobre o Álcool			X													X		2
2007	Política Nacional de Desenvolvimento Sustentável dos Povos e Comunidades Tradicionais	X		X							X				X	X	X		6
2007	Diretrizes para Política Federal de Saneamento Básico						X					X							2
2008	Política Nacional de Turismo														X	X			2
2009	Política Nacional de Aviação Civil								X	X									2

Year	Policy name	SDG																	Total
		1	2	3	4	5	6	7	8	9	10	11	12	13	14	15	16	17	
2009	Política Nacional de Desenvolvimento Sustentável da Aquicultura e da Pesca	X	X						X						X				4
2009	Política Nacional sobre Mudança do Clima													X				X	2
2009	Política Nacional para a População em Situação de Rua	X	X	X							X						X		5
2010	Política Nacional de Segurança Alimentar e Nutricional		X	X							X								3
2010	Política Nacional de Assistêencia Técnica e Extensão Rural para a Agricultura Familiarr e Reforma Agrária	X	X						X		X								4
2010	Política Nacional de Resíduos Sólidos									X		X	X						3
2010	Política Nacional de Segurança de Barragens						X												1
2011	Política Nacional de Incentivo ao Manejo Sustentado e ao Cultivo do Bambu									X						X			2
2011	Política Nacional de Segurança e Saúde no Trabalho			X					X	X									3
		SDG																	
2012	Política Nacional de Agroecologia e Produção Orgânica	X	X						X						X	X			5
2012	Política Nacional de Gestão Territorial e Ambiental de Terras Indigenas	X		X												X	X		4
2012	Política Nacional de Mobilidade Urbana										X	X							2
2012	Política Nacional de Proteção e Defesa Civil						X					X							2
2012	Política Nacional de Proteção dos Direitos da Pessoa com Transforno do Espectro Autista			X							X								2
2013	Política Nacional de Irrigação		X				X												2
2013	Política Nacional de Integração Lavoura-Pecuária-Floresta		X													X			2
2013	Política Nacional para os Trabalhadores Rurais Empregados	X							X										2
2014	Política Nacional de Atenção às Mulheres em Situação de Privação de Liberdade e Egressas do Sistema Prisional			X		X					X						X		4
2014	Política Nacional de Promoção da Saúde			X							X								2
2014	Política Nacional de Cultura Viva				X												X		2
2015	Estatuto da Metrópole											X							1
2015	Política de Educação para o Consumo Sustentável				X								X						2
2015	Política Nacional de Comabte à Desertificação e Mitigação dos Efeitos da Seca	X	X						X			X							4
2015	Política Nacional de Informação e Informática em Saúde			X															1

(continued)

Table 2.1 (continued)

Year	Policy name	SDG																	Total
		1	2	3	4	5	6	7	8	9	10	11	12	13	14	15	16	17	
2015	Política Nacional de Repressão ao Furto e Roubo de Veículos e Cargas									X							X		2
2016	Política Nacional de Formação dos Profissionais da Educação				X														1
2016	Política Nacional de Inteligência																X	X	2
2017	Política Nacional de Biocombustíveis							X											1
2017	Política Nacional de Inovação Tecnológica na Saúde			X															1
2017	Política Nacional de Recuperação da Vegetação Nativa															X			1
2017	Law de Migração			X	X				X	X	X	X					X	X	8
2018	Política Nacional de Avaliação e Exames da Educação Básica				X														1
2018	Política Nacional de Exportaçã e Importação de Produtos de Defesa																X	X	2
2018	Política Nacional de Incentivo à Produção de Cacau de Qualidade	X	X						X	X									4
2018	Política Nacional de Lawtura e Escrita				X														1
2018	Política Nacional de Pisos Mínimos do Transporte Rodoviário de Cargas								X	X									2
2018	Política Nacional de Segurança da Informação																X		1
2018	Política Nacional de Segurança de Infraestruturas Críticas									X							X	X	3
2018	Política Nacional de Segurança Pública e Defesa Social																X		1
2018	Política Nacional de Trabalho no âmbito do Sistema Prisonal								X								X		2
2019	Política Nacional de Alfabetização				X														1
2019	Política Nacionalde Busca de Pessoas Desaparecidas																X	X	2
2019	Política Nacional sobre Drogas			X													X		2
2019	Política Nacional de Desenvolvimento Regional	X															X		2
2019	Política Nacional da Erva-Mate								X	X									2
2019	Política Nacional de Incentivo à Ovinocaprinocultura		X							X									2
2019	Política Nacional de Prevenção da Automutilação e do Suicídio			X															1
Total		8	7	10	8	2	3	2	9	8	6	6	2	1	2	5	15	6	82

Table 2.2 National programs linked to the implementation of the SDGs. Source: Authors

Year of creation	National program	SDG																	T
		1	2	3	4	5	6	7	8	9	10	11	12	13	14	15	16	17	
1993	Benefício de Prestação Continuada	X	X	X							X								4
2003	Program a Luz para Todos	X	X	X	X		X	X			X							X	8
2004	Programa Bolsa Familia PBF	X	X	X	X	X			X		X			X					8
2005	Programa Nacional de Inclusão de Jovens (Projovem)	X	X		X	X			X		X	X						X	8
2005	Piano Nacional de Implementação da Convenção de Estocolmo sobre Poluentes Orgânicos Persistentes (NIP-POPs)		X	X			X			X			X		X	X		X	8
2005	Programa Universidade para Todos (Prouni)	X			X				X	X	X								5
2008	Fundo Amazônia						X							X		X	X	X	5
2011	Programa Bolsa Verde (PBV)	X	X		X				X		X		X			X			7
2011	Programa de Fomento às Atividades Produtivas Rurals	X	X			X			X		X			X					6
2011	Programa Nacional de Acesso ao Ensino Técnico e Emprego (Pronatec)	X			X				X	X	X						X		6
2013	Programa Cisternas	X	X	X		X	X				X			X			X		8
2013	Projeto Opções de Mitigação de Emissões de Gases de Efeito Estufa em Setores-Chave do Brasil		X					X	X	X		X		X		X		X	8
2013	Programa Nacional de Apoio ao Associativismo e ao Cooperativismo Social - Pronacoop Social	X							X		X		X				X		5
2016	Programa Criança Feliz (PCF)	X		X	X	X					X								5
2017	Rede Brasil Mulher			X	X	X			X		X						X		6

An interesting and methodologically divergent aspect between our study and SIOP (2018) is the grouping or macro-categorization procedure: the latter study started from the classic division of four sustainability dimensions (social, economic, environmental, and institutional), adopting an ex ante categorization.

In contrast, the present study proceeded to an ex post grouping, that is, without a previously formed categorization. As the analysis evolved, five macro-categories (or axes) were generated: *axis 1 for rights* (access, equality/equity, empowerment); *axis 2 for risks* (security, vulnerability, risk, and disaster management); *axis 3 for the market* (growth in terms of the market, development, productivity/technological efficiency); *axis 4 for policies* (public policies and management, social protection/services, global governance); and *axis 5 for the environment* (resilience/protection, ecosystems, management of recruits).

It is possible to establish a correspondence between axis 1 for rights with the social dimension of sustainability, axis 3 for the market with the economic dimension of sustainability, axis 5 for the environment with the environmental dimension of sustainability, and axis 4 for public policies with the institutional dimension of sustainability. The emergence of axis 2 for risks is noteworthy given that it has gained attention years after the concept of sustainability was formulated and is present in contemporary debates on social policies and vulnerability (Beck 2006). In the same way, this debate leads to the incorporation of the theme in the world agenda through the Hyogo Framework for Action: Building the Resilience of Nations and Communities to Disasters (2005–2015) (UNISDR 2005) and Sendai Framework for Disaster Risk Reduction (2015–2025) (UNISDR 2015). In fact, specifically on the effectiveness of the implementation of the Hyogo Framework for Action in Brazil, one can consult Branco and Saito (2017).

Another aspect that must be emphasized is the fact that these initiatives can suffer a severe setback due to changes in the current national government's way of conducting public policies. This is primarily due to the extinction of the National Social Participation Policy, instituted by Decree 8243 on May 23, 2014, and revoked by Decree 9759 on April 11, 2019. Nevertheless, it should be noted that the same decree also led to the end of the activities of the National Commission for the SDGs (CNODS) and its Thematic Chamber on Partnerships and Means of Implementation (CTPMI) on June 28, 2019. Three more Thematic Chambers were created but were never implemented due to a lack of constitutive membership.

Our analysis showed that the most promising initiatives with regard to meeting the largest number of SDGs correspond to the Cisterns Program (*Programa Cisternas*), Family Grant Program (*Programa Bolsa Família*), National Youth Inclusion Program (*Projovem*), and Light for All Program (*Programa Luz para Todos*), all in connection with eight different SDGs. It can be said that not only these initiatives but also others that have relationships with multiple SDGs in general operate in the axis of rights, which corresponds to the social pillar of sustainability. This means that the Brazilian Government was in line with the central SDG guideline, based on a social commitment to "leave no one behind."

Table 2.2 also presents the National Implementation Plan for the Stockholm Convention on Pollutants Organic Persistent (*Plano Nacional de Implementação da Convenção de Estocolmo sobre Poluentes Orgânicos Persistentes-NIP-POPs*) and the Greenhouse Gas Emissions Mitigation Options Project in Key Sectors of Brazil, connected with eight different SDGs, followed by the Green Grant Program (*Programa Bolsa Verde*), which has seven connections to SDGs. These initiatives' importance is recognized largely thanks to the government's monitoring of SDG implementation in its integrated planning and budgeting system (SIOP 2018), which noted repeated indications of the contribution of these programs to different SDGs.

In addition, the legal instruments for creating these initiatives are clear in their objectives and principles regarding the links with the SDGs. It has been observed that not only these initiatives but also others that have relationships with multiple SDGs generally operate along the axis of rights, which corresponds with the social pillar of sustainability. This means that the Brazilian state was in line with the central SDG guidelines, the social commitments to "leave no one behind" in each nation and to reduce global inequalities among nations, with more attention to the most vulnerable countries and peoples. The SDG with the greatest coverage by government programs and actions is SDG 10, followed by SDG 1.

What is also highlighted are those initiatives that are strongly linked to the maintenance of ecosystem services and the effort to avoid disruptions in the self-regulatory capacity of natural systems. This category includes the National Plan for the Implementation of the Stockholm Convention on Persistent Organic Pollutants (NIP-POPs) and the Project Options for Mitigation of Greenhouse Gas Emissions in Key Sectors of Brazil, as well as the Green Grant Program.

It is important to remember that in 2016, Brazil ratified the climate agreement signed in 2015 during COP21 and committed to reducing greenhouse gas emissions by 37%, to below 2005 levels, by 2025 in accordance with the Intended Nationally Determined Contributions (INDCs). A set of mechanisms were established to help achieve this, such as the ongoing Project Options for Mitigation of Greenhouse Gas Emissions in Key Sectors of Brazil and the National Plan for Adaptation to Climate Change (PNA) launched in 2016.

Thus, while there is concern regarding the end of activities of the National Commission for the SDGs (CNODS), it is imperative that the emerging initiatives that are most integrative and comprehensive regarding the themes of the SDGs be maintained and strengthened so that Brazil can meet the targets it is committed to. The importance of maintaining programs and projects with the greatest impact on the implementation of the SDGs will be addressed in the next section, which proposes to present in a didactic manner the importance of these initiatives at the national level.

4.2 Developing Educational Material About the Relationship Between Government Actions and the SDGs

As previously discussed, national policies in Brazil are implemented through plans, programs, and projects and are put into practice through initiatives. Every initiative corresponds to an action. Thus, to strengthen the understanding of relations between the most connected national policies and local SDGs at the level of initiatives, a didactic treatment was applied with the development of a qualitative model to communicate the chain of interdependencies and explain why a given policy is connected to all the mentioned SDGs.

The qualitative model is founded on scientific literacy, which is in turn grounded in a systemic view that can be facilitated by concept maps.

Based on these referential frameworks, a set of 18 cases of socioecological conflict (original problem) and the respective positive action (the proposed solution) was elaborated with, for each case, the support of photos and a description of the problem and the solution based on a concept map, as shown in Fig. 2.3, which presents one of the programs most connected to SDGs.

For each governmental initiative, the respective SDGs are highlighted, as is the justification for establishing the link. Each initiative is presented as a solution to a previously detected problem that was the main motivator for the emergence of the initiative. The underlying legal instrument is identified, and the justification for establishing connections with the SDGs is presented in the form of a concept map and explanatory text. The text in the box describes the solution to the presented problem and corresponds to the graphical description presented in the concept map. The concept maps allow the development of systemic reasoning about socioecological processes (Saito 2016, 2017) so that each program can be understood as an ongoing and dynamic complex process (Richmond 1993; Monat and Gannon 2015). This strategy converges with the idea presented by Norström et al. (2014) of embracing the perspective of an integrated socioecological system to acknowledge the key dynamics of such systems, including the multiple cross-interactions.

In this didactic material, the cross-interactions and the chain of interdependence are the core issues of the systemic view. By this way, we would like to promote literacy, and thus we doubly reinforce the presentation, both graphically and textually.

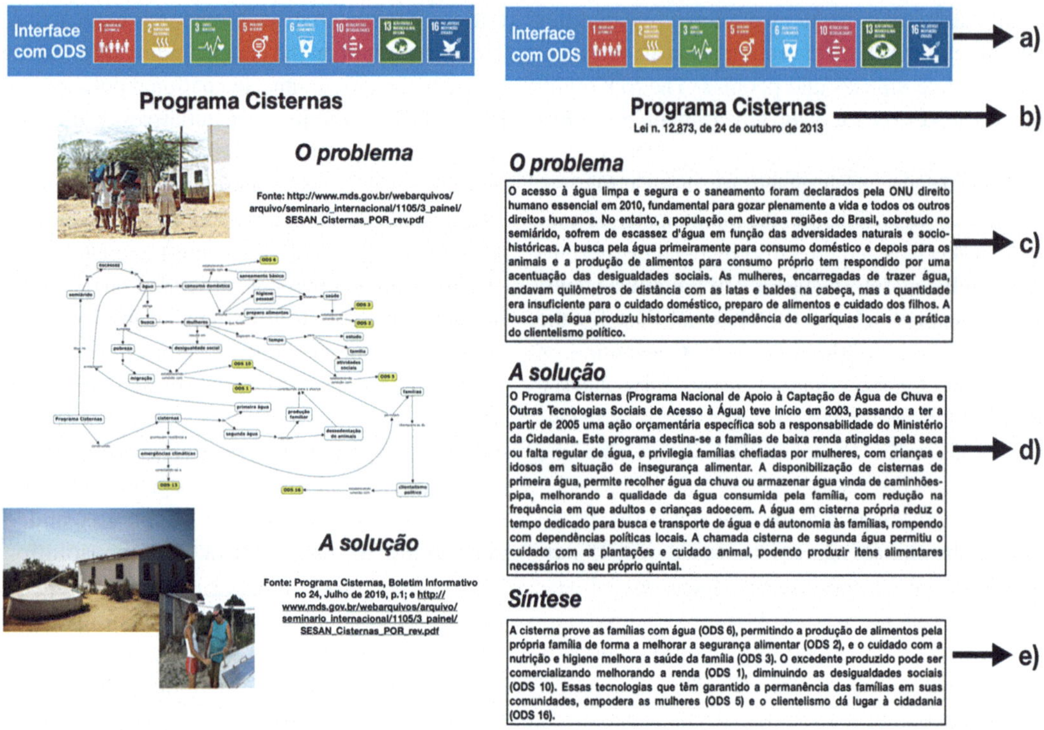

Fig. 2.3 Example of didactic material developed for the Cisterns Program and the structure of the portrait with transcripts. Source: Authors

Figure 2.3 can exemplify this didactic pattern through the portrait structure for the Cisterns Program (Brazil 2013): (a) Number of related SDGs. (b) Policy identification: The Cisterns Program, Law N° 12873 of 24 October 2013. (c) The problem: Access to clean and safe water and sanitation was declared by the UN to be an essential human right in 2010, fundamental to fully enjoying life and all other human rights. However, the population in several regions of Brazil, especially in the semiarid region, suffers from water scarcity due to natural and sociohistorical adversities. The search for water first for domestic consumption and then for animals and the production of food for own consumption have led to an increase in social inequalities. Women, charged with bringing water, walk miles away with cans and buckets on their heads, but the amount of water is insufficient for household care, food preparation, and child care. The search for water has historically produced dependence on local oligarchies and the practice of political clientelism. (d) The solution: The Cisterns Program (National Program to Support the Collection of Rainwater and Other Social Technologies for Access to Water) started in 2003, and from 2005 onwards, the program has represented a specific budgetary action under the responsibility of the Ministry of Citizenship. This program is aimed at low-income families affected by drought or regular lack of water and privileges families headed by women, with children and the elderly in a situation of food insecurity. The availability of cisterns of "first water" allows families to collect rainwater or store water from water trucks, which improves the quality of the water consumed by the family and reduces the frequency of illness among adults and children. Water in families' own cisterns reduces the time spent searching for and transporting water and gives families autonomy, breaking with local political dependencies. The so-called "second water" cistern allows food cultivation and animal care, enabling families to produce necessary food items in their own backyard. (e) Synthesis: The cistern provides families with water (SDG 6), allowing the family to produce food to improve food security (SDG 2), and the care for nutrition and hygiene improves the family's health (SDG 3). The surplus produced can be sold to improve income (SDG 1) and reduce social inequalities (SDG 10). These technologies that have enabled families to remain in their communities empower women (SDG 5), and political clientelism gives way to citizenship (SDG 16).

The overall purpose of this educational material is to reframe the debates on SDGs in a broader sense and go beyond the application of the indicators as an end point in itself (Mair et al. 2018) to think about SDGs' influence at local scale. At this scale, local people should develop full awareness and systemic view is capable to promote the understanding of how government policies and local initiatives can help achieving SDGs. Internalization and territorialization will be supported by the hierarchical understanding of nested system as part of an even larger system (Hieronymi 2013). Participatory processes will be benefited by the development of a creative way of thinking due to systemic view (Miller 1983; Hurd 1998; Laugksch 2000). Finally, systemic thinking can help develop rationality in social practices when combined with sociopolitical commitment in an emancipatory context (Bazin 1977; Berlinck and Saito 2010). The synthetic view of the material was presented in a language accessible to different social actors at different scales (from the national to the local levels), including suggestions and directions for its use as an instrument for mobilization and communication and as a progress-tracking tool.

5 Conclusions

The qualitative modeling adopted for analyzing SDG-public policy interactions and the development of the instructional material proved to be adequate to the proposed objectives. The improvement of the model should occur through transdisciplinary dialogues with both public and social institutions, in particular with decision makers directly working on the theme of integration and coherence between SDGs and public policies.

The national policies highlighted in association with the SDGs present a promising path in the adoption of integrated approaches to improve implementation results. Presenting cases by conflict situation and positive action, supported by scientific literacy and grounded in a systemic view facilitated by the concept maps, sheds light on local practical experiences related to the process of meeting global commitments. As didactic material, it can promote scientific literacy to resolve socioecological conflicts, and promote more equality conditions for the local and the community-based contributions, improving the 2030 Agenda as an inducer of participatory and democratic processes, with bottom-up and top-down decisions.

Thus, in collaborative settings, planning and managing public policies in accordance with social, economic, and environmental issues contribute to the implementation of more effective and sustainable solutions. The implementation challenges are complex, and the results reinforce the importance of maintaining and strengthening existing initiatives to enhance their contribution towards the achievement of the 2030 Agenda.

It has also been noted that national policy initiatives that are seen as priorities in terms of achieving SDGs and those that relate to multiple SDGs generally act within the framework of human rights, which corresponds to the social pillar of sustainability. This means that historically, different Brazilian Government mandates, in different periods of the country's history, have been in line with the SDG core guideline, based on a social commitment to "*leave no one behind*" and reduce global inequalities. This has been the "face" of the Brazilian state since the re-democratization process, with the exception of the present government.

This study hopes to contribute to the maintenance and strengthening of initiatives in the Brazilian context towards the global commitment to achieve the sustainable development goals. It is also expected to support discussions of how to implement and monitor SDGs at local scale, and how municipalities and local communities could pay attention to SDG' implementation and monitoring.

Acknowledgements The authors are grateful to INCT-ODISSEIA Project, "Observatory of Socio-environmental Dynamics," at Brasília University (UnB) for their technical support and to the National School of Public Administration (Escola Nacional de Administração Pública - Enap), Coordination for the Improvement of Higher Education Personnel (Capes), National Council for Scientific and Technological Development (CNPq), and Research Support Foundation of the Federal District (FAP-DF) for their financial support.

References

AAAS (1990). *Science for All Americans - Project 2061*. New York: Oxford University Press.

Bazin, M.J. (1977). O cientista como alfabetizador técnico. In S. Anderson, & M. Bazin (Eds.), *Ciência e In/Dependência, vol. 2* (pp. 94-98). Lisbon: Livros Horizonte.

Beck, U. (2006) Living in the world risk society. Economy and Society, 35(3), 329-345. https://doi.org/10.1080/03085140600844902.

Berlinck, C.N. & Saito, C.H. (2010). Action research for emancipation informed by Habermas and hierarchy of systems: Case study on environmental education and management of water resources in Brazil. *Systemic Practice Action Research, 23*(2), 143-156. https://doi.org/10.1007/s11213-009-9150-z

Branco, M.S.L.C., & Saito, C.H. (2017). Prevenção de desastres de origem hídrica no Brasil: avaliação do nível de implementação do Quadro de Ação de Hyogo. *Desenvolvimento e Meio Ambiente, 42*, 180-201. https://doi.org/10.5380/dma.v42i0.48754.

Brazil (1988). *Constitution of the Federative Republic of Brazil: Constitutional text of October 5, 1988, with the alterations introduced by Constitutional Amendments no. 1/1992 through 64/2010 and by Revision Constitutional Amendments no. 1/1994 through 6/1994*. Brasília: Chamber of Deputies.

Brazil (1999). *Law N° 9765, of April 27, 1999. Política Nacional de Educação Ambiental*. Brasília: Federative Republic of Brazil.

Brazil (2007). Decree N° 6040, of February 7, 2007. *Política Nacional de Desenvolvimento Sustentável dos Povos e Comunidades Tradicionais*. Brasília: Federative Republic of Brazil.

Brazil (2009). Decree N° 7053, of December 23, 2009. *Política Nacional para a População em Situação de Rua*. Brasília: Federative Republic of Brazil.

Brazil (2012). Decree N° 7794, of August 20, 2012. *Política Nacional de Agroecologia e Produção Orgânica*. Brasília: Federative Republic of Brazil.

Brazil (2013). Law N° 12873 of 24 October 2013. *Programa Nacional de Apoio à Captação de Água de Chuva e Outras Tecnologias Sociais de Acesso à Água - Programa Cisternas, entre outros*. Brasília: Federative Republic of Brazil.

CEPAL (2018). *Comisión Económica para América Latina y el Caribe. La ineficiencia de la desigualdade*. Santiago: CEPAL/ONU.

CNM (2017). *Guia para Integração dos ODS nos Municípios Brasileiros*. Brasília: Confederação Nacional de Município. https://www.cnm.org.br/cms/biblioteca/Guia_para_Integração_dos_ODS.2017.pdf. Accessed 01 April 2020.

CNODS (2017). *Comissão Nacional dos Objetivos do Desenvolvimento Sustentável – Plano de Ação 2017-2019*. Brasília: CNODS. http://www4.planalto.gov.br/ods/publicacoes/plano-de-acao-da-cnods-2017-2019. Accessed 01 April 2020.

Dedecca, C.S. (2014). O enfoque multidimensional da pobreza no Plano Brasil sem Miséria. *Cadernos de Estudos - Desenvolvimento Social em Debate, 19*, 22-35. https://aplicacoes.mds.gov.br/sagirmps/ferramentas/docs/cadernos%20de%20estudos_19_Final%20(1).pdf. Accessed 02 August 2020.

Ebbesson, J. (2010). The rule of law in governance of complex socio-ecological changes. *Global Environmental Change, 20*(3), 414-422. https://doi.org/10.1016/j.gloenvcha.2009.10.009.

Folledo, M. (2000) Raciocínio Sistêmico: uma boa forma de se pensar o meio ambiente. *Ambiente & Sociedade, 6-7*, 105-145. https://doi.org/10.1590/S1414-753X2000000100006.

Freire, P. (1970). *Pedagogy of the Oppressed*. 1st Edn. New York: Herder and Herder.

Fretwell, H. (2009). Environmental Education: the Science of Fear. *PercReports, 27*(3), 8-11. https://www.perc.org/wp-content/uploads/2018/01/PRfall_winter09.pdf. Accessed 01 April 2020.

Gallagher, R., & Appenzeller, T. (1999). Beyond reductionism. *Science 284*(5411), 79. https://doi.org/10.1126/science.284.5411.79

Gattie, D.K., Kellam, N.N., Turk, H.J. (2007). Informing ecological engineering through ecological network analysis, ecological modelling, and concepts of systems and engineering ecology. *Ecological Modelling, 208*(1), 25-40. https://doi.org/10.1016/j.ecolmodel.2007.04.027.

Gräber, W. (2000). Aiming for Scientific Literacy through self-regulated learning. In G. Stochel & I. Maciejowska (Eds.), *Interdisciplinary education - challenge of 21st century* (pp. 101–109). Krakow: FALL.

Griggs, D.J., Nilsson, M., Stevance, A., McCollum, D. (2017). *A Guide to SDG Interactions: from Science to Implementation*. Paris: International Council for Science (ICSU). https://council.science/wp-content/uploads/2017/05/SDGs-Guide-to-Interactions.pdf. Accessed 01 April 2020.

Hieronymi, A. (2013). Understanding Systems Science: A Visual and Integrative Approach. *Systems Research and Behavioral Science, 30*(5), 580-595. https://doi.org/10.1002/sres.2215.

Holbrook, J., & Rannikmae, M. (2009). The Meaning of Scientific Literacy. *International Journal of Environmental & Science Education, 4*(3), 275-288.

Hurd, P.D. (1998). Scientific literacy: New minds for a changing world. *Science Education, 82*(3), 407-416.. https://doi.org/10.1002/(SICI)1098-237X(199806)82:3%3C407::AID-SCE6%3E3.0.CO;2-G

Ing, D. (2013) Rethinking Systems Thinking: Learning and Coevolving with the World. *Systems Research and Behavioral Science, 30*(5), 527-547. https://doi.org/10.1002/sres.2229.

Laugksch, R.C. (2000). Scientific Literacy: A Conceptual Overview. *Science Education, 84*(1), 71-94. https://doi.org/10.1002/(SICI)1098-237X(200001)84:1%3C71::AID-SCE6%3E3.0.CO;2-C.

Le Blanc, D. (2015a). Towards Integration at Last? The Sustainable Development Goals as a Network of Targets. *Sustainable Development, 23*(3), 176-187. https://doi.org/10.1002/sd.1582.

Le Blanc, D. (2015b). *Towards Integration at Last? The Sustainable Development Goals as a Network of Targets. DESA Working Paper No. 141 ST/ESA/2015/DWP/141*. New York: United Nations Department of Economic and Social Affairs (DESA).

Leopold, L.B., Clarke, F.E., Hanshaw, B.B., Balsley, J.R. (1971). *A procedure for evaluating environmental impact.* Washington: US Geological Survey (Geological Survey Circular 645), https://doi.org/10.3133/cir645.

Liverman, D.M. (2018). Geographic perspectives on development goals: Constructive engagements and critical perspectives on the MDGs and the SDGs. *Dialogues in Human Geography, 8*(2), 168-185. https://doi.org/10.1177/2043820618780787

Lopes, J., Reis, R., Gaivizzo, L., Litre, G., Rodrigues-Filho, S., Saito, C.H. (2020). The Contribution of Community-Based Recycling Cooperatives to a Cluster of SDGs in Semi-arid Brazilian Peri-urban Settlements. In: G. Nhamo,. G. O. A. Odularu, V. Mjimba (Ed.). *Scaling up SDGs Implementation - Emerging Cases from State, Development and Private Sectors. Sustainable Development Goals Series*. 1ed (pp. 141-154). Johannesburg: Springer International Publishing. https://doi.org/10.1007/978-3-030-33216-7_10.

Mair, S., Jones, A., Ward, J., Christie, I., Druckman, A., Lyon, F. (2018). A Critical Review of the Role of Indicators in Implementing the Sustainable Development Goals. In: W. Leal Filho (Ed.) *Handbook of Sustainability Science and Research* (pp. 41-56). Berlin: Springer International Publishing. https://doi.org/10.1007/978-3-319-63007-6_3

Miller, J.D. (1983). Scientific Literacy: A Conceptual and Empirical Review. *Daedalus, 112*(2), 29-48.

Monat, J.P., & Gannon, T.F. (2015). What is systems thinking? A review of selected literature plus recommendations. *American Journal of Systems Science 4*(1), 11-26.

Nhamo G., Dube K., Chikodzi D. (2020) Sustainable Development Goals: Concept and Challenges of Global Development Goal Setting. In: R. Haring, I. Kickbusch, D. Ganten, M. Moeti (Ed.) *Handbook of Global Health*. Springer, Cham. https://doi.org/10.1007/978-3-030-05325-3_79-1.

Nhamo, G., & Mjimba, V. (2020). Scaling Up SDGs Implementation: Down the Road to Fast Approaching 2030. In: G. Nhamo,. G. O. A. Odularu, V. Mjimba (Ed.). *Scaling up SDGs Implementation - Emerging Cases from State, Development and Private Sectors. Sustainable Development Goals Series*. 1ed (pp. 3-19) Johannesburg: Springer International Publishing. https://doi.org/10.1007/978-3-030-33216-7_1.

Nilsson M, Griggs D, Visbeck M (2016) Map the interactions between sustainable development goals. *Nature, 534*, 320-322. https://doi.org/10.1038/534320a.

Norström, A.V., Dannenberg, A., McCarney, G., Milkoreit, M., Diekert, F., Engström, G., et al. (2014). Three necessary conditions for establishing effective Sustainable Development Goals in the Anthropocene. *Ecology and Society 19*(3), 8. https://doi.org/10.5751/ES-06602-190308.

Novak, J.D., & Canas, A.J. (2008). The theory underlying concept maps and how to construct and use them. *Technical Report IHMC Cmap Tools, 2006-01 Rev 01-2008*. Pensacola: Institute for Human and Machine Cognition. https://cmap.ihmc.us/docs/theory-of--concept-maps.php. Accessed 01 April 2020.

Patten, B.C., Fath, B.D., Choi, J.S., Bastianoni, S., Borrett, S.R., Brandt-Williams, S., et al. (2002) Complex Adaptive Hierarchical Systems In: R. Costanza, S.E. Jørgensen. *Understanding and Solving Environmental Problems in the 21st Century: Toward a New, Integrated Hard Problem Science* (pp. 41-94), Amsterdam: Elsevier Science. https://doi.org/10.1016/B978-008044111-5/50005-6.

Rasul, G., & Sharma, B. (2016). The nexus approach to water-energy-food security: an option for adaptation to climate change. *Climate Policy, 16*(6), 682-702. https://doi.org/10.1080/14693062.2015.1029865.

Richmond, B. (1993). Systems thinking: critical thinking skills for the 1990s and beyond. *System Dynamics Review 9*(2): 113-133. https://doi.org/10.1002/sdr.4260090203.

Rockström, J., Steffen, W., Noone, K., Persson, Å., Chapin, F.S., Lambin, E.F., et al. (2009). A safe operating space for humanity. *Nature, 461*(7263), 472-475. https://doi.org/10.1038/461472a.

Sachs, J.D. (2012). From Millennium Development Goals to sustainable development goals. *Lancet, 379*(9832), 2206-2211. https://doi.org/10.1016/S0140-6736(12)60685-0.

Saito, C.H. (2013). Environmental education and biodiversity concern: beyond the ecological literacy. *American Journal of Agricultural and Biological Sciences, 8*(1), 12-27. https://doi.org/10.3844/ajabssp.2013.12.27.

Saito, C.H. (2016). Concept Map for Environmental Education Planning: Capacitation of Volunteers for the FIFA Football World Cup in Brazil. *Journal of Education for Sustainable Development, 10*(2), 289-308. 10.1177%2F0973408216651944.

Saito, C.H. (2017). Concept Map to Visualize Opposite Perspectives of Rapa Nui History as a Whole.

Journal of Historical Archaeology & Anthropological Sciences, 1(5), 182-189. https://doi.org/10.15406/jhaas.2017.01.00029.
Saito, C.H., & Bastos, F.P. (2018). Tributo a Maurice Bazin: um farol da prática para a liberdade. *Alexandria, 11*(2), 279-305. https://doi.org/10.5007/1982-5153.2018v11n2p279.
Santos, I.A., Berlinck, C.N., Araujo, S.C.S., Steinke, E.T., Steinke, V.A., Pianta, T.F., et al. (2005). The Centrality of the mediation concept in the participatory management of water. *Canadian Journal of Environmental Education, Lakehead University, Ontario, 10*(1), 180-194. Available at: https://cjee.lakeheadu.ca/article/view/186/31.
SIOP (2018). *Agenda ODS no SIOP. Relatório Anual de Monitoramento (ano-base 2017)*. Brasília: Sistema Integrado de Planejamento e Orçamento. Ministério do Planejamento, Desenvolvimento e Gestão DF. https://www.gov.br/economia/pt-br/arquivos/planejamento/arquivos-e-imagens/secretarias/arquivo/spi-1/ppa-2016-2019/agendas-ods_volume-i-ods.pdf. Accessed 01 April 2020.
Stafford-Smith, M., Griggs, D., Gaffney, O., Ullah, F., Reyers, B., Kanie, N. et al. (2017) Integration: the key to implementing the Sustainable Development Goals. *Sustainability Science 12*(6), 911-919. https://doi.org/10.1007/s11625-016-0383-3.
Thalmeinerova, D., Fábrega, J.R., Guan, Y., Janusz-Pawletta, B., Kileshye-Onema, J.M., Saito, C.H. (2017). *IWRM Toolbox Teaching Manual*. Stockholm: Global Water Partnership, http://www.gwp.org/globalassets/global/toolbox/references/iwrm_teaching_manual.pdf. Accessed 01 April 2020.
Tosun, J., & Leininger, J. (2017). Governing the Interlinkages between the Sustainable Development Goals: Approaches to Attain Policy Integration *Global Challenges, 1*(9), 1700036. https://doi.org/10.1002/gch2.201700036.
UN (2014). *The road to dignity by 2030: Ending poverty, transforming all lives and protecting the planet: Synthesis report of the Secretary-General on the post-2015 sustainable development agenda.* New York: United Nations General Assembly A/69/700. 4 December 2014. https://www.un.org/disabilities/documents/reports/SG_Synthesis_Report_Road_to_Dignity_by_2030.pdf. Accessed 01 April 2020.
UN (2015). *Transforming our world: the 2030 Agenda for Sustainable Development.* New York: United Nations General Assembly A/RES/70/1. 25 September 2015.https://documents-dds-ny.un.org/doc/UNDOC/GEN/N15/291/89/PDF/N1529189.pdf?OpenElement. Accessed 01 April 2020.
UN (2018). *Working Together: Integration, institutions and the Sustainable Development Goals, World Public Sector Report 2018.* New York: United Nations, Department of Economic and Social Affairs. https://doi.org/10.3326/pse.42.4.5
UN-Water (2016). *Water and Sanitation Interlinkages across the 2030 Agenda for Sustainable Development.* Geneva: UN-Water. https://www.unwater.org/app/uploads/2016/08/Water-and-Sanitation-Interlinkages.pdf. Accessed 01 April 2020.
UNESCO (1978). *Intergovernmental Conference on Environmental Education - Tbilisi (USSR).* Paris: UNESCO. https://unesdoc.unesco.org/ark:/48223/pf0000032763. Accessed 01 April 2020.
UNISDR (2005). *Hyogo framework for action 2005-2015: Building the resilience of nations and communities to disasters.* Geneva: UNISDR. https://www.unisdr.org/2005/wcdr/intergover/official-doc/L-docs/Hyogo-framework-for-action-english.pdf.
UNISDR (2015). *Sendai Framework for Disaster Risk Reduction*. Geneve: UNISDR. https://www.unisdr.org/files/43291_sendaiframeworkfordrren.pdf.
Weitz, N., Carlsen, H., Nilsson, M., Skånberg, K. (2018) Towards systemic and contextual priority setting for implementing the 2030 Agenda. Sustainability Science, 13(2), 531-548. https://doi.org/10.1007/s11625-017-0470-0.
Weitz, N., Nilsson, M., & Davis, M. (2014). A Nexus Approach to the Post-2015 Agenda: Formulating Integrated Water, Energy, and Food SDGs. SAIS Review of International Affairs 34(2), 37-50. https://doi.org/10.1353/sais.2014.0022.
Waage, J., Yap, C., Bell, S., Levy, C., Mace, G., Pegram, T., et al. (2015) Governing the UN Sustainable Development Goals: interactions, infrastructures, and institutions *The Lancet Global Health, 3(*5): e251-e252. https://doi.org/10.1016/S2214-109X(15)70112-9.
Weymouth, R., & Hartz-Karp, J. (2018). Principles for Integrating the Implementation of the Sustainable Development Goals in Cities. *Urban Science, 2*(3), 77 https://doi.org/10.3390/urbansci2030077
Zamignan, G. (2018). *Gestão Integrada de Recursos Hídricos: Desenvolvendo Capacidades para a Construção de Visão Sistêmica sobre Gestão das Águas.* Brasília: Universidade de Brasília, Centro de Desenvolvimento Sustentável UnB.

Localising SDGs in Rural Uganda: Learning Active Citizenship Through the *Saemaul Undong* Model

3

Karembe F. Ahimbisibwe and Tiina Kontinen

Abstract

The Sustainable Development Goals (SDGs) are premised on the principles of 'leaving no one behind' and transformative development. Achieving the goals requires active citizens that are engaged in community development and claiming their rights. The chapter explores the ways in which a local NGO uses *Saemaul Undong* (SMU), a Korean community development model, to localise holistic achievement of a number of SDGs. Drawing on theories of the travel of global ideas in institutional sociology and based on participatory research including in-depth interviews, focus group discussions and participation in community activities, the chapter analyses how SMU's three pillars of self-help, diligence and cooperation were domesticated and translated in a local community in western Uganda. Findings show how the pillars were translated into practices of active citizenship such as hard work, responsibility and enhanced participation, which contributed to the improvement of livelihoods and to general efforts of local realisation of SDGs. The process successfully promoted active citizenship as community development. As a consequence, the chapter recommends that localisation of SDGs needs to emphasise the promotion of active citizenship to support their holistic achievement in the spirit of enhancing inclusive development.

Keywords

Active citizenship · Domestication · Localisation · SDGs · Saemaul Undong · Uganda

1 Introduction

The localisation of the Sustainable Development Goals (SDGs) to achieve transformative and inclusive development where 'no one is left behind' has been presented as the raison d'être of *Agenda 2030*. For instance, while the *Agenda* is described as being of 'unprecedented scope and significance for the entire world', the process of its making 'paid particular attention to the voices of the poorest and most vulnerable' (United Nations General Assembly 2015, p. 3) to highlight its localisation intent as 'an Agenda of the people, by the people and for the people' (United Nations General Assembly 2015, p. 12). Conceptualised as a process of 'taking deliberate efforts to make the aspirations of the SDGs become real to communities, households and

K. F. Ahimbisibwe (✉) · T. Kontinen
Department of Social Sciences and Philosophy, University of Jyväskylä, Jyväskylä, Finland
e-mail: tiina.t.kontinen@jyu.fi

G. Nhamo et al. (eds.), *Sustainable Development Goals for Society Vol. 1*, Sustainable Development Goals Series, https://doi.org/10.1007/978-3-030-70948-8_3

individuals, especially those who are at risk of falling further behind' (Steiner 2017), localisation requires multilevel stakeholders and local communities to participate in 'defining, implementing, and monitoring at the local level, strategies aimed at achieving global, national and sub-national goals and targets' (United Cities and Local Governments 2019, p. 16). Overall, citizens' participation in decision-making about their everyday lives is seen as a cornerstone in achieving the SDGs (Menon and Hartz-Karp 2019; Sriskandarajah 2018), especially in Africa where countries are gravely constrained with insufficient resources to implement national and global development agendas (Nhamo 2017).

However, different ideas on how to promote the necessary 'activeness' circulate in the international development discourses. On the one hand, models such as citizens' engagement (Gaventa and Barrett 2012) and social accountability (Hickey and King 2016; Sriskandarajah 2018) emphasise citizens' activeness in claiming their rights and keeping governments accountable. On the other hand, models of community development and empowerment emphasise local participation manifested in the active role of ordinary citizens to identify and address their own problems as well as help and learn from one another in mastering their shared destiny (Ibrahim 2006; Menon and Hartz-Karp 2019; Nhamo 2017).

Non-governmental organisations (NGOs) are often explicitly committed to the localisation of SDGs and supporting general community transformation through initiatives that promote active citizenship. Conceptualising active citizenship as 'constructed, learned and performed in practices taking place in communities involved in joint activities with an aim of taking care of shared issues' (Holma and Kontinen 2020a, p. 25), this chapter scrutinises a community-based development model, *Saemaul Undong* (SMU) being used by an NGO in rural Uganda to promote, domesticate and customise SDGs at a local level. The model, originating from South Korea (hereafter Korea), is widely acknowledged for heralding the rapid transformation of (mainly rural) Korea in the 1970s (Douglass 2013; Eom 2011; Park 2019; Yang 2017). SMU is anchored in three principles: self-help, diligence and cooperation (Park 2019; UNDP 2015; Yang 2017). Originally, it focused on challenging peasants to embrace change and break away from dependency, and to stop backward peasantry practices and features of 'hunger, poverty, idleness, gambling, alcohol drinking, low agricultural productivity, and lack of energy' (Iqbal and Milon 2017, p. 70). The model has been marketed and promoted by the Korea International Cooperation Agency (KOICA), Korean NGOs, global personalities such as the former UN Secretary-General Ban Ki-moon and other international organisations, as a possible panacea for reducing rural poverty in developing countries. Consequently, in several African countries, there are experimentations of the model being perceived to resonate with the *Agenda 2030* mission of inclusive and transformative development.

The Community Volunteer Initiative for Development (COVOID) was founded in 2003 as an indigenous grassroots NGO with a focus on empowering and strengthening the capacities of the community to support the rights and needs of children (COVOID 2016). The NGO is headquartered in Rubirizi district, in south-western Uganda (approximately 365 km from Kampala), and it mainly focuses on building community capacity to 'ensure that the child is very safe'[1]. Like many rural areas in sub-Saharan Africa, Rubirizi is largely agrarian with 79% of its 129,149 inhabitants eking out a living from subsistence agriculture (Republic of Uganda [RoU] 2017). However, unlike other districts in more excluded regions of the country, Rubirizi falls in the western region, which has been socially and politically stable with a comparably lower poverty index of 6.8% compared to the national average of 21.4% (RoU 2017). The district is also

[1] In an interview with the founder, who is also the executive director, he revealed that although the NGO keeps changing and responding to many global and national development forces, COVOID remains focused on ensuring the well-being of children, a vision he emphasised can only be realised if the general community, especially women, overcome poverty and are able to provide the basic necessities of life.

religiously homogeneous, with about 75% of the population belonging to the Roman Catholic denomination. The village of Nyakahama, perched on the outskirts of the sprawling Rubirizi town council, is a hilly community covered in leafy green banana plantations, a staple and cash crop in western Uganda. However, the village still shares some of the characteristics of rural communities, such as poverty, social exclusion, land shortage and fragmentation (RoU 2013, 2020). In an attempt to address these challenges at the grassroots level, in 2015 COVOID started to pilot the SMU model in order to localise the achievement of the following SDGs: 1. Poverty. 2. Hunger. 3. Health and well-being. 4. Inclusive education. 5. Gender equality. 6. Clean water and sanitation. 10. Reduced inequalities. 12. Responsible consumption and production (COVOID 2016). The use of this localisation model was inspired mainly through personal networks and random encounters, such as reading a local newspaper article about how SMU was transforming the village of Busanza in western Uganda. Later, on a trip to Bangkok, Thailand, the executive director met some Koreans who shared with him how the model transformed Korea from a poor country into a developed one (COVOID 2016).

Theoretically, the chapter draws on institutional sociology and its conceptualisation of the travel of global models as processes of translation (Czarniawska and Joerges 1996) and domestication (Alasuutari 2009, 2015; Alasuutari and Qadir 2013). These conceptualisations contend that instead of assuming that blueprint models designed in one location will be diffused and adopted in another, the analysis should pay attention on how they change during the travel. The focus should be on the ways in which models are adapted when perceived as responses to some existing societal needs and rhymed with the current processes in any particular location. Based on a qualitative case study in Nyakahama village in western Uganda, in this chapter we ask how SMU's three pillars of diligence, self-help and cooperation were translated and domesticated in the encounter between an NGO and a community in a particular case of localising SDGs. We examine the kinds of practices of active citizenship that emerged, and show how the model rhymed with the contextual needs of the local population and resulted in steps taken towards realisation of SDGs in the community. We further point out how the model and its pragmatic principles and values fitted well in the country's neoliberal narrative that challenges citizens to take responsibility of developing themselves and families out of poverty taking advantage of the prevailing peace and stability, rather than holding government accountable for services and citizenship rights (Makara 2020).

This chapter is organised in five sections. First, a literature review on SMU and the conceptualisations of translation and domestication of global ideas are provided. The second section describes the study context and methodology, while the third section presents empirical findings on how the three pillars of SMU were translated into actual practice in the community. The fourth and fifth sections, respectively, reflect on the tensions concerning active citizenship inherent in the SMU, and conclude by highlighting the importance of contextual analysis in engendering inclusive development that leaves no one behind.

2 Literature Review

Saemaul Undong (translated as 'New Community/Village Movement') was launched in the 1970s during the regime of Korean President Park Chung-hee as a rural development model to challenge peasants to embrace change and break away from dependency. The model, premised on the pillars of self-help, diligence and cooperation, was implemented through a 'carrot-and-stick' approach that combined elements of government support and villagers' self-help (Han 2012, p. 10) with competition, punishment, training and mindset change (Engel 2017; Odularu 2009). Increasingly, SMU is being presented by KOICA as an ideal model for 'participatory rural poverty alleviation and a blueprint for poverty alleviation with many developing countries expressing interest of replicating it in the hope that it will help them reproduce Korea's exceptional growth' (Jeong 2017, p. 160), and experimented by several international development agencies. For instance, the UNDP has developed a *Saemaul Initiative Towards Inclusive and Sustainable New Communities* as a guide to nations on how to localise and make the SDGs reach the poorest and most marginalised people who have the least resources and remain furthest behind (UNDP 2015).

In this chapter, the journey of the SMU model from 1970s Korea to a local community in contemporary Uganda is conceptualised by drawing on the institutional sociology of the global travel of models, ideas and policies, which argues that we should not perceive such travels as global diffusion or adoption of ready-made blueprint models. Rather, the global spread of ideas is understood as a process of translation (Czarniawska and Joerges 1996) and domestication (Alasuutari 2009, 2015; Alasuutari and Qadir 2013). In these accounts, nations and people are not passive adopters of ready-made models, but active creators who translate external, globally circulating ideas for their own use (Czarniawska and Joerges 1996) and domesticate global models through active reconfiguration to match local conditions and needs (Alasuutari 2009). Global ideas travel easily if they are perceived as responses to existing societal needs, if they are seen as a fashion to be followed, or if they are practiced by successful peers who should be imitated (Czarniawska and Joerges 1996). The promoters of SMU present Korea as a successful peer, a miracle of development to be mimicked, and the SMU model as a way to overcome livelihood challenges typically experienced by many rural communities. In the process of domestication (Alasuutari 2009), models are not just adopted, but are turned into actual practices embedded in certain local conditions, actors' own interests and already existing processes. The processes of translation, domestication or customisation result in the model gaining meanings that are different from the original blueprint, but more consistent with the particular community.

Consequently, we contend that SMU has increasingly gained popularity in global development because it challenges peasants to embrace change and break away from dependency through hard work and self-help while speaking to many contexts, including the African indigenous value systems (Avoseh 2001) and existing political contexts. According to Park (2019), Rwanda, South Africa, Uganda and Ethiopia are among the African countries that are most actively experimenting with the SMU model. However, in these countries, the model has mainly been promoted by Korean development actors, including Korean NGOs, as in the case of Rwanda (Nauta and Lee 2017). Thus, these experiments have remained largely scattered and, as our case illustrates, random and voluntary.

SMU formally entered Uganda in 2009 when two village projects were established in central districts of first Wakiso and later Mpigi with the collaboration of Korea *Saemaul Undong* Centre, KOICA and UNDP. The model was enthusiastically welcomed by the Ugandan state, perhaps partly due to its ideology that strengthened and rhymed with the narrative of what Buire and Staeheli (2017, p. 174) conceptualise as 'individualised, depoliticised and neoliberal subjects who work to enhance self-sufficiency' to meet their needs without pressuring governments for providing services. Moreover, Uganda's history of self-help spirit that is strongly etched into traditional community beliefs and practices (see Twesigye et al. 2019) further provided a fertile ground for SMU. Senior government officials visited Korea to benchmark the SMU model. For example, in May 2013, President Museveni visited the *Saemaul Undong* Centre in Sungham, and applauded the centre for the opportunity the model provided for Uganda to refocus on the country's lost communal traditions (Park 2019). In the same vein, in 2015 Vice President Edward Ssekandi told a convention organised by the Uganda *Saemaul Undong* Centre in Kampala that SMU's emphasis on diligence, self-help and cooperation was the 'best approach' to overcome Uganda's dependency syndrome problems (Park 2019, p. 331). While the politicians have seemingly been impressed by the transformative potential of SMU, the model has not been mainstreamed in the Ugandan Government policies and blueprints as an official national development approach. Instead, the *Saemaul Undong* Centre, KOICA and UNDP have been at the helm of promoting the SMU development model in Uganda with a broad goal of 'transforming communities with a long-term shared vision of a better life for all, and an infectious enthusiasm for local development, sustained by volunteerism at the community level' (Park 2019, p. 326). As our case shows, SMU villages continue to spring up in scattered and random ways in different locations of the country.

The notion of domestication as active reconfiguration and adaption of models to local conditions and needs (Alasuutari 2009) guides our analysis of SMU's promotion of active citizenship. The pillars of SMU emphasise self-reliant and collective citizenship manifested in joint efforts to address challenges related to, for instance, poverty. In bringing new kinds of practices under the banner of diligence, self-reliance and cooperation, the SMU model might potentially trigger learning and reformulation of citizenship practices (Holma et al. 2018; Holma and Kontinen 2020b). These new practices could provide a more enabling institutional environment for the emergence of disruptive innovations to significantly improve local livelihoods (Adegbile and Sarpong 2018), in accordance with SDG9 that focuses on industry and innovations.

Such promotion of active citizenship as community self-reliance centred around improving material livelihoods differs from the notion of active citizenship envisaged by the human rights-based approach common in civil society and donor discourses, which espouse ideas of good governance, accountability and democracy (Dagnino 2007; Gaventa and Barrett 2012; Gaynor 2011; Sriskandarajah 2018). For instance, in the context of the achievement of SDGs, Sriskandarajah (2018, p.1) calls for political bite and mounting pressure on governments through an 'accountability revolution' where 'citizens will hold governments accountable to the promises they made in 2015 … to deliver a more just and sustainable world by 2030'. In contrast, SMU emphasises hard-working and self-reliant citizens with a changed mindset and the mentality of 'we can do' (Doucette and Müller 2016), occasionally interspersed with religious-laced slogans such as 'God helps those who help themselves' (Jwa 2018, p. 197). In this vein, SMU is based on a particular idea of active citizenship, where rather than pressuring the state for solutions or paying attention to realisation of rights and democratisation (Jeon 2019), citizens are challenged through reward and punishment to actively engage in voluntary community activities, find solutions to their problems and collectively aspire for transformation. Therefore, it becomes interesting to explore how NGOs are applying the ideals of SMU to spur the community into mindsets and practices of self-reliance in the Ugandan neoliberal and hybrid regime that combines elements of freedom, authoritarianism, state withdrawal, patronage, elite corruption and stability (see Kalinaki 2020; Makara 2020; Mwenda and Tangri 2005; Tripp 2004).

3 Research Design and Methods

The research material was collected through a participatory methodology that combined in-depth interviews (n = 20), group discussions (n = 2) and active participation in everyday community activities and functions such as village meetings, communal work, funerals and religious ceremonies. The first author stayed in the community for an extended period of 3 months (from June to August 2019), and was therefore able to often take evening and weekend casual walks in the community and spontaneously engage in informal conversations with different categories of community members in, for example, bars, places of worship, homesteads, markets, restaurants and gardens. The use of the multiple methods was in line with the principles of participatory research, which emphasises the promotion of mutual learning, knowledge co-construction and empowering potential of the research process (Ahimbisibwe et al. 2020). The extended stay in the community enabled close observation, built close rapport with community members through informal discussion and helped in the triangulation of information with different methods. All of these actions contributed to the management of bias or, in the vocabulary of qualitative and participatory research, to the increase of this research's rigor and trustworthiness (Galdas 2017; Guba and Lincoln 2005). A qualitative thematic analysis with deductive orientation was conducted. SMU's three principles of self-help, diligence and cooperation were used as an analytical framework. The material was organised under these three themes particularly from the point of view of active citizenship practices, and a rich description of the content of each theme was produced.

4 Findings

In this section, we present the empirical findings, and show how SMU's principles of diligence, self-help and cooperation were domesticated and translated into specific citizenship practices in the interaction between COVOID and the community.

4.1 Diligence as Hard Work and Frugality

The domestication of the SMU model in resonance with the everyday life and experiences of the community could be seen in, among other things, the terminology in the Runyankore language used by the community. *Saemaul Undong* was domesticated to *Samwiri Odongo*[2] and the whole ideology became known as *Enkora y'aba Korea* ('Work ethos of Koreans'). The principle of diligence was translated as *okukora n'omutima*, which literally means 'putting your heart into what you are doing' or 'working tirelessly to achieve what ordinarily seems insurmountable'. COVOID used diligence to inculcate a work ethic that encouraged community members to work hard and practise frugality. The main message articulated in COVOID's training was that if residents did as SMU taught them to do, they would end poverty and develop as the Koreans did in the 1970s.

In addition, and perhaps in line with the implicit goal of engendering financial inclusion to achieve SDG 1 (no poverty), COVOID integrated the SMU model into the already existing practice and culture of communal saving, the Village Savings and Loan Associations (VSLAs). The NGO claimed to have trained community members in habits and practices of diligence such as frugality, joint planning and saving at the household level, target setting and healthy competition. These were in contrast to the existing practices of household conflicts and neglect, wastage, and reckless spending and jealousy.

> In the VSLAs, we have been training them to compete in savings not spending. Then after saving enough, you can compete positively. Now I can say that in the village and as you may have seen during your research visits, households are now competing to build a good house roofed with Sembule[3] iron sheets. They are moving away from the traditional white iron sheets that are associated with poverty. (Interview with executive director, August 16, 2019)

Thus, the integration of the SMU model with the existing joint community practices, such as an ingrained village culture of saving, further strengthened its domestication and created new momentum for local citizens to believe in their abilities to end their poverty. In different interactions, several community members revealed how SMU had taught them to be frugal and avoid a culture of reckless spending. Probed on what had changed, they responded: *'Tukaba turiira eryo … Hati titwakiriira eryo'*, (meaning: 'we used to spend [eat] money [recklessly] as if we would not live the next day, but that habit has stopped'). At an individual level, community members narrated how a switch from consumption of expen-

[2]When the SMU model was introduced, the locals could not easily pronounce the words *Saemaul Undong*, and instead localised it to Samwiri (Samuel) Odongo. Samuel, pronounced *Samwiri*, is a popular Christian/biblical name in Uganda, while *Odongo* is a popular name in Acholi, one of the dominant ethnic groups in northern Uganda.

[3]In Uganda, Sembule iron sheets are high-gauge and high-end pre-painted coloured roofing materials often used by rich people and, therefore, in many communities they are a mark of social class. This is in contrast to the low-gauge white iron sheets predominantly used by low-income earners.

sive fish to home-grown sources of sauce such as beans and green vegetables freed some money for saving in the village saving association. Taken together, the dietary change prompted by frugal habits inadvertently promoted the notions of good health, and responsible and sustainable consumption emphasised by SDGs 2 and 12, respectively. At the community level, the village meeting on 5 March 2017 resolved that each household was to make a monthly contribution of 2000 Uganda shillings (approx. €0.50) which would be deposited into the village account that the same meeting resolved to open in a saving and credit cooperative (SACCO) run by COVOID. This practice was evident during community meetings attended by the first author during fieldwork.

To encourage the emerging spirit of diligence, COVOID adopted SMU's dual-implementation strategy of 'motivation based on carrots and competition', along with training and fostering 'missionaries' (Odularu 2009, p. 156). For example, the NGO provided the community with basic farming tools and items such as wheelbarrows, pangas, tarpaulins and gumboots to use in joint community activities. It also periodically organised inter-household and intra-cluster competition on aspects such as the building of relatively better and permanent homes, sanitation, participation in community projects, projects generating household income and children's education. The winning households and clusters were further incentivised with additional rewards to boost the SMU spirit. At a national level, 'Nyakahama SMU village participated in the inter-SMU village competitions[4] held in 2016 at Kampiringisa SMU village in central Uganda, and emerged as the winner among the three new entrants and third overall of the eight villages that participated in the competition' (FGD with Nyakahama community members, 12 July 2019). For this performance, the village was later visited by officials from the Korean implementing agencies and UNDP and received an assortment of agricultural equipment and a huge signpost declaring Nyakahama village to be a '*Saemaul Undong* Ambassador'.

In the community, the strengthened work ethic was further evident in increased commitment and renewed voluntary participation in joint activities such as opening up and maintaining community roads, attending meetings and households' efforts to improve their material conditions. For instance, a random walk through the village revealed a noticeable trend in households shifting from old traditional mud-and-wattle houses to newly constructed, more modern permanent housing units, some of which were connected to solar energy and digital television. Also visible were households at different stages of preparedness to construct new houses, including cleared spaces and kilns in their compounds and the incremental purchase, accumulation and storage of building materials such as metal sheets and cement. In addition, most households maintained a backyard vegetable garden, a standard drying rack and some livestock, especially goats, pigs or chickens, as supplementary income for the family as a demonstration of the diligence demanded by the SMU model. Although not all the homesteads in the community were at the same level and quality, these local initiatives point to the village's journey and determination to meet SDGs related to equality (SDG 10), sanitation (SDG 6) and sustainable communities (SDG 11).

4.2 Self-Help as Responsible, Useful and Confident Community Membership

Translated as *okweyamba/okwekwatiramu* in the Runyankore language ('helping oneself' or 'getting personally involved'), self-help entails the recognition that it is the individual's responsibility to find solutions to one's own predicament without depending on others. In community practice, it is premised on the understanding that poor people with a shared fate of depravity and marginality can overcome their hardships through

[4]The SMU village competition is organised and sponsored by the Korean Government through the Korea International Cooperation Agency (KOICA), United Nations Development Programme (UNDP) and Government of Uganda.

selfless collectivisation and aggregation of individual efforts (Avoseh 2001; Ibrahim 2006). SMU's idea of the need for communities to 'help themselves' is easily domesticated in African landscapes, where communal self-help has been valued throughout nation-building processes. It also resonates well with the traditional practice and spirit of *bulungi bwansi* ('community service'): the obligation of every member to provide voluntary contribution in times of shared challenges, crises and joyous moments, such as funerals, marriages and, increasingly, poverty.

Moreover, COVOID's training messages focused on encouraging citizens to embrace the attitude of *Nitubaasa* ('We can do') and habits of voluntarism, self-reliance and inclusiveness. The pillar of self-help emphasised a certain attitude towards the government. COVOID trained communities that 'with the SMU model you do not wait for the government, you start addressing your problems, then government can find you where you have reached' (Interview with executive director, 13 August 2019). In addition, it showed that 'people should identify their problems and then work on those problems by themselves' (Interview with SMU model community facilitator, 12 August 2019).

These messages not only reinforced the pre-existing common self-organising initiatives and practices but also led to the emergence of new forms of self-help among citizens. Hence, monthly communal work on issues such as maintaining village roads, joint projects of brickmaking, mandatory monthly meetings and savings, individual household income generation projects and community vigilance started to take root in the community in addition to the pre-existing ones that revolved especially around burial and wedding ceremonies. Minutes from several monthly village meetings show that community members became more active and started to discuss and suggest solutions to the agreed problems. For example, a village meeting on 29 April 2019 tasked the chairperson of the Local Council I (LCI) together with SMU village committee to move around the village and identify homesteads that still had mud-and-wattle houses. The meeting resolved that the community would, starting from July 2019, begin to construct modern houses for them. There were several other cases of self-help that the village continued to deliberate upon and to find solutions for, such as improving participation in funeral arrangements, group farming and organising village households into ten-home clusters (*mayumba ikumi*).

4.3 Cooperation as Active Participation in Community Affairs

Mutual cooperation in the SMU model means working together to achieve a shared goal (Kim 2015). In the local language of Runyankore, the pillar of cooperation was translated as *okukwatanisa*, which simply means working or identifying with others at all times. It emphasises the idea of an active citizen who, in concert with others, contributes selflessly towards the common good of society. Cooperation, manifested through identifying and working with others, is a prerequisite for belonging and identity, while non-cooperation is generally treated as laziness, aloofness and bad neighbourliness, and may attract ridicule and isolation. In the implementation of the SMU model, cooperation involved infusing the spirit of collective action to ensure that each household participates and benefits in joint village projects and activities. The cooperation espoused by the SMU model was an inclusive one, distinct from the selective participation often mobilised by conventional NGO project approaches to target beneficiaries, often a small section of the community, and achieve time-bound predetermined outcomes.

> The SMU model challenges all village members to work together without leaving anyone behind whether it is communal work or during funerals and other community ceremonies. We have also started working jointly on our gardens through the *mayumba ikumi* arrangement. (Interview with vice chairperson, SMU committee, 16 August 2019)

Furthermore, discussions with several community members emphasised that good citizenship manifested in working together for mutual benefit. Good cooperation was said to be cross-cutting, from households to the immediate neighbourhood through to the entire community:

> According to me, good citizenship is when there is mutual cooperation between spouses at the household level. For example, when there is that cooperation, the family will be able to educate children, be welcoming and peaceful, and generally have development projects. But this cooperation at the family should extend to the community. So, to me a good citizen must also cooperate with community members in all activities such as *Samwiri Odongo*, saving groups and burials.
> (Interview with former leader SMU committee, 16 July 2019)

At the community level, COVOID emphasised the importance of regular community meetings and helped in the establishment of SMU village committees that worked hand in hand with existing village political structures in spearheading the mobilisation efforts of the model. The understanding of the NGO was that such strategies would foster more cohesive and participatory decision-making as well as increase the community voice for advocacy, all of which somehow contributed to the SDGs' overarching goal of inclusive development that leaves no one behind. For example, a female participant talked about how the SMU model had reawakened the practice of attending meetings, especially among men:

> Meetings used to be attended by women in this village, but since the introduction of *Samwiri Odongo,* men have increasingly appreciated the importance of meetings and are now more involved in community projects. (Female participant interview, 23 July 2019)

Although the meetings still have only average attendance, the open, free and dialogic deliberations often focus on finding solutions to the common problems in the village. For example, during a community meeting on 29 July 2019, the poor attendance of meetings was highlighted as one of the challenges facing the implementation of the SMU model. In several other meetings, several by-laws were passed and fines ranging from 10,000 Uganda shillings (UgX) (approximately €2.50) to 30,000 UgX (approximately €7.50) were suggested to be imposed for uncooperative and deviant behaviour in the village. The listed deviant behaviours included drunkenness, failure to attend meetings, failure to send children to school, failure to participate in communal work, *okwonesa* (meaning a failure to restrain ones' animals from destroying neighbours' crops), operating local bars at restricted times and defaulting on burial contributions. While we were not able to establish the level of enforcement or compliance with these fines, the decision illustrates well the 'stick' element of the SMU model in enhancing cooperative behaviour.

5 Discussion

In this section, we reflect on the emerging active citizenship practices being emphasised and learned through the SMU model, and the ways in which these localised practices potentially contributed to achieving SDGs at the village level. We posit that the SMU can revitalise the spirit of communal development, important in mobilising for inclusive grassroots transformation and poverty eradication envisaged by the SDGs. At the same time, however, it falls short in promoting the organised state-level implementation necessary to overcome critical citizenship constraints that limit the overall achievements of the SDGs in any country. In this regard, we identified some persistent tensions related to the kind of 'active citizenship' promoted through SMU, which should require further attention and analysis.

First, we identified a tension between communalist and individualist citizenship. The SDGs' rallying call to leave no one behind encapsulated in the localisation crusade presupposes a highly inclusive, homogeneous, democratic and egalitarian context (Menon & Hartz-Karp 2019; Weber 2017), where communities are supposed to share similar ideals of work ethics, aspirations and problems, further strengthened through models such as SMU. In reality, while it is broadly true that citizens in rural communities such as Nyakahama generally have a shared fate of poverty and marginalisation, these are experienced differently at personal and household levels. Factors such as gender, age and socioeconomic status influence citizenship and the enactment of practices of self-help, diligence and cooperation. Whereas SMU relies on communal values as the drivers of self-reliant, cooperative and frugal citizenship, it simultaneously offers a counter-narrative of individuals that should take care of their own affairs, inadvertently promoting practices of selfishness and individualised accumulation. In our case, this tension manifested, for instance, in the practice of each household having its own backyard garden of green vegetables, while picking vegetables from a neighbour's garden was punished according to the by-laws. It also showed as poor attendance of meetings by community members, especially males engaged in labour mobility outside the village to ensure household income.

Second, there are continuous tensions between the positive transformative potential of the self-help ethos promoted by SMU and its obvious shortcomings when it comes to transformations in structural constraints for citizenship, such as poverty and inequality. Critical scholars have suggested how notions of self-reliance and self-help being promoted by the United Nations to localise SDGs are just subtle attempts to depoliticise development, dampen dissent and replace state accountability with community self-reliance (Buire & Staeheli 2017; Gaynor 2011; Telleria 2018; Weber 2015, 2017). In the same vein, critical observations have pointed to the SMU model's tendency to support authoritarianism and disregard issues such as human rights and democracy (Kim et al. 2011; Jeong 2017). Thus, these critical points of view maintain that transformative development should be able to establish a link between issues such as poverty, market, freedom and democracy in order to have a chance to end global exclusion (Telleria 2018).

In the attempts to localise SDGs in contexts such as Uganda, the SMU model conveniently fits the popular public narratives on holding communities responsible for their own development. The SMU's principle of rural development based on communal ethos resonates with the emphasis on citizens assuming responsibility for their own and national development advocated by the state (RoU 1995, Objective. XXIX). Thus, such models can weaken poor citizens' agency to demand equality from an unfair system (Telleria 2018) characterised by the political economy of neoliberal state withdrawal (Wiegratz et al. 2018) and semi-authoritarian rule based on a clientelist relationship between state political elites and ordinary citizenry in Uganda (Alava et al. 2020; Kalinaki 2020; Muhumuza 2009; Mwenda 2007; Titeca 2006).

Yet, from the localised perspective of incremental learning and change in citizenship practices (Holma et al., 2018), the domestication of the three pillars of SMU promoted individual and collective enthusiasm to try to surmount the problems of poverty and marginalisation. By engaging with collective hard work combined with notions of self-help, the community members improved their material conditions, and boosted the confidence and assertiveness they later exercised in demanding services such as clean water and road-grading equipment from the district offices. The changes related to construction of new permanent houses, growing a culture of saving, education of children and advocacy for clean running water resulting from joint efforts, combined with regular meetings that offered dialogical spaces for citizenship expression, were significant for local CONTRIBUTIONS to achieving the SDGs. The instances of communities learning to solve their own problems and 'not waiting for government' may appear as a depoliticised means to shield duty bearers from responsibility, but at the same time, they are pragmatic ways of dealing with the prevailing contexts in this very moment.

6 Conclusion

The chapter showed how a Ugandan NGO used the three pillars of SMU to inculcate active citizenship in a rural community in its effort to localise SDGs. The pillars of the SMU model were translated, domesticated and locally customised: diligence into practices of hard work and frugality; self-help into responsibility, usefulness and confidence; and cooperation into the willingness of the community to jointly participate in activities that addressed different issues of their shared destiny. Each pillar spurred joint and regular community activities, which contributed to changes in resonance with inclusive development central to SDGs. Moreover, by addressing joint problems of immediate concern, the community incrementally learned to be active in shared affairs, including helping the most vulnerable in the spirit of leaving no one behind. They also gained capacity to take steps to realise some of the SDGs, such as reducing poverty and hunger while achieving decent accommodation, social inclusion, improved water and sanitation, local infrastructure and responsible consumption.

The analysis showed how development models travel globally in multiple ways, ranging from a high-level campaign by global institutions and national governments to civil society networks and, as in our case, through random individual encounters by NGO leaders. Consequently, in line with the concepts of domestication and translation, the models are continuously interpreted, domesticated, customised and contextualised in ways consistent with the conditions, interests, needs and circumstances of the given actors. Therefore, instead of expecting a model such as SMU implemented in Uganda in the late 2010s to produce outcomes similar to those it produced in Korea in the 1970s, we should conduct nuanced analyses of what kinds of changes the model can promote in different contexts.

The activeness promoted in domesticating the SMU model emphasised more of community members' initiatives to solve their own problems than increasing citizens' engagement in claiming their rights related to the achievement of various SGDs and access to public services. Hence, the smooth domestication of the model in the encounter between the NGO and the local community, we suggest, resulted from the way it provided opportunities to address immediate problems, and rhymed with existing local practices of active citizenship manifested in individual initiatives and not being antagonistic towards either local or national values, prevalent political system or power holders.

As a consequence, we contend that despite the apparent success of the NGO in using SMU to promote active citizenship, the impact remains limited and scattered across time and space. Domestication and localisation of SDGs in selected villages do not create the requisite citizen momentum and capacity to address asymmetrical and historical power structures and systemic inequalities that characterise citizen-state relations in most developing countries. It is not possible to speculate what the consequences for national development would be, if, beyond the rhetoric of endorsement by politicians, SMU was implemented nationwide in contemporary Uganda.

Finally, we suggest that in the efforts to localise SDGs of, for and by the people, any model that generally promotes holistic learning and practices of active citizenship has the potential to enable marginalised people to work together to reduce the constraints of poverty. Thus, models can, to some extent, help achieve the inclusive ideal of leaving no one behind particularly at community level. However, in each case, the potential realises in different ways, and detailed analysis of the contextualised potentialities and limitations to promote transformative development is continuously needed.

References

Adegbile, A., & Sarpong, D. (2018). Disruptive innovation at the base-of-the-pyramid. *Critical Perspectives on International Business, 14*(2/3), 111–138. https://doi.org/10.1108/cpoib-11-2016-0053

Ahimbisibwe, K. F., Ndidde, A. N., & Kontinen, T. (2020). Participatory methodology in exploring citizenship: A critical learning process. In K. Holma & T. Kontinen (Eds.), *Practices of Citizenship in East Africa: Perspectives from Philosophical Pragmatism* (pp. 159–175). London: Routledge.

Alasuutari, P. (2009). The domestication of worldwide policy models. *Ethnologia Europaea, 39*(1), 66–74.

Alasuutari, P. (2015). The discursive side of new institutionalism. *Cultural Sociology, 9*(2), 162–184. https://doi.org/10.1177/1749975514561805

Alasuutari, P., & Qadir, A. (Eds.). (2013). *National policy-making: Domestication of global trends*. London: Routledge.

Alava, H., Bananuka, T. H., Ahimbisibwe, K. F., & Kontinen, T. (2020). Contextualizing citizenship in Uganda. In K. Holma & T. Kontinen (Eds.), *Practices of Citizenship in East Africa: Perspectives from Philosophical Pragmatism* (pp. 57–72). London: Routledge.

Avoseh, M. B. (2001). Learning to be active citizens: Lessons of traditional Africa for lifelong learning.

International Journal of Lifelong Education, 20(6), 479–486. https://doi.org/10.1080/02601370110088454

Buire, C., & Staeheli, L. A. (2017). Contesting the 'Active' in active citizenship: youth activism in Cape Town, South Africa. *Space and Polity, 21*(2), 173–190. https://doi.org/10.1080/13562576.2017.1339374

COVOID. (2016). *COVOID piloting the Saemaul Undong (Korean Model of Development) in Rubirizi District, South Western Uganda* (Newsletter).

Czarniawska, B., & Joerges, B. (1996). Travel of Ideas. In B. Czarniawska & G. Sevón (Eds.), *Translating Organizational Change* (pp. 13–48). Berlin: de Gruyter.

Dagnino, E. (2007). Citizenship: a perverse confluence. *Development in practice, 17*(4–5), 549–556. https://doi.org/10.1080/09614520701469534

Doucette, J., & Müller, A. R. (2016). Exporting the Saemaul spirit: South Korea's Knowledge Sharing Program and the 'rendering technical' of Korean development. *Geoforum, 75*, 29–39.

Douglass, M. (2013). *The Saemaul Undong: South Korea's rural development miracle in historical perspective*. ARI Working Paper No. 197. Asia Research Institute, National University of Singapore.

Engel, B. A. (2017). Father and Daughter's Saemaul Undong: True Replication or Mere Glorification? In K. J. Ayhan (Ed.), *Korea's Soft Power and Public Diplomacy* (pp. 57–74). Seoul: Hangang Network.

Eom, S.-J. (2011). The Rural Saemaul Undong revisited from the perspective of good governance. *Korean Journal of Policy Studies, 26*(2), 17–43.

Galdas, P. (2017). Revisiting bias in qualitative research: Reflections on its relationship with funding and impact. *International Journal of Qualitative Methods, 16*, 1–2. https://doi.org/10.1177/1609406917748992

Gaventa, J., & Barrett, G. (2012). Mapping the outcomes of citizen engagement. *World development, 40*(12), 2399–2410. https://doi.org/10.1016/j.worlddev.2012.05.014

Gaynor, N. (2011). In-Active citizenship and the depoliticization of community development in Ireland. *Community Development Journal, 46*(1), 27–41.

Guba, E. G., & Lincoln, Y. S. (2005). Paradigmatic controversies, contradictions, and emerging confluences. In N. K. Denzin & Y. S. Lincoln (Eds.), *The sage handbook of qualitative research* (pp. 163–188). Thousand Oaks: Sage Publications.

Han, D.-H. (2012). *The successful cases of the Korea's Saemaul Undong (New community movement)*. Ministry of Strategy and Finance, Republic of Korea.

Hickey, S., & King, S. (2016). Understanding social accountability: Politics, power and building new social contracts. *The Journal of Development Studies, 52*(8), 1225–1240. https://doi.org/10.1080/00220388.2015.1134778

Holma, K., & Kontinen, T. (2020a). Practices and habits of citizenship and learning. In K. Holma & T. Kontinen (Eds.), *Practices of citizenship in East Africa: Perspectives from philosophical pragmatism* (pp. 15–28). London: Routledge.

Holma, K., & Kontinen, T. (Eds.). (2020b). *Practices of Citizenship in East Africa: Perspectives from Philosophical Pragmatism*. London: Routledge.

Holma, K., Kontinen, T., & Blanken-Webb, J. (2018). Growth into citizenship: Framework for conceptualizing learning in NGO interventions in sub-Saharan Africa. *Adult Education Quarterly, 68*(3), 215–234. https://doi.org/10.1177/0741713618768561

Ibrahim, S. S. (2006). From individual to collective capabilities: the capability approach as a conceptual framework for self-help. *Journal of human development, 7*(3), 397–416. https://doi.org/10.1080/14649880600815982

Iqbal, M. H., & Milon, S. R. H. (2017). Ways forward to rural development of Bangladesh: lesson from south Korean Saemaul Undong movement. *Asian Journal of Social Sciences & Humanities, 6*(3), 70–80.

Jeong, H. (2017). Globalizing a rural past: The conjunction of international development aid and South Korea's dictatorial legacy. *Geoforum, 86*, 160–168.

Jwa, S.-H. (2018). Understanding Korea's Saemaul Undong: Theory, Evidence, and Implication. *Seoul Journal of Economics, 31*(2), 195–236.

Kalinaki, D. (2020, July 23). Opposition is dying, and that is not a good thing for democracy. *Daily Monitor*. Kampala, Uganda. https://www.monitor.co.ug/OpEd/columnists/DanielKalinaki/Opposition-is-dying%2D%2Dand-that-is-not-a-good-thing-for-democracy/878782-5598004-t0v6qmz/index.html. Accessed 25 July 2020

Kim, D. K. (2015). A Schumpeterian Analysis of the Saemaul Undong Movement in 1970s South Korea within the CDD Framework. *Seoul Journal of Economics, 28*, 415–453.

Kim, T., Kwon, H.-J., Lee, J., & Yi, I. (2011). Poverty, Inequality, and Democracy: 'Mixed Governance' and Welfare in South Korea. *Journal of Democracy, 22*(3), 120–134. https://doi.org/10.1353/jod.2011.0037

Makara, S. (2020). Presidential Term Limits in Uganda: Do Elections Provide an Avenue for Alternate Power Succession? In J. R. Mangala (Ed.), *The Politics of Challenging Presidential Term Limits in Africa* (pp. 105–137). Cham: Springer Nature.

Menon, S., & Hartz-Karp, J. (2019). Linking Traditional 'Organic' and 'Induced' Public Participation with Deliberative Democracy: Experiments in Pune, India. *Journal of Education for Sustainable Development, 13*(2), 193–214. https://doi.org/10.1177/0973408219874959

Muhumuza, W. (2009). From fundamental change to no change: The NRM and democratization in Uganda. *Les Cahiers d'Afrique de l'Est/The East African Review*, (41), 21–42.

Mwenda, A. M. (2007). Personalizing power in Uganda. *Journal of democracy, 18*(3), 23–37. https://doi.org/10.1353/jod.2007.0048

Mwenda, A. M., & Tangri, R. (2005). Patronage politics, donor reforms, and regime consolidation in Uganda. *African affairs, 104*(416), 449–467. https://doi.org/10.1093/afraf/adi030

Nauta, W., & Lee, T. (2017). South Korean civic actors in Rwanda: Promoting rural development and an emerging civil society in contexts controlled by the state. In A. Graff & A. Hashim (Eds.), *African-Asian Encounters: new cooperations and dependencies* (pp. 183–215). Leiden: Amsterdam University Press.

Nhamo, G. (2017). New global sustainable development agenda: A focus on Africa. *Sustainable Development, 25*(3), 227–241. https://doi.org/10.1002/sd.1648

Odularu, G. O. A. (2009). Enhancing the Rural Development Policy Space: Will the Saemaul Movement be Adaptable to Africa? *Wook Chae, President*, 2025151.

Park, J.-D. (2019). *Re-Inventing Africa's Development: Linking Africa to the Korean Development Model.* Cham Switzerland: Springer Nature.

RoU. (1995). *Constitution of the Republic of Uganda.* Kampala: Uganda Printing and Publishing Corporation.

RoU. (2013). *Uganda Vision 2040.* Kampala: National Planning Authority.

RoU. (2017). *The National Population and Housing Census 2014 – Area Specific Profile Series, Rubirizi district.* Kampala: Uganda Bureau of Statistics.

RoU. (2020). *Third National Development Plan (NDPIII) 2020/21 – 2024/25.* Kampala: National Planning Authority.

Sriskandarajah, D. (2018). Toward an Accountability Revolution? Citizen Participation and the SDGs. In H. Kharas, H. Kato, J. W. McArthur, & R. M. Desai (Eds.), *From Summits to Solutions: Innovations in Implementing the Sustainable Development Goals.* Brookings Institution Press.

Steiner, A. (2017, October 12). *Localizing the Implementation of the SDGs.* Speech presented in World Bank Group MC Building, Washington DC.

Telleria, J. (2018). Can we 'transform our world' without affecting international power relations? A political analysis of the United Nations development agenda. *Globalizations, 15*(5), 655–669. https://doi.org/10.1080/14747731.2018.1467744

Titeca, K. (2006). Political patronage and political values: The developmental role of political patronage and its impact on shaping political values in rural Uganda. *Afrika Focus, 19*(1–2), 43–67. https://doi.org/10.21825/af.v19i1.5414

Tripp, A. M. (2004). The changing face of authoritarianism in Africa: The case of Uganda. *Indiana University Press, 50*(3), 3–26. https://doi.org/10.1353/at.2004.0035

Twesigye, J., Twikirize, J. M., Luwangula, R., & Kitimbo, S. (2019). Building Resilience through Indigenous Mechanisms: The Case of Bataka Groups in Western Uganda. In J. M. Twikirize & H. Spitzer (Eds.), *Social Work Practice in Africa: Indigenous and Innovative Approaches* (pp. 145–160). Kampala, Uganda: Fountain Publishers.

UNDP. (2015). Saemaul Initiative Towards Inclusive and Sustainable New Communities: Implementation Guidance. file:///C:/Users/user/Downloads/Implementation%2520Guidance%2520WEB%2520Nov.pdf. Accessed 16 June 2020

United Cities and Local Governments. (2019). *Towards the Localization of the SDGs* (Local and Regional Governments' Report to the 2019 HLPF No. 3). Barcelona: United Cities and Local Governments.

United Nations General Assembly. (2015). Transforming our world: the 2030 Agenda for Sustainable Development, 21 October 2015, A/RES/70/1. https://www.refworld.org/docid/57b6e3e44.html. Accessed 18 August 2020

Weber, H. (2015). Reproducing inequalities through development: the MDGs and the politics of method. *Globalizations, 12*(4), 660–676.

Weber, H. (2017). Politics of 'leaving no one behind': contesting the 2030 Sustainable Development Goals agenda. *Globalizations, 14*(3), 399–414. https://doi.org/10.1080/14747731.2016.1275404

Yang, Y. (2017). Saemaul Undong revisited: a case of state–society dynamics in social capital mobilisation, focusing on the role of local leaders in South Korea of the 1970s. *Journal of International Development, 29*(7), 993–1010. https://doi.org/10.1002/jid.3048

Ending Extreme Poverty (SDG 1) in Chegutu District of Zimbabwe: An Analysis of Tsungirirai Welfare Organisation's Interventions

4

Itai Kabonga, Ernest Dube, Cowen Dziva, and Nyasha Chaminuka

Abstract

The study analyses the contribution of Tsungirirai Welfare Organisation's (TWO) interventions in ending extreme poverty (SDG 1) in Chegutu district, Zimbabwe. The study was qualitative in nature and data were collected using in-depth interviews, focus group discussions (FGDs) and documentary analysis. In a bid to reduce extreme poverty, TWO is implementing activities that include economic strengthening, vocational training, healthcare provision and educational assistance to poor households. Internal Savings and Lending Schemes (ISALS) have increased financial income, allowing households to improve on indicators of health, education and food and nutrition. Households' capability to acquire physical assets like wheelbarrows, scotch carts and ploughs has improved because of household income-generating projects. Vocational training has reduced unemployment of youths. The chapter argues that the TWO has contributed significantly to the reduction of extreme poverty although some challenges remain in the district and country at large. To address TWO's limited coverage, the study recommends partnerships with other NGOs and the private sector.

Keywords

SDG 1 · Tsungirirai Welfare Organisation · Poverty · Extreme poverty · NGOs

I. Kabonga (✉)
Department of Humanities and Social Sciences, Zimbabwe Ezekiel Guti University, Bindura, Zimbabwe

E. Dube
Department of Development Sciences, Marondera University of Agricultural Sciences and Technology, Marondera, Zimbabwe

C. Dziva
Nehanda Centre for Gender and Cultural Studies, Great Zimbabwe University, Masvingo, Zimbabwe

N. Chaminuka
Department of Development Studies, Zimbabwe Open University, Gweru, Zimbabwe

1 Introduction

The Sustainable Development Goals (SDGs) represent a set of 17 common goals to meet urgent global environment, economic, social and political challenges by the year 2030. The SDGs substituted the Millennium Development Goals (MDGs) in 2015 "with the objective of producing a set of common goals that meet urgent global environmental, economic, and political challenges by 2030" (Fofana et al. 2019, p. 4). Amongst other imperatives, the SDGs took off as

G. Nhamo et al. (eds.), *Sustainable Development Goals for Society Vol. 1*, Sustainable Development Goals Series, https://doi.org/10.1007/978-3-030-70948-8_4

a global effort to address the indignity of poverty. SDGs focus on issues of wider and global implication relating to social, economic and environmental sustainability as well as pertinent aspects of peace, justice and effective institutions (Ankeroye et al. 2018). Mugadziwa (2012) sees the SDGs as an attempt to balance the three dimensions of development—economic, social and environment—in a manner that ensures sustainable use of resources to benefit the present and future generations.

This chapter focuses on the implementation of SDG 1—to end extreme poverty by 2030—by a non-governmental organisation (NGO), Tsungirirai Welfare Organisation (TWO) in Chegutu district. Poverty ranks as one of the major problems affecting the Chegutu district. Statistics show that in Chegutu district 72.4% of the people are living in poverty (Zimbabwe Poverty Atlas Map 2015, p. 4). Extreme poverty in the district is reflected by food shortages, homelessness, lack of clothing, sickness due to lack of medical attention and lack of income at the household level. Poverty at the household level in Chegutu has also led to an increase in child prostitution and teenage pregnancies. Many people are dying of avoidable diseases that can be easily treated. The growing poverty level in Chegutu district is related to at least two dynamics that occurred in the broader trajectory of Zimbabwe as a country. These include the neoliberal stance that was adopted by the country in the 1990s. Contextualising it to Chegutu, the neoliberal stance adopted by the government in the form of Economic Structural Adjustment Programme (ESAP) resulted in retrenchments and increased difficulties as the society could not cope with the increased withdrawal of the government in social spending. Compounding the situation was the Fast Track Land Reform Programme that resulted in the plunder and looting of the once viable farms like Hopedale. The land reform programme saw farmworkers plunged into poverty gravitating towards what Mowawa (2013) calls "sex panning", an indication of growing levels of poverty. With the coming in of the "Zimbabwean Crisis"—a shorthand for complex state failure (Murisa 2010, p. 3)—the general fall of the economy increased poverty across Zimbabwe and Chegutu district was no exception. Magure (2015) posits that as a result of economic challenges, most people in the district are in the informal economy doing informal trading, gold panning and prostitution as livelihood strategies.

TWO is tackling poverty through several approaches such as investing in education, health provision, social provision and economic strengthening to poor households. The chapter argues that TWO has significantly contributed to ending extreme poverty although some challenges remain in the district and the country at large. The organisation in its attempt to fight poverty is cooperating with other NGOs and the government, significantly aligning with one of the SDGs' core principles of partnership. To achieve the SDGs, there is need for collaboration between governments, civil society organisations (CSOs) and the private sector (Adjei et al. 2012). The fight against poverty is no easy battle as the organisation is facing numerous challenges such as donor dependency, shifting donor requirements and effects of structural economic conditions in Zimbabwe.

While SDGs are global goals that nations must achieve by 2030, there is an absence of micro studies on how NGOs are contributing to the ending of extreme poverty in the context of sustainable development. The academia, policy crafters and development practitioners are in the dark regarding the strategies and activities used by local level NGOs to end extreme poverty. CEEWeb for Biodiversity (2017) argues that NGOs play an important role in the implementation of SDGs. To achieve goal number 1, NGOs' roles include suggesting improvements to policies aimed at reducing poverty, critical analysis of public policies, holding governments and the private sector accountable through position papers, practical implementation of poverty reduction projects and involvement of the people in such projects (CEEWeb for Biodiversity 2017; Hege and Demailly 2018; Schaefer et al. 2018). Up to now, NGOs are doing the most in advancing the SDGs, whereas governments are hesitant and making the poorest contribution in advancing the SDGs (Globe-Scan Sustainability Survey 2019).

The objectives of the study are (1) to analyse activities implemented by TWO to reduce extreme poverty, (2) to analyse the contribution of the activities to ending extreme poverty and attainment of SDG 1 and (3) to document challenges faced by NGOs in attaining SDG 1. The chapter is organised into different sections. The first section is the introduction that details the focus of the study. This is followed by an empirical literature review that traces the contribution of NGOs to ending extreme poverty in the context of SDGs and a section on the conceptualisation of key terms. This is followed by the methodology section. Lastly is the presentation and discussion of findings using a thematic approach.

2 Literature Review

2.1 NGOs' Contribution to Ending Extreme Poverty by 2030

Poverty levels in Africa, Zimbabwe included, have worsened of late. Whereas over 1 billion people are living in poverty in the world (Suharko 2007), there is consensus that poverty is increasing at a faster pace in Africa compared to other continents (Kange'the and Chivanga 2015). The rhetoric of "Africa Arising" has been debunked by the fact that more than half of the extreme poor live in sub-Saharan Africa. The number of poor in the region increased by 9 million, with 413 million people living on less than US$1.90 a day in 2015, more than all the other regions combined. If the trend continues, by 2030, nearly nine out of ten extreme poor will be in sub-Saharan Africa (World Bank 2019). This reality paints a gloomy picture in terms of the desire to achieve goal number 1. On the other hand, this reality calls upon the involvement of all stakeholders in fighting poverty, NGOs included. In Zimbabwe, the involvement of NGOs cannot be overemphasised given the threat of poverty.

Though there have been attempts to reduce poverty using the SDG framework, Zimbabwe is facing poverty problems of great proportion. Nhapi (2019) blames Zimbabwe's prolonged dictatorship and associated disorder as having disrupted social well-being. United Nations Children Fund (UNICEF) (2011) claims that about 250,000 households, inclusive of approximately 700,000 children in Zimbabwe, live in extreme poverty. Poverty in Zimbabwe is believed to be higher in rural areas than in urban areas (Machingura and Nicolai 2018). Even the rapidly growing urban slums are not any better. Income insecurity has worsened the situation, particularly in rural areas though the urban zones are scarcely any better. Nhapi (2019) argues that based on the Total Consumption Poverty Line (TCPL) 62.6% of households in Zimbabwe are poor, whereas 16.2% of households are extremely poor. Rural poverty is driven chiefly by low income from farming activities and "very low prices for the goods sold at the market" (Nhapi 2019, p. 159). Even the political changes predicated in the ouster of Robert Mugabe and the coming in of the new president have seen little prospects in terms of poverty reduction. The new regime's approach to economic development is criticised by Bond (2019, p. 1) because of "a more brutal fiscal policy plus an even tighter state squeeze on hard currency". Given this reality, at least for now, the drive to end extreme poverty in all its forms is in serious jeopardy. This reality is even manifested at the global level where "of the world's people, 1 in 12 is living in extreme poverty. One in nine go hungry. Half lack essential healthcare. Half are not covered by social protection. One in five children is not attending school" (Manuel et al. 2019, p. 2). However, despite the pressures and weaknesses of Zimbabwe's drive towards ending extreme poverty, it is important to document the NGOs' strategies in the fight to reduce extreme poverty.

NGOs in Zimbabwe can play important roles in the implementation of SDG 1. While advocacy is not foreign to the functions of NGOs, NGOs are expected to hold governments accountable since governments are signatories to the SDGs (CEEWeb for Biodiversity 2017). This is important because without the pressure from NGOs, governments are less likely to prioritise SDGs

(Hege and Demailly 2018). Thus, the role of NGOs also involves monitoring of governments' incorporation of SDGs in the country's development policies. To successfully play the advocacy role, Hege and Demailly (2018) recommend that NGOs should form alliances and strategic partnerships with other NGOs. This can be through joint position papers, conferences and progress reports on the achievement of SDGs. NGOs are not only mandated with holding governments accountable, but they can also hold the private sector accountable through indirectly pushing for government action or directly aiming at the profit sector. The private sector has a role in achieving the SDGs, particularly through their voluntary commitments. Part of the NGOs' role is, therefore, to monitor that the private sector delivers on the principles of commitment and transparency in the processes.

Partnership as a function of NGOs in delivering the SDGs is not a new principle since it is part of NGOs' day-to-day activities (Lewis 2010). This involves NGOs establishing coalitions amongst themselves and across sectors, resulting in partnerships with the government and the private sector (Hege and Demailly 2018). Partnerships help NGOs to broaden their voice and position. However, for a partnership to produce the desired results, the silo mentality or particularism inherent in the NGO sector requires decimation to facilitate the sharing of ideas, knowledge and experiences (Hege and Demailly n.d.). Examples of partnerships like SDG Europe Watch and Alliance SUD are good practices of holding governments accountable in the implementation of SDGs. Adjei et al. (2012) submits that a close collaboration between governments and non-state actors is a workable framework to achieve poverty reduction. To achieve SDG 1, partnerships are imperative since it is one of the core principles of SDGs (Schaefer et al. 2018).

A direct role of NGOs in the achievement of SDG 1, which resonates with what TWO is doing in Chegutu district, is practical project implementation. Practical implementation involves NGOs carrying out projects, or supporting communities or other organisations in carrying out projects (Hege and Demailly 2018). Projects by NGOs need not necessarily be SDG projects, but NGOs rather strive to use the goals and targets in pre- and post-evaluation of projects to ensure that they are well aligned with the SDGs, positively contributing to some and not negatively affecting others (CEEWeb for Biodiversity 2017). Bangladesh Rural Advanced Committee (BRAC)'s practical projects to end extreme poverty in Bangladesh include supporting agriculture, microfinance and food security programmes (Hassan and Forhad 2013), whereas in Zimbabwe practical projects by NGOs to end extreme poverty include savings groups, provision of inputs to farmers, income-generating projects and provision of social services (Chimire and Chitongo 2018; Kabonga 2016). SOS Children's Villages (n.d.) submits that the organisation's activities support the achievement of goal 1 through building the capacity and resilience of families in order to break the cycle of poverty. Elsewhere, NGOs such as Plan International, World Vision, Save the Children and CARE International have over the years been building traction and strong foundations around goal 1 of ending extreme poverty.

2.2 Conceptualising Non-governmental Organisations (NGOs)

The concept of NGOs by far embodies contestations as there are disagreements on what NGOs are. The simplest definition of NGOs is that they are organisations that operate independent of governments and they do not exist for profit-making (Lewis and Kanji 2009; Lewis 2010; Willetts 2002). Even though the above authors view NGOs as independent of the government, the reality is that there are NGOs that are fronted and funded by the government called government-controlled NGOs (GONGOs). Hence, Antonio (2015) concluded that not all NGOs are non-governmental.

An attempt to understand NGOs shows that there are a diversity of organisations that fall within the category of NGOs, from small community-based organisations to large international NGOs (Banks and Hulme 2012). Within this pendulum, it is abundantly clear that NGOs can be formal or informal, registered or unregistered (Dar 2014). For Salamon and Anheier (1992) NGOs can be conceptualised using a three-pronged approach. First, NGOs can be understood from a legal perspective. This view argues that NGOs are a legal entity since they are formed and guided by a certain legal provision. In Zimbabwe, NGOs are guided by the Private Voluntary Organisations Act (17:05). Second, the financial and economic perspective notes that NGOs are those organisations whose funding is less than 50% from the government and finally the structural and operational perspective defines NGOs from a characterisation perspective. NGOs are formal, implying that they are institutionalised; private, meaning they run independently of the government; non-profit-making, meaning NGOs do not exist to make a surplus; and voluntary, meaning that there is some form of free participation. For Willetts (2002) these boundaries are nebulous as one can find NGOs generating profits from lending out properties and sale of products from their projects.

Lewis (2010) locates three functions of NGOs. NGOs are implementers (they implement development projects), catalysts (they speed up development) and innovators (they come up with new approaches to development). Activities by NGOs to reduce extreme poverty are located within the implementers' function.

2.3 Conceptualising Poverty

Conceptualising poverty remains contested in social sciences. While there are many definitions of poverty, many scholars converge on the idea that poverty entails deprivation of basic human needs (Dube 2019). Argued from a basic needs approach, poverty is a shortage of important basic needs: water, food, clothing, shelter and a clean environment. Dominant definitions of poverty in the past tended to view poverty as a lack of income, wealth, assets and material possessions (Julius and Bawane 2011). According to Dube (2019, p. 50), "contemporary thinking recognises poverty as a multidimensional concept that includes lack of access to education, lack of access to health care and infrastructure, the denial of opportunities and choices to take independent decisions, denial of the respect of others, and even remoteness to political power". This view of poverty is also supported by Adeyemi et al. (2009) who viewed the multidimensionality of poverty as insecurity, lack of material well-being, lack of freedom of choice, low self-confidence, psychological distress, social isolation and not believing in oneself. There are many "poverties" in society; hence, one can refer for instance to education poverty, health poverty and infrastructure poverty. The authors agree with the United Nations Department of Economic and Social Affairs (2005, p. 14), that "while poverty has many dimensions, its two fundamental aspects are the lack of economic power owing to low incomes and assets, and the lack of socio-political power, as reflected in the limited access to social services, opportunities and information and often in the denial of human rights and the practice of discrimination". Though there are many debates on what poverty is and what it is not, there is consensus that poverty is unwanted in human societies, hence the emphasis on SDG 1 of ending extreme poverty in its forms everywhere.

2.4 Theoretical Framework: Sustainable Development Goals

This study is guided by the Sustainable Development Goals framework. The Sustainable Development Goals (SDGs) comprise 17 goals as opposed to the 8 MDGs that ended in 2015, but the targets set for the SDGs were developed in a more complementary manner, highlighting the interconnected nature of the goals and calling all countries to action (CEEWeb for Biodiversity 2017; Ankeroye et al. 2018). The SDGs represent the 2030 agenda for sustainable development and were adopted by 193 members of the United Nations (UN) in September 2015. The goals act as a guideline for national and international decision-making until 2030. Aligned to the SDGs

Table 4.1 The 17 Sustainable Development Goals

SDG goal	Narration of the goal
SDG 1	No poverty
SDG 2	No hunger
SDG 3	Good health and well-being
SDG 4	Quality education
SDG 5	Gender equality
SDG 6	Clean water and sanitation
SDG 7	Affordable and clean energy
SDG 8	Decent work and growth
SDG 9	Industry, innovation and infrastructure
SDG 10	Reduce inequality
SDG 11	Sustainable cities and communities
SDG 12	Responsible consumption and production
SDG 13	Climate action
SDG 14	Life under water
SDG 15	Life land
SDG 16	Life on land
SDG 17	Partnerships for the goals

Source: Nour (2017)

are 169 targets that outline specific actions that governments and other bodies including NGOs should take to realise the SDGs (Jaiyesimi 2016). Unlike the MDGs the SDGs are more holistic and collaborative, integrating environmental, economic and governance aspects. The SDGs represent a cross-cutting approach to ensure that economic, social and environmental problems are addressed together (Schaefer et al. 2018). People, planet, prosperity, peace and partnerships are the five core dimensions of the SDGs. In many aspects, the SDGs are the quintessence of the problems that the world faces today. The 17 goals are shown in Table 4.1.

3 Materials and Methods

In this study, the qualitative research design was adopted as the strategy of inquiry. According to Jackson et al. (2007) qualitative research relies on non-numeric data, usually in the form of words. Qualitative research is underpinned by a desire to understand a phenomenon in depth from the perspectives of those with lived experiences or who would have witnessed an experience. This study was premised on understanding in depth from the perspective of beneficiaries and officials the contribution of TWO to attaining SDG 1. The advantage of the qualitative research approach is that it allowed probing in areas that required clarity. This was the case during the in-depth interviews conducted with NGO officials and TWO's beneficiaries.

The study used a case study design to unpack the contribution of TWO to achieving SDG 1 of ending extreme poverty in the Chegutu district. Cresswell (2003) explains that when using a case study, the researcher explores a programme, an event, an activity, a process or one or more individuals in depth. This case study was bounded by activity and time. In terms of time, the researchers focused on projects currently being implemented whereas in terms of activities the research focused on various activities constituting projects that bring socio-economic development to Chegutu district as part of initiatives to end extreme poverty. The use of a case study design enabled the researchers to use several data collection methods (Jackson et al. 2007). Case studies are normally used when there are theoretical inadequacies in an area. A case study in this instance was applicable because currently there is a lack of academic inquiry into the contribution of NGOs to the achievement of goal number 1 in Chegutu district.

In the study, the researchers used purposive sampling because it allowed the inclusion of respondents useful to the study. The researchers sampled 15 community members benefitting from NGO projects. These were sampled from TWO database of beneficiaries. These community members participated in in-depth interviews and FGDs were conducted. The respondents were considered information rich and of value to the study. Purposive sampling allowed the exclusion of elements of less value to the study.

The study used several data collection methods. These include in-depth interviews in which respondents shared their views, opinions and lived experiences. The in-depth interviews were held with 15 beneficiaries of TWO activities and 2 staff members involved in poverty reduction programming.

Two FGDs comprising eight participants each were conducted. In all the FGDs the males were four and the females were four. There was a need to balance gender so that no gender category dominates in the discussion. All the participants were adults with their ages ranging from 20 years to 40 years. Education-wise, the respondents had varying educational level attainment. Those that had the lowest education attainment had a primary level with highest education attainment being tertiary level. The interaction within the groups enabled the respondents to influence each other with their responses.

Data were also gathered through documentary review. The review of literature mainly journals, books, book chapters and conference papers shaped the orientation of the study. The researchers reviewed the organisation's documents such as narrative reports, strategic plans and meeting minutes. These documents outlined the focus of TWO, achievements and challenges. Data used in this study were collected from 2017 to 2018 during the period in which one of the researchers worked in the Chegutu district.

The study employed triangulation to increase the dependability of the research. Different data collection methods were used and these are in-depth interviews, FGDs and documentary reviews. Triangulation ensured corroboration allowing the researchers to move away from relying on a single data point. Interview techniques such as probing were used by the researchers to ensure rigour. The researchers probed on areas that required clarity. The credibility of the study was enhanced through the use of member checking. The interpreted findings were taken to respondents to confirm the findings before the writing of the paper.

4 Presentation and Discussion of Results

4.1 TWO Activities to Reduce Extreme Poverty

TWO activities to end extreme poverty are discussed below. The organisation is using economic strengthening, vocational training, educational support and healthcare support to end extreme poverty. After the presentation of activities used by TWO to fight extreme poverty, a deeper analysis of the contribution of the activities to poverty reduction follows.

4.1.1 Economic Strengthening

To counter extreme poverty emanating from lack of income, TWO is using Internal Savings and Lending Schemes (ISALS) to fight off extreme poverty. According to the respondents, the organisation since 2015 has trained over 6000 people emanating from over 5000 households. The trained household representatives are making monthly savings and spearheading household income-generating projects like gardening, buying and selling and small shops (tuckshops). Explaining the organisation's ISALS approach, respondent 1 said "... we have realized that many aspects of poverty emanate from lack of income. Consequently, our organization is using the ISALS approach to ensure that households have access to financial income". Some studies in Zimbabwe concur with the use of the ISALS methodology to fight poverty (Machokoto 2014; Zimunya 2015). In Ghana, Adjei et al. (2012) found out that a strategy used to reduce poverty is expanding the economic activities of the poor. Many NGOs in Zimbabwe such as Care International, World Vision, Plan International, GOAL Zimbabwe and International Rescue Committee (IRC) prioritise the ISALS approach in fighting poverty (Muchenje 2018). It is important to note that TWO's economic strengthening activities are integrated with other activities. For instance, the households being assisted with school fees participate in ISALS. The savings made on education are supposed to be used to start ISALS savings. From participation in ISALS, the household representatives are expected to start household income-generating activities based on a market analysis done with assistance from TWO. This integrated approach shown in Fig. 4.1 resonates with the SDG framework where several challenges are addressed together (Schaefer et al. 2018).

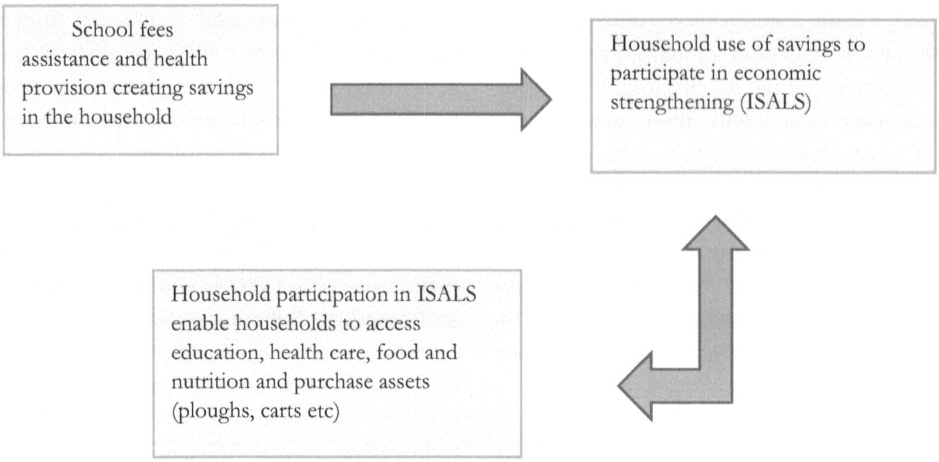

Fig. 4.1 Integration of TWO economic strengthening activities. Source: Authors

4.1.2 Vocational Training for Youth

As a result of poverty, over half of in-depth interview respondents mentioned that many young people in Chegutu district are resorting to antisocial behaviours like engaging in prostitution, alcohol abuse and drug abuse. "Many of our young people because of poverty end up abusing drugs, involved in theft and many other social ills", reiterated respondent 3. To counter poverty amongst the youth, TWO is spearheading vocational training. The organisation is sending young people to Mashayamombe Training Centre and Chibero College to pursue farming, hairdressing, dressmaking, welding, carpentry and cosmetics. Literature shows that vocational training is not affordable for many households no wonder non-state actors like TWO chip in (Pindiriri and Muhoyi 2011). Respondent 3 who is an official of TWO highlighted that "… we are sending young people normally 16 years to 24 years who have just completed secondary education to vocational training centres to enable them to be able to take care of their needs be it financial or otherwise after the completion of the training". Thus far the organisation has managed to send 25 young people to vocational centres. The organisation is focusing on those who have completed ordinary level studies. Studies have shown that as a result of poverty, this age group is keen to experiment with drugs and prostitution endangering their lives. This age group in many societies marks the transition into adulthood. Thus, the need for self-reliance is important. As the SDGs prioritise inclusivity, vocational training offers the opportunity to include youth in mainstream development.

4.1.3 Education Provision

TWO is fighting education poverty through education provision. According to narrations from the respondents, the organisation is assisting orphans and vulnerable children with school fees payment and stationery support. Many studies agree that education is the pathway to success (Julius and Bawane 2011; Palmer et al. 2007). For Mihai et al. (2015, p. 856) "in the nowadays economy, to complete a post-secondary education can make the difference between a life of poverty and a secure economic future". Adjei's et al. (2012) study shows that the provision of educational assistance to needy households is a common strategy used by NGOs to fight education poverty in poor households. The organisation's educational support fights poverty on two fronts. Firstly, paying school fees for children drawn from struggling households allows the households to make savings and channel the funds to other household needs. In the context of TWO, the savings made on school fees are supposed to be channelled towards ISALS and household income-generating projects. Respondent 1, an official of TWO reiterated:

"Educational assistance allows households to make savings and direct the money saved to other households needs. Savings made must be channelled to ISALS to enable households to generate the financial income needed to afford household basics such as food and health care". Secondly, there are better individual prospects for an educated individual (Julius and Bawane 2011). TWO's activities are in line with the SDGs, which seek to eliminate poverty and provide quality education by 2030.

4.1.4 Health Provision

TWO health provision activities in Chegutu district include HIV risk assessment, Adolescent Girls Empowerment Program (AGEP) and Early Childhood Stimulation (ECS). Respondents stated that in the recent past before the decentralisation of anti-retroviral therapy (ART), the organisation used to provide transport to children to enable them to access drugs. Respondents further narrated that these activities, just like in supporting education, allow households to make savings on funds that could otherwise be spent on health care. For sustainable community development to take place, health communities are an essential component; hence, the SDG framework places "People" at the centre of development. Provision of health is not a new activity for NGOs. In Africa health activities by NGOs include sexual reproductive health (SRH), family planning, fighting HIV and AIDS, supporting immunisation programmes and distribution of mosquito nets (Adjei et al. 2012).

4.2 Analysis of the Contribution of TWO Activities in Ending Extreme Poverty

Some households participating in household economic strengthening activities like ISALS reported that ISALS has ensured improved access to financial income. During an FGD, respondent 9 said that "... because of loans and share outs I am getting from ISALS, my financial position has improved. I purchased assets like ploughs, scotch carts, axes and wheelbarrow thanks to ISALS". This shows that economic strengthening like ISALS plays an important role in asset accumulation. Zimunya's (2015) study in Zimbabwe shows that participants of village saving groups accumulated durable assets like scotch carts, ploughs and other assets. Poverty has been, in many instances, conceptualised as lack of assets. Other respondents in the study reported the ability to send children to school, purchase food and pay for medical care. For instance respondent 15 submitted that "... ever since I started participating in ISALS, I have managed to provide decent meals to my children and pay for their school fees" and yet other respondents cited the ability to purchase food for household consumption eliminating food poverty. "... Chegutu district has been experiencing severe droughts for the past years, thus food shortage is a reality for many households. As a family, we have survived because the income from ISALS is helping us to buy maize and other food requirements for the family", said respondent 6. Studies in Africa such as those of Anyako et al. (2007), Harelimana (2018), and Mwansakilwa et al. (2016) show that participation in savings groups leads to improved access to food, ability to pay for school fees and medical needs for the household. In a study by Zimunya (2015), households in savings groups escaped malnutrition as they were able to afford a balanced diet. These indicators on food, education and health care are important because they denote the presence of poverty or lack of it. A similar study on village savings also discovered that "members invested in the business, which had increased households' disposable incomes and enabled them to continue building their savings base and expanding or diversifying their business" (Anyako et al. 2007, p. 15). To ensure the sustainability of well-being, households are engaging in income-generating activities (see above paragraph) as an offshoot of ISALS. ISALS were found to be helping many households reduce many "poverties" they face like food poverty, education poverty and asset poverty.

To argue that all households have experienced poverty reduction as a result of participating in ISALS belies reality. This is because the extremely poor, even after making savings on education and health care, are unable to participate in ISALS nor to establish IGAs. Even when coerced by TWO to join ISALS as per programme design, the extreme poor are not able to join or when they join, they find it difficult and become inactive. This approach instead of eradicating poverty has increased the vulnerability of the poor because when the extremely poor join ISALS instead of actively using loans borrowed from groups, they again borrow to service the loan, impoverishing themselves further.

The youths that have managed to start small businesses after vocational training have managed to create a niche for themselves particularly in rural areas. An interviewed respondent 11 reiterated that "… I was fortunate to get capital from a relative and I am running a small dressmaking business. Many of my clients are parents who place an order for uniforms. I cannot complain, as the money I am getting is better than nothing. I can afford to take care of myself as well as contributing to the welfare of our household". Asked how the vocational training has enabled them to escape extreme poverty many youths framed their responses in terms of employment creation vis-à-vis growing unemployment in the country. Respondent 15 submitted that "… I can say that at least I am getting some few dollars from my farming business. I can buy food, clothing and airtime for myself. It is better than doing nothing given high unemployment levels amongst the youth in the country". Vocational training has only worked for the youths who accessed capital after their training. Prosperity, a key component of the SDGs, has been witnessed in some youths. However, those youths who finished their training and lack access to capital continue to wallow in poverty. This finding concurs with a study by Pindiriri and Muhoyi (2011) who argue that young people who went through vocational training have difficulty setting up businesses largely due to lack of capital. In addition, SNV's (2009) study propounded that vocational training is failing to produce desired results in Zimbabwe due to lack of toolkit provision for the trainees. With rapid deindustrialisation since 2000, getting employment is even a bigger challenge. This reveals a gap in TWO programming; the youths that finish vocational training must be linked to sources of capital to enable them to put into practice their vocational training skills.

The causality relationship of education and poverty is clear—educated people earn more than uneducated people in wage employment (Palmer et al. 2007). Mihai et al. (2015, p. 857) concur that "lack of education perpetuates poverty and poverty limit access to education". A longitudinal study is required, unlike this study that only covered the period from 2017 to 2018. This is because in the long term, education enhances skills and the productivity of poor households in as much as it also increases the wage level and overall welfare of the population (Julius and Bawane 2011). Having children attend school is important because "inadequate education can be considered a form of poverty" (Julius and Bawane 2011, p. 72). In Chegutu, the savings made on education and health care have enabled households to participate in ISALS which improved access to income, food and nutrition and enabled households to acquire assets such as wheelbarrows. However, for labour-constrained households and extremely poor households even with savings made on education and health care, participation in ISALS, a gateway to poverty reduction, is an overwhelming task. TWO health and education activities, despite contributing to poverty reduction, are contributing to achieving SDGs on health and education (Manuel et al. 2019), a reflection of the holistic and integrated nature of the SDGs.

4.3 Challenges for the Attainment of SDG 1

4.3.1 Donor Dependency

The greatest criticism levelled against NGOs in development work is their over-dependency on donors. Brown and Kalegaonkar (2002) submit that organisations that depend on international funders are forced to pursue the interests and priorities of the funders. Concurring is Moyo (2001) who argued that NGOs end up doing something completely new outside their vision and aims. TWO is dependent on Hospice Palliative

Association of Zimbabwe (HOSPAZ) and USAID for funding. This means that the organisation's approaches are determined by its funders. Low-level solutions to poverty horned over the years are superseded by donor requirements. In the context of SDGs, the funding arrangements must be flexible, to avoid creating donor dependency, which is characteristic of current donor-NGO funding arrangements.

Related to donor dependency is the particularism of the NGOs. Particularism refers to a scenario where NGOs cannot do something outside their work plans, budgets and proposals. This finding agrees with Brown and Kalegaonkar (2002, p. 235) who found out that "NGO particularism can also be a sectoral weakness when NGOs fail to respond to interests outside their narrowly defined constituency". As a result of restricted focus, TWO has failed to respond to poverty using other well-known tools of fighting poverty, that is, advocacy and being a watchdog over the Government of Zimbabwe. Failure to watchdog over mismanagement of the country by NGOs, TWO included, explains why economic structural problems that dwarf local level achievements (see below) persist. To deal with poverty TWO must move away from concentrating only on service provision and engage in advocating for pro-poor policies. At present, because of particularism determined by proposals, the organisation cannot do that.

4.3.2 Economic Structural Challenges

Since the year 2000, Zimbabwe has been experiencing economic problems of great proportion (Murisa 2010). These problems include inflation, unemployment, liquidity problems and poor economic performance (Bond 2019). These economic problems have affected TWO's efforts towards poverty reduction in two ways. Firstly, the operating environment for the organisation is difficult. Due to liquidity problems, obtaining cash from banks to spearhead activities is a difficult task. This has delayed the implementation of poverty reduction activities. Shortages of basic requirements like fuel have also made Zimbabwe a difficult operating environment. This finding converges with Brown and Kalegaonkar's (2002) finding that material shortages pose a difficult choice for NGOs, hence the dependence on donors which creates further problems.

Secondly, the macroeconomic difficulties tend to erode poverty reduction gains made at the household level. Other studies are instructive in that poverty in the villages transcends households to encompass the entire community, where inadequate stock of physical and social infrastructure within the communities provides a weak foundation to support individual households that are trying to make strides out of endemic poverty (Adjei et al. 2012). Macroeconomic conditions in Zimbabwe do not support households' efforts to escape from poverty. Inflation is the biggest contributor to the erosion of gains made in the attempt to lift households from poverty. This finding reveals that efforts by NGOs alone may not end poverty without enabling macroeconomic conditions. Attaining SDG 1 ought to be a holistic drive, underlined by both state (macro) and NGO (micro) efforts. The NGOs' drive to attain prosperity for the poor is being severely undermined by broader economic constraints like inflation.

4.3.3 Limited Coverage

While TWO is playing important roles in ending extreme poverty, limited coverage is limiting its impact. Because of donor requirements, the organisation has targets, which means that some of the poor households that could benefit from the organisation activities are excluded. For instance, in Chegutu district there are so many young people being affected by poverty and in need of vocational competencies. So far, the organisation has only assisted about 25 young people with vocational training. Added to that, the education assistance being offered by TWO suffers from limited coverage. With over 16,000 OVC in the district (National AIDS Council 2018 statistics), the organisation's assistance covers slightly above 1.5% of OVC. Limited coverage by NGOs is also reflected in their "piecemeal style" (Kabonga 2016, p. 91) where they cover certain wards in the district without saturating the district. Thus, with this approach, a universal reduction of poverty in the Chegutu district remains a daunting task.

5 Conclusion

There are limited studies that assess the contribution of NGOs in attaining SDGs in Zimbabwe, more so, on how NGOs are contributing to ending extreme poverty. This knowledge gap underlines the focus of the study on the contribution of TWO to end extreme poverty. Given the reality of poverty in the Chegutu district, TWO is responding to the poverty dynamics mainly through project implementation. Activities implemented to end extreme poverty include vocational training, education assistance, health provision and economic strengthening through the ISALS component. Vocational training has worked for the youth who managed to get capital after skill training. Income generated from ISALS has helped households to pay school fees for children and pay for medical care and accumulation of household assets. Assistance rendered on education and health care, besides contributing to the attainment of SDG goals on education and health, is reducing poverty through relieving households so that they can direct savings to needy areas. The chapter argues that TWO is making a significant contribution to ending extreme poverty though there are both internal and external challenges hindering progress.

Externally, the macroeconomic constraints in Zimbabwe tend to erode benefits made at the local (micro) level. Donor dependency as another challenge that confines the organisation's poverty reduction strategies to the dictates of the donors superseding micro level strategies perfected over time. Given the limitations of TWO, the study recommends the following:

- To address limited coverage inherent in many NGO operations, NGOs must partner with the private sector. Such partnerships are important in broadening coverage. Such partnerships are bound to work since the private sector has obligations to contribute to the attainment of goal 1.
- A weakness observed with TWO's contribution to attaining goal 1 is confining itself to project implementation, neglecting a greater role of advocacy. A greater role that NGOs should adopt is holding the government accountable. The Government of Zimbabwe (GoZ) has over the years exhibited a hesitant approach in prioritisation of SDG implementation.
- GoZ must implement favourable policies to achieve the SDGs. Currently, macro-level policies like the tight fiscal policy of the current government tend to erode the strides achieved at the local level by NGOs.

References

Adeyemi, L.S., Ijaiya, G.T., & Raheem, U. (2009). Determinants of poverty in Sub-Saharan Africa. *An International Multidisciplinary Journal. 3*(2), 162-177.

Adjei, P.O.W., Agyemang, S., & Afriyie, K. (2012). Non-Governmental Organization's poverty reduction in Northern Ghana: Perspectives of beneficiaries on strategies impact and challenges. *Journal of poverty alleviation and international development*, *3*(2), 47-73.

Ankeroye, T.O., Nygard, H.M., & Eyo, A. (2018). Towards implementation of Sustainable Development Goals (SDGs) in developing nations. *International Area Studies Review. 21*(1) 3-8.

Antonio, G. (2015). *Doing development in extraordinary environments: Exploring the impact of NGOs in Zimbabwe*. University of Denver, Denver, United States of America.

Anyako, E et al (2007) V*illage and Loans Associations: Experiences from Zanzibar*, Nairobi: Decentralised Financial Services.

Banks, N., & Hulme, D. (2012). *The role of NGOs and civil society in development and poverty reduction* (BWPI Working Paper 171). Manchester: The Brooks World Poverty Institute.

Bond, P. (2019). *In Zimbabwe, capitalist crisis +Ultra-neoliberal policy='Mugabesque' authoritarianism*; available at https://www.sahistory.org.za/archive/zimbabwe-capitalist-crisis-ultra-neoliberal-policy-mugabesque-authoritarianism-patrick-bond.

Brown, L.D., & Kalegaonkar, A .(2002). Support organisation and the evolution of the NGO sector. *Profit and Voluntary Sector Quarterly, 31*(2) 231-258.

CEEWeb for Biodiversity. (2017). *Uniting for the 2030 agenda for Sustainable Development; Best practises for NGO involvement*. Budapest: CEEWeb for Biodiversity.

Chimire, C., & Chitongo, L. (2018). Non-Governmental Organisations and sustainable development: The case of Action Faim community garden in Gutu Ward 10, Zimbabwe. *European Journal of Research in Social Sciences. 6*(6)-12-24.

Cresswell, J.W. (2003). *Research design: Qualitative, quantitative and mixed methods approaches*. Thousand Oaks: Sage.

Dar, F. (2014). Emerging role of NGOs in the world's socio-political affairs. *International Journal of Peace and Development Studies, 6*(1), 1-9. doi https://doi.org/10.5897/IJPDS204.0202.

Dube, E. (2019). Rarity or reality of some theories of poverty in explaining the poverty in Africa. *The Journal of Rural and Community Development, 14*(3), 48–60.

Hege, E., & Demailly, D. (2018). *NGO mobilisation around the SDGs*. Paris: IDDRI

Hege, E., & Demailly, D. (n.d.). *How do NGOs mobilise around the SDGs and what the ways forward*. Paris: IDDRI

Fofana, I., Mabugu, M.C., & Mabugu, R. (2019). *South Africa milestones to achieving the Sustainable Development Goals on poverty and hunger*. (IFPRI Discussion Paper 01731). Washington: International Food Policy Research Institute.

Harelimana, J.B. (2018). The contribution of village savings and loans associations to household welfare in Rwanda. *Global Journal of Human Sciences, 18*(4), 1-11.

Hassan, A., & Forhad, A. (2013). The roles of NGOs in the sustainable development in Bangladesh. *Present Environment and Sustainable Development, 7*(2) 60–72.

Jackson, R., Darlene, K., & Camara, K. (2007). What is qualitative research? *Qualitative Research Reports in Communication. 8*(1), 21-28.

Jaiyesimi, R. (2016). The challenges of implementing Sustainable Development Goals in Africa: The way forward. *Africa Journal of Reproductive Health*, 20(3), 13-18.

Julius, M.K., & Bawane, J. (2011) Education and poverty: Relationship and concerns, A case Study of Kenya. *Problems of education in the 21ST Century. 21*.72-85.

Kange'the, S.M., & Chivanga, S. (2015). Exploring the benefits and gaps inherent in non-governmental organization (NGO) in their quest to effectuate poverty reduction: Examples from African countries. *Journal of Human Ecology, 51*(1.2), 131-140.

Kabonga, I. (2016). Development in quandary, shortcomings and inadequacies of Non-Governmental Organisations: A perspective from the field. *Research Journal of Humanities and Social Sciences, 7*(2) 89-94.

Lewis, D. (2010). *Nongovernmental organizations, definition and history*. Retrieved from https://doi.org/10.1007/978-0-387-93996-4_3

Lewis, D., & Kanji, N. (2009). *Non-governmental organizations and development*. New York: Routledge.

Machokoto, A.G.V. (2014). *Assessing the village savings and loans programme as a sustainable cost-effective rural finance in Zimbabwe: Case of Midlands Province* (Unpublished Dissertation), Harare: University of Zimbabwe.

Machingura, F., & Nicolai, S. (2018). *Contextualising the SDGs to leave no one behind in health: A Case study from Zimbabwe*: Washington: ODI.

Magure B .(2015). Interpreting urban informality in Chegutu, Zimbabwe. *Journal of Asian and African Studies. 50*(6), 650–666. https://doi.org/10.1177/0021909614535568

Manuel, M., Manea, S., Samman, E., & Evans, M. (2019). *Financing the end of extreme poverty*. Washington. ODI.

Mowawa, S. (2013). *Political economy of crisis, mining and accumulation in Zimbabwe: Evidence from Chegutu Mhondoro area*. (Unpublished Doctoral Thesis), University of KwaZulu Natal, Durban, South Africa.

Mihai, M., Titan, E., & Manea, D. (2015). Education and poverty. *Procedia Economics and Finance, 32* (1) 855 – 860.

Moyo, B.(2001). International foundations, agenda setting and non-profit sector in South Africa. *AJIA, 4*(1) 93-118.

Muchenje, R.(2018). *Advent of internal savings and lending schemes: A sustainable solution to Women Empowerment Thirst in Shamva*. (Unpublished Dissertation). Bindura: Zimbabwe Ezekiel Guti University.

Mugadziwa, D.L.(2012). Accelerating towards the achievement of Sustainable Development Goals in Zimbabwe: A youth centric perspective. Retrieved from: https://www.google.com/search?q=Mugadziwa%2C+D.L.(2012).+Acceralating+Towards+the+achievement+of+Sustainable+Development+Goals+in+Zimbabwe%3A+A+youth+centric+perspective.&rlz=1C1GCEU_enZW884ZW884&oq=Mugadziwa%2C+D.L.(2012).+Acceralating+Towards+the+achievement+of+Sustainable+Development+Goals+in+Zimbabwe%3A+A+youth+centric+perspective.&aqs=chrome..69i57.295759646j0j7&sourceid=chrome&ie=UTF-8

Murisa, T. (2010). *Social development in Zimbabwe: Development Foundation for Zimbabwe Discussion Paper*. Harare: Development Foundation of Zimbabwe.

Mwansakilwa, C., Tembo, G., Zulu, M.M., & Wamulume, M. (2016). Village savings and loans associations and household welfare: Evidence from Eastern and Western Zambia. *Journal of Agriculture and Resource Economies 12*(1), 85-97.

Nhapi, T.G. (2019). Socioeconomic Barriers to Universal Health Coverage in Zimbabwe: Present Issues and Pathways Toward Progress. *Journal of Developing Societies, 35*(1):153–174.

Nour, M.M.A. (2017). Quest for effectiveness local government institutions to implement sustainable development goals in Bangladesh. *Journal of Administrative Studies, 10*(1), 9-20.

Palmer, R., Wedgehood, R., & Hayman, R. (2007). *Education out of poverty: A Synthesis Report on Ghana, India, Kenya, Rwanda, Tanzania and South Africa*. Edinburgh: Centre for African Studies.

Pindiriri C & Muhoyi, E (2011) An assessment of vocational training and youth employment in Zimbabwe. *Zambezia: Journal of Social Sciences, 38* (1),77-94.

Schaefer, A., Muller, A., & Alfonso-Gallegos, L. (2018). The role of development partners. In W. Fourie (ed), *Implementing the Sustainable Development Goals in South Africa* (pp 28-37*)*. Pretoria: South African SDG Hub.

Salamon, G., & Anheier, H. (1992). In search of the non-profit sector: In search of definitions. *Voluntas, 13*(2), 125–152.

SNV-Netherlands Development Organization. (2009). *Scoping study on the provision of post-primary school technical/vocational education in Zimbabwe*. Harare: SNV.

SOS Children's Villages. (n.d.). *Care for children is care for development*. SOS Children's Villages.

Suharko, J. (2007). *The role of NGOs in rural poverty reduction: The Case of Indonesia and India*. Japan: Nagoya University.

The Globe-Scan Sustainability Survey. (2019). *Evaluating progress on the SDGs*. Globe Scan.

United Nations Children Fund (UNICEF). (2011). *Annual Report (Zimbabwe*). UNICEF

United Nations Department of Economic and Social Affairs (2005). *The inequality predicament: Report on the World social situation 2005*. New York, NY: United Nations. Retrieved June 02, 2018 from https://www.un.org/esa/socdev/rwss/docs/2005/rwss05.pdf

Willetts, P. (2002). *What is a non-governmental organization?* Oxford: Eolss Publishers Oxford.

World Bank. (2019), *Understanding poverty*. Retrieved from https://www.worldbank.org/en/topic/poverty/overview

Zimbabwe Poverty Atlas. (2015). *Small area poverty estimation: Statistics for poverty eradication*. Harare: ZIMSTAT

Zimunya, A.A (2015). *The impact of village savings and lending association on rural livelihoods: A case of Vheneke village district* (Unpublished Dissertation). Gweru: Midlands State University.

Programs to Support Sustainable Development Goals: Contributions from Social, Inclusive, and Transformative Innovation

5

Eliana Villa, Daniel Cardona Valencia, and Alejandro Valencia-Arias

Abstract

This chapter analyzes some social initiatives developed in Colombia related to the Third Innovation Policy Framework, which focuses on social, basic, inclusive, frugal, and/or transformative innovation practices in the Sustainable Development Goal (SDG) implementation. A literature review approach that allows to identify innovative social programs was adopted as the research method. Subsequently, the initiatives are analyzed based on established categories, which draw on the economical, the social, and the environmental dimension of sustainable development. From this categorization, the identification of elements of analysis (principles) that are stipulated in the Green Paper 2030: Science and Innovation Policy for Transformation in Colombia is carried out. The investigation looks at participation, learning, and experimentation, as well as interdisciplinary and anticipation of results and effects. The findings allow the characterizing of key players involved in the initiatives and the identification of innovative forms that are being used to adapt, locate, and implement the changes necessary to achieve the SDGs. The findings also allow the proposal of new policies, plans, programs, practices, and projects that can be adapted in similar contexts.

Keywords

SDGs · Colombia · Social programs · Innovation · Transformative

1 Introduction and Background

The 17 intertwined Sustainable Development Goals (SDGs) are global development goals agreed by the United Nations member countries to improve living conditions around the world. These SDGs bring together social, economic, and environmental issues, integrating various sectors of society and requiring the coordination of joint actions between public organizations, businesses, civil society, and academia at both the national

E. Villa (✉)
Instituto Tecnológico Metropolitano ITM, Universidad Nacional de Colombia, Bogotá, Colombia
e-mail: elianavilla@itm.edu.co

D. Cardona Valencia
Instituto Tecnológico Metropolitano ITM, Medellín, Colombia
e-mail: danielcardona@itm.edu.com

A. Valencia-Arias
Fundación Universitaria Católica del Norte, Medellín, Colombia

G. Nhamo et al. (eds.), *Sustainable Development Goals for Society Vol. 1*, Sustainable Development Goals Series, https://doi.org/10.1007/978-3-030-70948-8_5

and international levels (United Nations 2015). The SDGs integrate society, not only from the perspective of the beneficiary, but also from the planning and implementation point of view. The 2030 Agenda for Sustainable Development is the plan that countries around the world should and have been implementing to advance efforts to comply with the SDGs. It requires the implementation of better strategies, public policies, business, and government participation, where science, technology, and innovation (STI) can have a fundamental role to meet the challenges of the globalized society (Chavarro et al. 2017).

Sustainable development cannot be separated from science, technology, and innovation, even less so in today's increasingly globalized and interconnected world. Advances in STI can not only be applied to SDGs that require specific, state-of-the-art knowledge to achieve their goals; they must also be integrated into all contexts and reinvented. According to the Scientific Advisory Board (SAB) of the UN Secretary-General (2014), the role of STI for sustainable development is crucial to achieving the necessary advances, since it can integrate different types of knowledge, educate more people about science, and formulate solutions to everyday problems. It also offers mechanisms to break down political, national, cultural, and even mental barriers. Whether through technology transfer or creation of proprietary technology, these mechanisms can greatly involve the specific needs of citizens to meet the different goals of the SDGs innovatively. Social innovation is a great tool for this inclusion process.

From 2013 to 2015, as part of the process of analysis and consolidation of the SDGs, a global survey of citizens was carried out to find out opinions on the different SDG issues and to prioritize sustainable development issues. This consultation process would then provide a global vision regarding their importance, but also an overview of the visions of each country. According to the results of the survey during which 9.5 million participants chose their priorities concerning development, it was discovered that the five main priorities at a global level were improvement of education, good health service, better job opportunities, honest and sensitive government, and access to healthy and nutritious food (My World 2015). In Colombia, similar results were obtained and education emerged as the main priority. Integrating the protection of forests, rivers, and oceans was another priority and this can be related to environmental consciousness in Colombia linked to having one of the biggest and most important ecological systems in the world.

In Colombia, a series of programs have been incorporated that aim to link human welfare, environmental conservation and preservation, and economic growth of the different regions with a social component. This positioning allows the opportune implementation of SDGs and a series of actions and projects directed at the advancement of the agenda 2030 to take place. These goals can be linked to the expectations for addressing aspects that require attention to improve the life of Colombians, such as discrimination, hunger, environmental pollution, and even poorly structured development plans that have affected progress in various sectors of society. This brings into consideration what the United Nations, and therefore all member nations, declared as the main objective, which is to ensure the protection of people and the planet and assure prosperity (Klasen 2018), the so-called triple bottom line from a business perspective.

It is important to highlight that Colombia is a country located in South America, with important natural resources and a privileged geographical position because of its access to the Pacific and Atlantic oceans. It is a country full of natural, cultural, and social wealth and with people recognized for their resilience and positivity. However, throughout its history, the country has also lived through long and harsh internal conflicts that have generated administrative difficulties and significant numbers of forced internal displacements and economic inequality; therefore, there is evident socioeconomic vulnerability (Sierra-Puentes and Correa-Chica 2019). Given such situations, actions in the direction of sustainable development, in search for poverty reduction and equitable growth as proposed in the SDGs and the 2030 Agenda for Sustainable Development, become relevant.

Within the structuring of the 17 SDGs, there are three main foundations: the social, the economic, and the environmental aspects. Within this study, we emphasize three of these goals, which are part of the social foundation, namely ending poverty—ending fundamental needs in all forms worldwide (SDG1); zero hunger—ending starvation, achieving food security and improved nutrition, and promoting sustainable agriculture (SDG2); and quality education—ensuring inclusive, equitable, and quality education and promoting lifelong learning opportunities for all (SDG4) (United Nations 2015). Overall, to meet the SDGs, the national government has advanced the creation of the High-Level Interagency Commission for the Enrollment and Effective Implementation of the Post-2015 Development Agenda and its Sustainable Development Goals, as well as the initiative of the Center for Sustainable Development Goals for Latin America and the Caribbean, based at the University of Los Andes (Chavarro et al. 2017). Both are entities that are responsible for the process of generating ideas, projects, strategies, and plans that can be implemented, monitored, and consolidated within the framework of the SDGs in the regions of the country, emphasizing the characteristics of the territory, forming leaders, and allying science, technology, and innovation as great tools within this global project.

In recent decades, innovation policies and frameworks have been developed in the constant search for economic development, social interest, and solutions to problems related to the capitals of countries (Echeverría 2008a; Kaplinsky 2013; Lichtensztejn 2001; Sabal 2002). For many years, innovation has been seen as a generator of economic development and a positive contributor to growth, since it generates higher levels of productivity; improves the relationship between the state and companies; and strengthens exports, foreign trade, generation of employment, and addressing of social and environmental challenges (Hernández and Pérez 2016).

However, for some years now, the need has been established for an approach aimed at solving specific problems in society as a living and mobile actor which not only receives but also contributes and builds within the interventions, as a path to equity and inclusion (Fressoli 2015; García-Mosquera et al. 2018; Villa et al. 2017). These visions propose a scope of the processes of science, technology, and innovation (STI), as a process of inclusive and human nature, capable of delivering solutions from the plurality of specific demands, of social, environmental, economic, and sustainable character with the complicity of base entities (Cardona 2020; Chataway et al. 2017; Fressoli 2015; Ramírez 2016; Villa et al. 2017).

This new concept shares a similar relationship with social innovation, which is a dynamic and collective element that focuses on the improvement of living conditions in society, through the participation of different actors. This concept is based on the "social turn" that has been given to innovation, which no longer obeys only the implementation of technology, but also efforts since the social with profound and improved benefits arising from the integration of citizens in development processes (Ezponda and Malillos 2011).

In this aspect, the role that society has within these processes of change and innovation is of vital importance in the path of adopting measures to achieve the SDGs, which precisely seek to provide equality and dignity to people as the main axis of development. Therefore, and considering that the SDGs are established from a global scope, it becomes necessary to think about the diversity of scenarios and particularities depending on the context of each country and its regions, which makes it necessary to have a broad vision, which is at the same time precise to identify problems so as to advance towards an inclusive and equitable development. Once again, it is important to have different actors that allow for collective action for the social change that is expected to be achieved, allowing for the formation of the best alliance, where formal knowledge is integrated with the reality of the population (Montgomery 2016).

In the same sense, Gutiérrez et al. (2019) indicate that the processes in which citizens have participation can be considered as elements that improve the impact of the initiatives, since they allow an adequate identification of the problems and consequently an active proposal of the possible alternative solutions that are more coherent with the context, making that the knowledge, the science, and the technologies can be involved simultaneously, contributing much more to the improvement of the quality of life of the citizens from the experience generated in the day to day of the beneficiaries.

This chapter therefore investigates some social initiatives developed in Colombia related to the Third Innovation Policy Framework. The framework focuses on social, basic, inclusive, frugal, and/or transformative innovation practices in the SDG implementation. The rest of the chapter comes as follows: the next section focuses on the materials and methods. This is followed by the presentation and discussion of results, before the conclusion.

2 Materials and Methods

The research is based on a literature review done to identify innovative social programs at the national, departmental, and/or municipal level, leveraged on local and municipal development plans. These initiatives are analyzed based on categories established from the three dimensions of sustainable development, namely economic, social, and environmental. Based on this categorization, the identification of elements of analysis that are stipulated in the Green Paper 2030: Science and Innovation Policy for Transformation in Colombia is carried out, namely directionality, participation, learning and experimentation, interdisciplinarity, and anticipation of results and effects.

3 Presentation and Discussion of Findings

Over the years, guidelines for innovation frameworks have been identified, seeking to establish relationships between key actors for development, such as the state, the university, and the private sector, based on a series of responsibilities and guidelines. Models focused on research and development (R&D) (Chataway et al. 2017; Schot and Steinmueller 2016) and on innovation systems (IS) (Dutrénit 2013). Fundamentally, interest was in competitiveness and economic development especially where society was not included as an active and fundamental part of the innovation processes (Chataway et al. 2017; Schot and Steinmueller 2016).

When talking about processes involving innovation, research, science, and technology, universities and the education sector in general have played the role of generators of knowledge, as well as trainers of future leaders and, more specifically in sustainable development, as drivers of change and observers of the fulfillment of goals. In this way, academia can generate models of sustainable practices for society, educate the population in skills to confront those great social challenges, as well as in research to provide solutions to those challenges, thus achieving interaction with local actors (Stephens et al. 2008; Zilahy and Huisingh 2009). However, not only these efforts do depend on the academia, but it is also important to involve government, industry, and especially civil society, who can participate in building community networks, building development plans, and creating frugal innovation. Thus, these processes can be seen in a much fairer way by having the participation of citizens, and knowing their points of view, which provides more detailed information about the context.

For some years now, innovation frameworks have begun to include the need to highlight society and to discuss innovation from the perspective of social inclusion and transformation. Authors such as Fressoli (2015) or Hernández and Pérez (2016) propose to include innovation also in informality with the inclusion of a diversity of participants as tools to create unexpected solutions to complex problems from different practices, at different levels and with views that include solutions generated from the same actors that are reflected in the base of the pyramid and that may have a different view of the same problem. The purpose of inclusive innovation is to promote development and transformation in the face of inequity and (unfair) distribution of wealth with a view that also encompasses excluded groups (De Beckker et al. 2019; Dutrenit 2017; GBM 2018).

According to Echeverría (2008b), traditional innovation frameworks propose that the concept of innovation refers to success in a market measured from a purely economic perspective, an inclusive social innovation that seeks learning and application of everyday life to policymaking processes. The role of society in innovation has been exposed in different experiments and examples such as those of Riggs and Hippel (1994) show that innovation could come from any actor in the productive chain such as users, distributors, and suppliers of supplies and not only from manufacturers and/or producers of good services. In addition to the information and telecommunications society, the relevance of social innovation practices becomes fundamental to support the producers of knowledge. Although in the end innovations are manifested in the markets, many of them are born and have their prior emergence in a clear social, cultural, artistic, or local domain. The indicators of social innovation can detect or anticipate possible innovations in a canonical sense of the term (OECD/Eurostat 2018).

Furthermore, within the nontraditional processes, such as the incorporation of innovation in sustainable development, there are options on how to integrate the community in a more participatory way in transformation and contextual changes that also articulate with the goals of the SDGs. Inclusive innovation, for example, is a type of innovation that targets not only affordability but also localization in underdeveloped countries, with the aim of reducing poverty (Mortazavi, et al. 2020). There are grassroots innovations as well which take advantage of the knowledge that is outside academia and enhance the creativity of the beneficiary populations to contribute to the transformation processes. Finally, there is open innovation, which is facilitated by digital technologies, and which seeks the creation of collaborative solutions under the philosophy of transdisciplinary knowledge sharing (Chavarro et al. 2017). These are some of the ways innovation can become an instrument that broadly aids the transformation processes led by the members of the society itself.

The STI policy formation processes, as mechanisms to address the challenges posed by SDGs, need to be supported by academia, business, state, and civil society to have a transformative and context-specific approach. In particular, STI policy in developing countries, in order to comply with the 2030 agenda, requires the mobilization of a variety of actors, including citizens, through participatory processes that enable the transition to sustainable development (Walsh et al. 2020). This participation among the different actors seeks to give a better focus to the innovation systems that allow to respond to the systemic failures, which are characterized by the fact that the actors of the national, regional, sectoral, and/or technological systems do not cooperate or have weak links that make difficult the learning and increase of their capacities to generate innovation (Weber and Rohracher 2012). According to Villa et al. (2020) the spaces in which the different actors converge around the processes of innovation and the construction of STI policy are of great importance since they allow knowing the capacities of the different actors, as well as exposing their realities and the search for consensus to achieve the proposed objective.

Society not only contributes from the generation of innovation within a transformative approach, but also generates spaces of visibility, discussion, deliberation, and reflection to socio-technical, environmental, and economic changes in a sustainable way. This society does not start the process from scratch since some processes and initiatives are already being carried out in the territories but may not have been supported or prioritized by local or national policies. It is in the society where adverse aspects are experienced, such as the conflicts which, from the transformative innovation, are understood as an opportunity to reason and validate different points of views with which narratives are constructed (Colciencias & SPRU-University of Sussex 2018).

3.1 Transformative Innovation Framework

The Transformative Innovation Framework (TIF) aims at achieving socio-technical changes, linking interactively the actors of the system where not only the state, the academia, and the private sector but also the society stand out in active participation (Fig. 5.1), to look for specific solutions. The policies form an inclusive approach, which generates development through alternative ways and which recognizes the communities not only as recipients but also as fundamental generators of intentions (Chataway et al. 2017; Schot and Steinmueller 2016; Villa et al. 2017).

Innovation has a growing directionality towards being a fundamental mechanism for the development and economic growth of nations and contributes from a position that generates both challenges and solutions (Chataway et al. 2017). The transformative concept refers to the role that the actors of the innovation framework have in terms of productivity and competitiveness, but from joint learning, local knowledge, and perspective of living conditions (Amaro-Rosales and de Gortari-Rabiela 2016).

Inclusive innovation as set out in this framework is more oriented towards the challenges and needs of excluded groups, defined as all individuals and communities that do not have similar contexts, access to goods and services, and scope of administrative and/or political decisions (Hernández and Pérez 2016). These require partnership, generation of production and knowledge networks, transmission of ideas, and governance in decisions from their inclusion in ICTs (Ramírez 2016).

Society is included in this framework to ensure that all partners and constituent parts of the country's problems are addressed. This means involving individuals, families, and communities (traditional beneficiaries) in the design and implementation of each part of the solution, taking into account their perceptions, ideas, opinions, and feedback; that is, there must be innovation not only in the solution but also in the

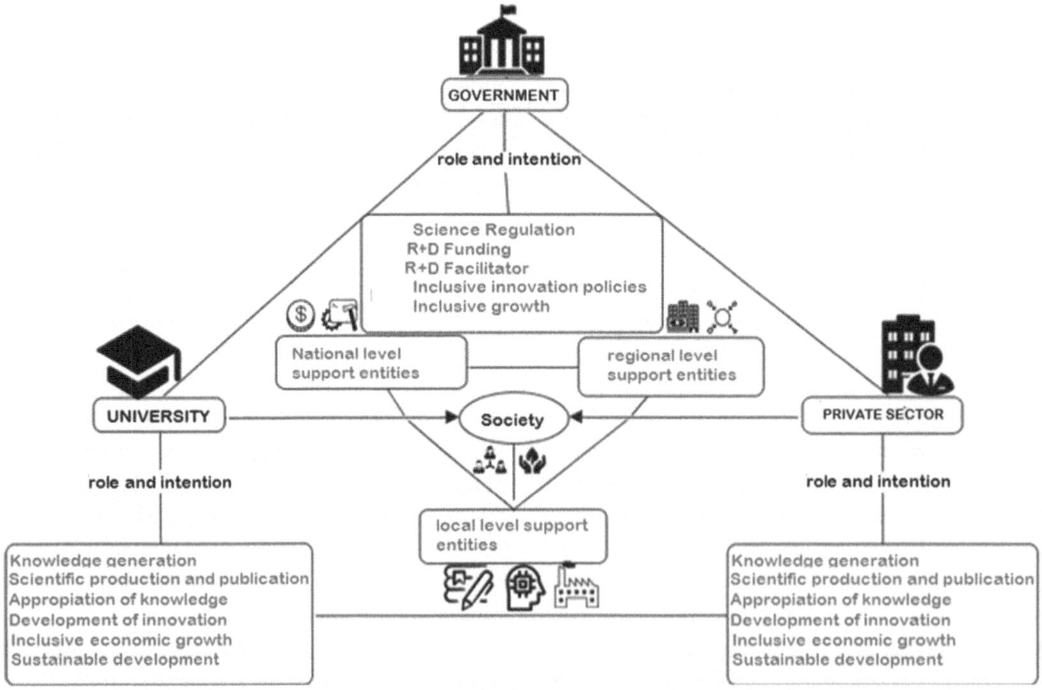

Fig. 5.1 Transformative Innovation Framework. Source: Adapted from Dutrénit (2013), Schot and Steinmueller (2016) and Chataway et al. (2017)

process through which that solution is created, financed, and implemented (Guaipatín and Humphreys 2014; Schot and Steinmueller 2018).

3.2 Examples of Social Intervention as a Factor of Change in the World

Over the years there has been the development of innovation frameworks in Colombia. Framework 1 is associated with research and development (R&D) and with a clear division of labor (Schot and Steinmueller 2016). Framework 2 is associated with an innovation system focused on actors and their relationships (Dutrénit 2013) and the more recent commitment to a framework of social and transformative innovation (Chataway et al. 2017). The successful contribution of STI to the SDGs has been highlighted based on the efforts not only of academia, but also of civil society, industry, and governments.

However, at this level of reflection on policy frameworks for development, communities are vital to the ownership of these goals, joining in with ideas and actions that contribute to this common achievement. Proof of this are some social programs that have emerged from the communities themselves, with leaders who intervene with ideas that move people with a common goal, thus promoting solutions to the SDG-related problems of these communities.

For example, Project H Design recently adjusted to the name "Girls Garage" (according to its gender focus). It is a proposal that was born in rural Bertie in North Carolina, United States, in 2008. This small town, the poorest in the state, had problems in the education system; its economy was based only on agriculture and many of its inhabitants who had the opportunity left to go to big cities. This program, led by women and born from the community, integrates design as a tool for education at all levels, teaching youth design, creativity, and vocational building skills by giving them the creative, technical, and leadership tools needed to achieve a positive and lasting change in their lives and communities. This program is aligned with SDGs 4 (quality education) and 11 (sustainable cities and communities), as it has improved the education that youth receive and promoted new economic and social dynamics within the community. By 2020 they had developed more than 120 projects and trained hundreds of girls (Girls-Garage 2020).

Another example of social actions is that of the Indian Shaffi Mather, who with his idea called 1298 for Ambulance has contributed to improving the lives of many inhabitants of India. This idea focuses on emergency care through ambulance service in regions with little or no access to this service. It is a community service independent from government entities or health bodies, oriented towards SDGs 3 (health and welfare) and 10 (reduction of inequalities). It provides opportunities for emergency medical assistance, helping to reduce the number of deaths in the city of Mumbai, promoting a crowdsourcing system where people with good economic status and willingness to support the service pay, while the poor pay little or nothing, and get help from other aid funds and investors (Acumen 2020).

These and many other ideas that have been born from the frugality of the communities give an account of the orientations that should be taken into account when it comes to action policies. The communities are the ones that know their needs best and within them the priorities, and it is precisely on these priorities that these social initiatives work. It is important to emphasize that the search for initiatives highlights in a very marked way the role of women as fundamental actors in families and communities that are always thinking about better socioeconomic conditions for their own and in the constant search for better conditions for the next generation.

According to Basu et al. (2013) frugally oriented initiatives born from society reflect a response to the call for sustainability from the same social, environmental, and economic needs around the world as citizens with products and services, which because they are generated from the same base are considered appropriate, adaptable, affordable, and accessible. These proposals for actions, in their great majority, are generated by entities that are classified as foundations, non-governmental organizations (NGOs), and small businesses of the triple bottom line type (businesses with a social, economic, and environmental dimension) and that seek to create value, generate income, and reach beneficiaries and that seek to generate an impact in addition to a sustainable activity (Landoni and Trabucchi 2020).

3.3 Social Initiatives in Colombia

Based on the analysis of social innovation witnessed in different contexts of the world and the orientations that these initiatives must become sustainable ideas, the guidelines proposed by Krlev et al. (2014) and McNally et al. (2020) identify elements of social innovation and social entrepreneurship from the identification of sustainable, economic, and social aspects and environmental and transformative problems and with levels of social innovation based on the capacity to identify and interpret social problems from the dynamics of the environment; the capacity to plan new products, services, processes, or methods aimed at solving problems and social inclusion; the capacity to interpret the impact, evaluate it, and learn about it; and the capacity for governance, understood as the mechanisms for public participation.

To classify these initiatives in the country, we have as reference the six elements proposed by the Transformative Innovation Policy Consortium (TIPC) (2019). They establish a basis on CIT guidelines (Table 5.1) that social innovation initiatives must be directed at solving the problems and needs of territories based on their particularities, but with a global scope and direction and taking into account the use of diverse knowledge and local know-how.

Table 5.2 shows a list of frugal and socially based initiatives born and developed in the communities. Three fundamental aspects are highlighted: *Framework Conditions of Development* refers to the conditions in which the initiative is generated, its policies, social climate, actors, interactions, and its physical and financial structure. *SDGs to which the activity is oriented* relate to the general development characteristics of the initiative seen from the SDG approach and the classification of the interventions. Finally, *Specific Products and Results Obtained* are the results of the innovation activity, to those products or services that meet a social or human need which include education, employment, environment, health and care, housing, social capital, networks, and political or community participation.

Table 5.1 Elements of Transformative Innovation Policy

Elements	Definition
Directionality	The policy has a clear approach to considering the non-neutrality of technology, which reviews a wide range of options and addresses the social and environmental problems they could cause
Social objective	The initiative focuses on the major social challenges expressed in the SDGs
Systemic level impact	It focuses on high-impact change at the socio-technical system level
Learning and reflection	The initiative allows for "second-order" learning
Conflict and consensus	Differences of opinion among stakeholders are recognized and encouraged
Inclusion	It requires the inclusion of civil society actors and/or end users in its design

Source: Transformative Innovation Policy Consortium (TIPC 2019)

The United Nations goal of ending global poverty (SDG2) by 2030 will require that high economic growth rates are maintained and that growth is inclusive and shared across society so that poverty reduction is maximized. High and sustained growth occurs through structural change; however, structural change is associated with growing disparities between rich and poor. In contrast, inclusive growth is best achieved with decreasing inequality to maximize poverty reduction (ESRC GCRF 2016).

From these premises, it is evident that the fundamental questions for achieving sustainable development are related to the need to find an economic development model that guarantees rapid economic growth and structural change that achieves an expanding share of national income for the poor. Additionally, it is fundamental that,

Table 5.2 Social initiatives identified in Colombia. Source: Authors' own construction

Entrepreneurial activity	Framework conditions for development	SDGs to which the action is directed	Field-specific products and results
Light for the Wayuu communities: Spinoff of Barefoot College https://www.barefootcollege.org	Under the foundation model, with a voluntary character, the aim is that rural communities use their cultural heritage and full of culture and adapt to new technologies Barefoot College generates innovation by allowing semiliterate indigenous women over 35 years to be trained and educated to pass on knowledge to their children, grandchildren, and the community; learning technical knowledge in solar engineering to achieve and bring light to their homes and generate sustainability in these communities to have, among members of the same community, knowledge of technologies, and to develop community self-management of the use of light, payment for services, and creation of fund savings for the purchase of new equipment	This initiative encourages gender equity and the use of renewable energy as a strategy to reduce inequalities through affordable and clean energy systems for sustainable communities It involves the community in the manual construction of their energy systems	• Generation of clean and renewable energy • Generation of financial saving strategies
Power Pal Pueblo: Empowerment of territories https://www.powerpalpueblo.com/	Nonprofit organization with the support of international consultants and volunteers works under a welfare model with the joint work of families to paint their homes, collect garbage, and strengthen the culture, recreation, and urban agriculture for the quality of life of rural and vulnerable populations	Through its environmental care campaigns, the NGO encourages the care of water sources and basic sanitation, as well as responsible production and consumption in the communities, generating dignified conditions for housing This entity stimulates the community on the improvement of housing with the contribution of the people of the community, who contribute their labor to the transformation of spaces	• Improving housing conditions in rural communities • Health services, improved drinking of water, better communications, and the Internet • Production and trade of products
Saldarriaga Concha: for academic inclusion https://www.saldarriagaconcha.org/la-fundacion/	A nonprofit family foundation that focuses on developing projects for people with disabilities and older adults. They design projects to transform public policies and human behavior	This initiative seeks to reduce social exclusion and improve the generation of quality formal education for children and young people with disabilities through the generation of decent academic and employment opportunities It contributes to the community, using their effort and time for the training of the disabled and their families, who build future together	• Flexible teaching practices and teaching materials • Inclusive employment fund • Community gardens

(continued)

Table 5.2 (continued)

Entrepreneurial activity	Framework conditions for development	SDGs to which the action is directed	Field-specific products and results
Genesis y Corona Fundación https://www.genesis-foundation.org/	A foundation that seeks to reduce school dropout rates in children from low socioeconomic levels under the premise of improving the number of children who complete their basic studies	The initiative seeks quality education and reduction of academic inequalities in early childhood in vulnerable conditions	• Training in language and mathematical skills for children and teachers in marginal areas of Colombia
Fundación Visibles http://fundacionvisibles.org/es/	This initiative was born from students and professionals from different areas who seek to develop intervention projects, volunteer training, and research on social problems in the city of Medellín, with a special focus on making street inhabitants visible and generating educational and recreational spaces for children in vulnerable situations	The entity works on projects to reduce hunger and to end poverty, based on interventions with people who depend on begging The model is not purely assistance oriented; it seeks to get people to reflect and prepare themselves to continue with a productive life based on training and attitudes to life	• Educational activities in marginalized communities • Delivery of food and resocialization processes to street people
Picachu con Futuro https://www.picachoconfuturo.org/index.php/empresas-sociales/telecentro-picacho.html	Social initiative in a vulnerable area of the city of Medellín, where mothers, young athletes, community action councils, and social leaders organized themselves along lines of work to achieve the development of a marginal neighborhood together. Its objective is to promote scenarios of citizen participation where security is generated for women, a food bank, and a digital inclusion center	This initiative, which does not have any legal incorporation, seeks to reduce poverty and hunger in the areas of community cooperation, training, and sport as strategies for health and welfare and sustainability The community is part of the actions born and developed by the community itself	• Training in Information and Communication Technologies • Events to gather economic resources that are reinvested in the community • Food Bank

based on public policies to manage the compensation between structural change and inclusive growth, there are economic and institutional arrangements that regulate the socioeconomic compensations of structural transformations in the most equitable manner (ESRC GCRF 2016).

From the scientific evidence, it is suggested to address these compensations through:

(a) Economic incentives: Reforms have an impact on private incentives. Based on a highly distorted balance, where the incentive is not aligned with the allocation and efficient use of resources, reforms can have a major impact on poverty and inequality.
(b) Technological changes: Technological changes include the adaptation of better technologies, increased productivity through "learning by doing," and efforts that lead to technological innovations. Investing in R&D and STI activities is necessary.
(c) Physical and human capital accumulation and access to the capital market: With the right incentives, physical and human capital accumulation has a large impact on future income. Unequal access to the capital market and quality education has a profound implication on income inequality.
(d) Social transformations: The first three factors can have a profound impact on the household and community. These are low-frequency changes, however, within one or two generations; these changes have enormous implications, such as changing patterns of gender inequality, fertility, and demographic transitions. A key development associated with vestiges of social change in the direction of the SDGs is the rural-urban immigration, reducing overtime and employee participation in rural areas, transforming agriculture from subsistence family farming into large and efficient corporate production. However, this transition can increase urban poverty, impacting income inequality.
(e) Political changes: Economic expansion and the resulting changes in social organization and income distribution have been associated with political changes, where underrepresented groups (women, poor, indigenous, etc.) can gain influence. Similarly, new developing sectors can reduce the political influence of declining sectors. These changes can lead to more equal access to education, health services, safety nets, etc. (Aizenman et al. 2012).

Empirical evidence from the social sciences shows that the effects between structural change and inclusion can be mediated when technologies achieve new significance as societies build institutions in response to structural changes that threaten cohesion or identity (De Haan 2015). As mentioned above, the trade-offs between inclusion and structural change in Latin America are addressed from the desire of nations to achieve the SDGs.

4 Conclusions

From this work, it emerged that innovations of a social, inclusive, or transformative nature exist and are developed in communities. However, this innovation requires greater visibility, structure, and alignment with local and national policies. Many of these initiatives are empirical and do not have an administrative structure that makes them sustainable social enterprises. In the international sphere, the constitution of NGOs for the development of these activities is noteworthy, but in the local sphere, reference is made more to the constitution of foundations and not to associations, since this generates tax benefits associated with the local regulation that exempts foundations from payments. The innovative initiatives presented could have been approached from the traditional innovation paradigm, but the direction and intention change with the inclusive approach. The intention is to give specific directionality and intentionality to excluded communities, not served by the common business sectors, nor by the state. Additionally, there should be high- or low-technology solutions accessible and affordable to these communities. The intention is to achieve inclusion, social equality and equity, respect for the environment, and SDGs.

References

Acumen. (2020). 1298 for Ambulance. Retrieved from https://acumen.org/investment/1298.html/

Aizenman, J., Lee, M., & Park, D. (2012). *The Relationship between Structural Change and Inequality: A Conceptual Overview with Special Reference to Developing Asia*.

Amaro-Rosales, M., & de Gortari-Rabiela, R. (2016). Innovación inclusiva en el sector agrícola mexicano: los productores de café en Veracruz. *Economía Informa, 400*, 86–104. https://doi.org/10.1016/j.ecin.2016.09.006

Basu, R. R., Banerjee, P. M., & Sweeny, E. G. (2013). Frugal Innovation Core Competencies To Address Global Sustainability Introduction: A Call For Global Sustainability. *Journal of Management for Global Sustainability, 2*, 63–82. https://doi.org/10.13185/JM2013.01204

Cardona, D. (2020). Revisión bibliográfica sobre inclusión financiera como estrategia de recuperación y de crecimiento fintech. *Semestre Económico, 23*(55), 183–203.

Chataway, J., Daniels, C., Kanger, L., Ramirez, M., Schot, J., & Steinmueller, E. (2017). Developing and Enacting Innovation Policy. *Developing and Enacting Transformative Innovation Policy*, 18–21.

Chavarro, D., Vélez, M., Tovar, G., Montenegro, I., Hernández, A., & Olaya, A. (2017). Los Objetivos de Desarrollo Sostenible en Colombia y el aporte de la ciencia, la tecnología y la innovación. *Colciencias*. https://doi.org/10.13140/RG.2.2.31118.87368

Colciencias y SPRU-University of Sussex. (2018). *Orientaciones para la formulación de políticas regionales de innovación transformativa en Colombia*. Retrieved from http://tipconsortium.net/wp-content/uploads/2018/07/Orientaciones-formulacion-politicas_defweb.pdf

De Beckker, K., De Witte, K., & Van Campenhout, G. (2019). Identifying financially illiterate groups: An international comparison. *International Journal of Consumer Studies*, 1–20. https://doi.org/10.1111/ijcs.12534

De Haan, A. (2015). *Social inclusion and structural transformation: Concepts, measurements and trade-offs*.

Dutrenit, G. (2017). *Innovaciones inclusivas: un modelo basado en agentes*. (March).

Dutrénit, G. J. S. (2013). *Sistemas de Innovación para un Desarrollo Inclusivo "Experiencia de latinoamerica."*

Echeverría, J. (2008a). El manual de Oslo y la innovación social. *ARBOR, Pensamiento y Cultura, CLXXXIV*(732), 609–618. https://doi.org/10.3989/arbor.2008.i732.210

Echeverría, J. (2008b). El manual de Oslo y la innovación social. *Arbor*.

ESRC GCRF. (2016). Structural Change, Inequality and Inclusive Growth: Tensions and Trade-offs (Academic Panel) – EADI NORDIC 2017.

Ezponda, J. E., & Malillos, L. M. (2011). Cambio de paradigma en los estudios de innovación: El giro social de las políticas Europeas de innovación. *Arbor*. https://doi.org/10.3989/arbor.2011.752n6002

Fressoli, M. (2015). Movimientos de base y desarrollo sustentable : la construcción de caminos alternativos. *CIENCIA E INVESTIGACIÓN, 65*, 56–68.

García-mosquera, J., Villa Enciso, E., & Cardona. (2018). Innovación inclusiva. revisión a partir de redes de colaboración: Caso Colombia. *Universidad Nacional de México*.

GBM. (2018). Inclusión financiera. Retrieved from https://www.bancomundial.org/es/topic/financialinclusion/overview

Girls-Garage. (2020). Girls Garage. Retrieved from https://girlsgarage.org/

Guaipatín, C., & Humphreys, M. O. (2014). Innovación social en salud. *I, LAB Banco Interamericano de Desarrollo*.

Gutiérrez, A., Gómez, A. M., De Alba Bastarrechea, R., Del Pino Carazo, E., Villagrasa, A. M. C., García, B. G., … García, E. G. (2019). La participación ciudadana. In *Manual de uso del diputado*. https://doi.org/10.2307/j.ctvk3gp9b.41

Hernández, J. L. S., & Pérez, C. D. (2016). Innovación para el desarrollo inclusivo: Una propuesta para su análisis. *Economía Informa, 396*, 34–48. https://doi.org/10.1016/j.ecin.2016.01.002

Kaplinsky, R. (2013). *Innovation Knowledge Development EARLY 21 ST CENTURY IN LOW AND MIDDLE INCOME ECONOMIES*.

Klasen, S. (2018). Human Development Indices and Indicators: A Critical Evaluation. *2018 UNDP Human Development Report Office Background Paper*.

Krlev, G., Bund, E., & Mildenberger, G. (2014). Measuring What Matters - Indicators of Social Innovativeness on the National Level. *Information Systems Management*. https://doi.org/10.1080/10580530.2014.923265

Landoni, P., & Trabucchi, D. (2020). Sustainability Models for Social Innovation Projects: An Empirical Perspective. In *Research for Development*. https://doi.org/10.1007/978-3-030-31117-9_11

Lichtensztejn. (2001). Pensamiento económico que influyó en el desarrollo latinoamericano en la segunda mitad del siglo veinte. *Revistas Bancomext*, 92–99.

McNally, P., Apostolopoulos, N., & Al-Dajani, H. (2020). *Social Innovation in Refugee Support: Investigating Prerequisites Towards a Conceptual Framework*. https://doi.org/10.1007/978-3-030-26245-7_8

Montgomery, T. (2016). Are Social Innovation Paradigms Incommensurable? *Voluntas*. https://doi.org/10.1007/s11266-016-9688-1

Mortazavi, S., Eslami, M. H., Hajikhani, A., & Väätänen, J. (2020). Mapping inclusive innovation: A bibliometric study and literature review. *Journal of Business Research*. https://doi.org/10.1016/j.jbusres.2020.07.030

My World, the U. N. G. S. for a B. W. (2015). *My World Survey 2015*. Retrieved from https://myworld2030.org/

OECD/Eurostat. (2018). Oslo Manual 2018. In *Handbook of Innovation Indicators and Measurement*. https://doi.org/10.1787/9789264304604-en

Ramírez, M. (2016). Inclusive Innovation, Capacity Building and Social. *Palmas*, *37*, 151–158.

Riggs, W., & von Hippel, E. (1994). Incentives to innovate and the sources of innovation: the case of scientific instruments. *Research Policy*. https://doi.org/10.1016/0048-7333(94)90008-6

Sabal, J. (2002). *Financial Decisions in Emerging Markets* (Oxford Uni). New York.

Schot, J., & Steinmueller, W. E. (2016). *Framing Innovation Policy for Transformative Change: Innovation Policy 3.0*. (September), 0–26.

Schot, J., & Steinmueller, W. E. (2018). Three frames for innovation policy: R&D, systems of innovation and transformative change. *Research Policy*. https://doi.org/10.1016/j.respol.2018.08.011

Sierra-Puentes, M., & Correa-Chica, A. (2019). Realidad socioeconómica de personas en condición de desplazamiento forzado en Colombia: revisión sistemática. *Suma Psicológica*. https://doi.org/10.14349/sumapsi.2019.v26.n2.1

Stephens, J. C., Hernandez, M. E., Román, M., Graham, A. C., & Scholz, R. W. (2008). Higher education as a change agent for sustainability in different cultures and contexts. *International Journal of Sustainability in Higher Education*. https://doi.org/10.1108/14676370810885916

Transformative Innovation Policy Consortium (TIPC). (2019). Tip criteria.

UN Secretary-General. (2014). *The Crucial Role of Science for Sustainable Development and the Post-2015 Development Agenda. Preliminary reflection and comments*. Retrieved from https://www.leopoldina.org/fileadmin/redaktion/Internationales/2014_The_Crucial_Role_of_Science_for_the_Post-2015_Development_Agenda.pdf

United Nations. (2015). Transforming our World: The 2030 Agenda for Sustainable Development. New York: United Nations Secretariat.

Villa, E., Cardona Valencia, D., Valencia-Arias, A., Hormechea, K., & García, J. (2020). *Transformative Innovation Policy, SDGs, and the Colombian University*. https://doi.org/10.1007/978-3-030-26157-3_14

Villa, E., Hormecheas, K., & Robledo, J. (2017). *De la innovación competitiva a la innovación inclusiva: el rol de la universidad latinoamericana*.

Walsh, P. P., Murphy, E., & Horan, D. (2020). The role of science, technology and innovation in the UN 2030 agenda. *Technological Forecasting and Social Change*. https://doi.org/10.1016/j.techfore.2020.119957

Weber, K. M., & Rohracher, H. (2012). Legitimizing research, technology and innovation policies for transformative change. *Research Policy*. https://doi.org/10.1016/j.respol.2011.10.015

Zilahy, G., & Huisingh, D. (2009). The roles of academia in Regional Sustainability Initiatives. *Journal of Cleaner Production*. https://doi.org/10.1016/j.jclepro.2009.03.018

Financial Inclusion as a Complementary Strategy to Address the SDGs for Society

6

Daniel Cardona Valencia, Carola Calabuig, Eliana Villa, and Fray Betancur

Abstract

Based on the Sustainable Development Goals (SDGs), the different and changing functions of society have been examined as part of the Transformative Innovation Policy and the necessary actions from different perspectives, without reference to the role of finance. The objective of this chapter is to analyze the importance of financial inclusion as a complementary strategy to address sustainable development and as a leverage tool to reduce socioeconomic gaps. The methodology we implemented is based on a systematic literature review to establish the relationship between development and financial inclusion and the orientations that can be identified in policies in that area and their influence on the SDGs. Our findings (a) clarify the scope and objective of financial inclusion as a concept and the key pillars to address its absence, (b) establish a causal relationship between financial inclusion and development in the achievement of the SDGs, and (c) and propose policies, strategies, methodologies, and practices that can be applied and/or replicated in different contexts to contribute to the 2030 Agenda. The main contribution of this chapter is that it confirms the importance of financial inclusion in society as a fundamental factor in innovation frameworks and its contribution to SDGs.

Keywords

SDGs · Financial inclusion · Social innovation · Poverty · Financial development

1 Introduction

Since the establishment of the Millennium Development Goals (United Nations 2015) and the Sustainable Development Goals (SDGs) (United Nations Statistics Division 2016) the different and changing role of society has been discussed as part of the Transformative Innovation Policy.

D. Cardona. Valencia (✉)
Instituto Tecnológico Metropolitano ITM,
Universidad Politécnica de Valencia,
Medellín, Colombia
e-mail: danielcardona@itm.edu.com

C. Calabuig
Universidad Politécnica de Valencia, Escuela Técnica Superior de Ingeniería Industrial, Valencia, Spain
e-mail: cacator@dpi.upv.es

E. Villa · F. Betancur
Instituto Tecnológico Metropolitano ITM,
Universidad Nacional de Colombia,
Valencia, Colombiae-mail: elianavilla@itm.edu.co;
fraybetancur64534@ correo.itm.edu.co

G. Nhamo et al. (eds.), *Sustainable Development Goals for Society Vol. 1*, Sustainable Development Goals Series, https://doi.org/10.1007/978-3-030-70948-8_6

According to Schot and Steinmueller (2018), in order to talk about a transformative innovation policy and the active role of society in it, different discussions and frameworks were generated that evolved the concept of research and development (R&D) from economic growth, through the competitive approach and the creation and grouping of networks, to a framework linked to environmental and social challenges that demand a truly transformative change in society. The transformative concept has to be linked to social innovation (Fressoli 2015).

Transformation and growth are related to the need to include the community as a living and contributing player to social, economic, and environmental problems (Transformative Innovation Policy Consortium (TIPC) 2019). After the financial crisis of 2008, concepts such as Fintech and microfinance took hold and political actions and initiatives aimed at "financial inclusion," propoor markets, and microcredit financing began to be identified as part of international development and plans of some countries. However, there were constant doubts about the understanding of these concepts from a purely economic and capitalist perspective or from the vision of powerful strategies that can contribute to the dynamics of sustainability (Gabor and Brooks 2017).

This chapter presents a systematic literature review meant to establish the relationships between development, financial inclusion, and efforts that can be identified in different policies and their prevalence where the scope of the 2030 Agenda and the SDGs is concerned. Furthermore, we aim to identify different possible tools to support poverty reduction as well as methodologies, proposals, or strategies (widely known in the academic context) that promote education, growth, equity, and poverty reduction based on the participation in and access to financial services.

2 Literature Review

Although financial inclusion has made progress in the international context in recent years, no consensus has been reached on its definition. Some institutions have called it *bancarization* (i.e., the increasing penetration of banking services), a term that may be confusing because it could be understood as the use of and access to financial services offered exclusively by traditional banking, as in Ruiz Ramírez (2011) and Rueda et al. (2015). Nevertheless, some studies have tried to give the concept a universal scope bringing it closer to the financial sector in general and the larger population group it covers (Emara and Ayah 2019; GBM 2018; Lauer and Lyman 2015). According to Sarma (2008), financial inclusion (as opposed to financial exclusion) is defined as easy access, availability, and use of the financial system by all the actors in a community. Financial exclusion, in turn, is the lack of adequate and equitable access to financial services due to factors related to prices, market, socioeconomic conditions, perceptions, and self-exclusion (Sinclair 2001).

According to the Global Partnership for Financial Inclusion (GPFI) and the Consultative Group to Assist the Poor (CGAP), financial inclusion refers to "a situation in which all the working age adults have effective access to credits, savings, payments, and insurances provided by formal service providers" (CGAP 2011, p. 8). This definition acknowledges different actors beyond traditional banking that increasingly influence the system. It defines formal institutions as "providers of financial services that have a legal status, which includes organizations (in some countries, even individuals) that meet a wide variety of mandatory characteristics and are subject to different levels and types of external supervision" (World Bank 2018).

The National Council for Financial Inclusion, created in Mexico in October 2011 (CNIF 2016, p. 3), defines financial inclusion as the access to and use of formal financial services under an appropriate regulation that guarantees consumer protection programs and promotes financial education in order for individuals in all the population segments to improve their resource management skills. The Organization for Economic Co-operation and Development, through the International Network on Financial Education, proposes a different description of financial inclusion, more from the perspective of those who receive it than that of the organizations that offer it, but limited to regulated products (OECD/INFE 2013, p. 71). It is the process of disseminating accessible, timely, and adequate knowledge about a wide variety of regulated financial products and services and the expansion of their use to all the segments of society by implementing customized, existing, and innovative approaches, including financial awareness and education in order to promote financial well-being and economic and social inclusion.

Along the same lines, the Central Bank of Brazil complements this idea by adding an orientation toward "a contribution to the quality of life of those who conduct financial operations" (Banco Central do Brasil 2015, p. 19). The United Nations Economic Commission for Latin America and the Caribbean (UNECLAC) defines financial inclusion as a tool that fosters opportunity creation and strengthens the achievement of the Sustainable Development Goals (SDGs) based on poverty and inequality reduction, work, and economic growth which are in turn based on access to financial services with adequate durations for society at large, including vulnerable groups (Aldasoro 2020; CEPAL 2019). Allen et al. (2016) showed that greater financial inclusion is associated with lower account fees (costs associated with the operation and use of services such as bank accounts), more proximity to financial intermediaries, stronger legal rights, and more stable political environments.

Financial inclusion processes should be combined with efforts to design adequate and accessible financial products that promote responsible use. Dermirgüç-Kunt et al. (2008) typified the profiles of financial exclusion (Fig. 6.1) according to willingness, necessity, cultural, religious, and access reasons. In this classification, we should highlight the groups that present voluntary exclusion; they are the communities and households closest to vulnerability conditions. Enabling effective access is only the first step because the products should also be remarkably affordable, convenient, simple, and safe.

According to the database Global Findex (2018), by the end of 2017, there were 1.7 billion unbanked adults, 40% of whom were in the poorest households (Niger, Central African Republic, Chad, Southern Sudan, Burundi, Mali, Eritrea, Burkina Faso, Sierra Leone, Mozambique, and the Democratic Republic of Congo). The World Bank defines the poorest households in developing countries as those that earn less than US$1.90 a day adjusted by the purchasing power parity (Demirguc-Kunt et al. 2018). Their report also stresses the fact that 20% of the unbanked adults refer to a lack of documentation on the topic and distrust in the financial system as their reasons for not considering financial institutions. Nowadays, in addition to a tool for savings and the generalization of the financial system, financial inclusion prepares individuals for unexpected impacts on their income and to smooth out crises, and it is an instrument for economic projection that should be adapted to ICTs (Demirguc-Kunt et al. 2018).

3 Methodology

The methodology of this work is based on a systematic review of literature and computer graphics in order to validate correlations between development indicators and financial inclusion. Subsequently a methodological approach was developed, made up of several processes that allowed the identification of strategies to achieve the SDGs. These processes included:

- Bibliometric study and journalistic follow-up to reports from the United Nations, ECLAC, World Bank, and collaborating agencies: These bibliometric studies consist of evaluating databases with indicators of different natures that seek to determine impact, evolution, state of the art, and trends (Sancho, 1990).
- Banking and NGO consultations regarding SDGs: Follow-up on high-impact documents, reading their conclusions and scopes to determine methodologies, proposals, and challenges in the field.

Based on this methodology, a context is established between financial inclusion and ODS and

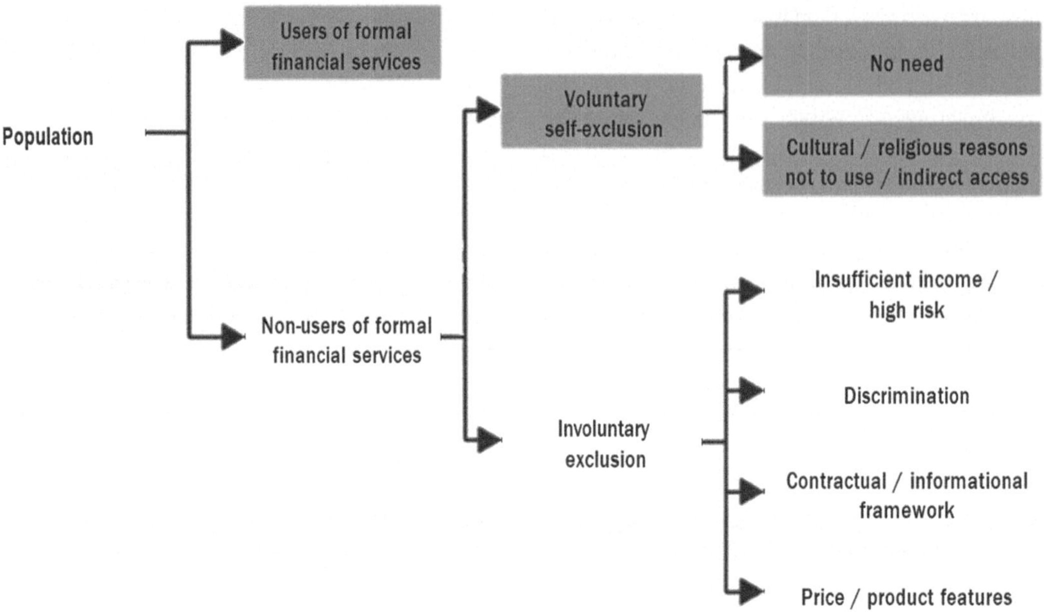

Fig. 6.1 Classification of the use of financial services. Source: Aslı Demirgüç-Kunt et al. (2008).

strategies and actions for inclusion to be recommended are also identified.

4 Presentation and Discussion of Results

The gross domestic products of countries have rapidly grown in recent decades because of the development of standardized products and information technologies, despite constant economic crises. With a rising life expectancy, retirement plans and social welfare have grown, as well as the number of jobs and skills required to maintain an adequate quality of life in the face of the said technological developments and more complex financial products (Annamaria Lusardi 2019). In this situation, financial inclusion becomes important as a fundamental part of economic development and poverty reduction (Cámara and David 2015).

According to Sarma (2008) and Oz-Yalaman (2019), the levels of human development and financial inclusion of countries are correlated, and development means financial inclusion because generalized growth and equality are not possible if part of the population is outside the financial system. King and Levine (1993) used data from 80 countries from 1960 to 1989 and presented econometric evidence that indicates that the financial system promotes economic growth. They found that financial development is a good predictor of long-term growth and is strongly correlated with future economic growth rates, the accumulation of physical capital, and an increase in economic efficiency.

Economic and financial stability is aimed at the generation of competitiveness and efficiency, but instability does not necessarily mean crisis (Nasir et al. 2015). Financial instability refers to the condition of a financial system that can continue operating but is compromised by the savings allowance, investment opportunities, and payment processing in its economy. Therefore, financial inclusion would be a key tool to reduce economic deceleration (Padoa-Schioppa 2003). Financial education should also be inherent to countries' projections because it aims at making daily life easier and enabling households, companies, and countries to establish objectives for different terms, manage liquidity, and respond to unexpected emergencies (World Bank 2018).

Sarma and Pais (2011), examining the relationship between financial inclusion and develop-

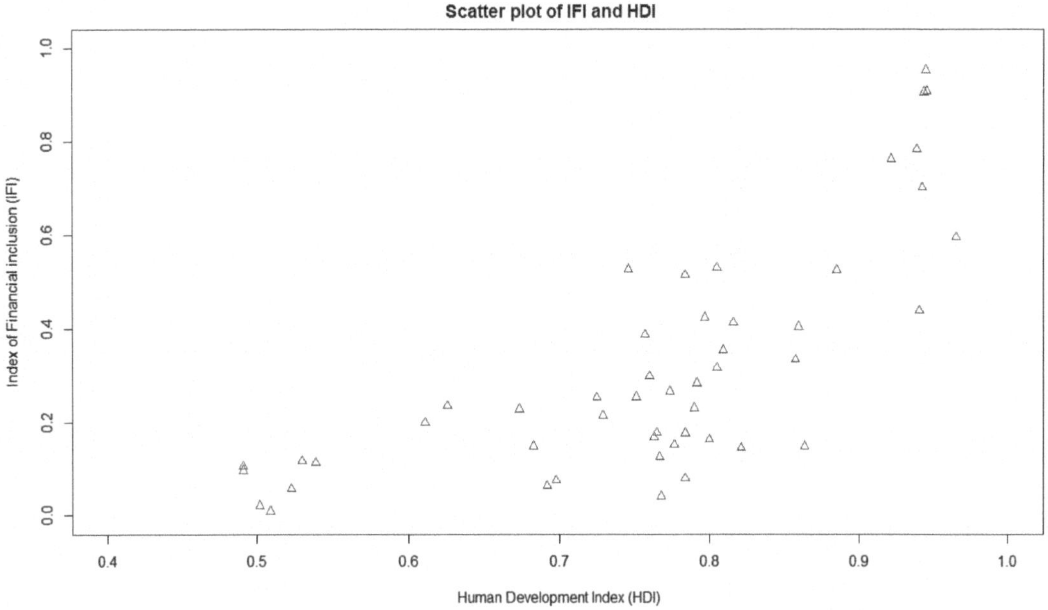

Fig. 6.2 Scatterplot of Financial Inclusion and Human Development Index. Source: Sarma and Pais (2011)

ment, found that the countries with a high and medium financial inclusion index were classified by the United Nations as having high human development (Fig. 6.2). That is, they could establish a positive and growing relationship between (1) the top countries in terms of healthy life expectancy, access to education, and dignified life and (2) the number of nationals that had a savings account, their level of indebtedness, and money management. Income inequality, financial education for adults, urbanization, and connectivity are decisive factors for this relationship.

This positive correlation between developed countries and financial inclusion is constant in more recent studies. For example, Achugamonu et al. (2020) based on the World Bank database examined African countries in the last 10 years and determined that nations with high exclusion also present low financial stability and high unemployment rates, especially in rural areas that have difficulties to access information systems. Likewise, Sha'ban et al. (2020) developed a multidimensional financial inclusion index for a global sample of 95 countries over 12 years. They reported findings that indicate that financial inclusion is a variable that changes over time (Fig. 6.3) and is positively and significantly associated with the per capita GDP, employment, banking competition, human development, government integrity, and Internet use.

4.1 Financial Inclusion and SDGs

On September 25th, 2015, the General Assembly of the United Nations adopted a set of strategies projected 15 years in the future called the Sustainable Development Goals (SDGs) as part of the 2030 Agenda for people, planet, and prosperity often termed the triple bottom line in development studies. Such goals are fundamental instruments that involve leaders and governments around the world to face the challenges of modern society. The effective implementation of the Agenda requires the commitment of governments, the private sector, and civil society. According to the UN, the 2030 Agenda was created by people, for the people. For this reason, the civil society plays the role of not only overseer and guarantor of the implementation of the SDGs, but also a transcendental one in the incidence, support, impact, and development of initiatives in favor of the SDGs (United Nations 2015).

Fig. 6.3 Time trend of the financial inclusion index of three income groups. Source: Sha'ban et al. (2020).

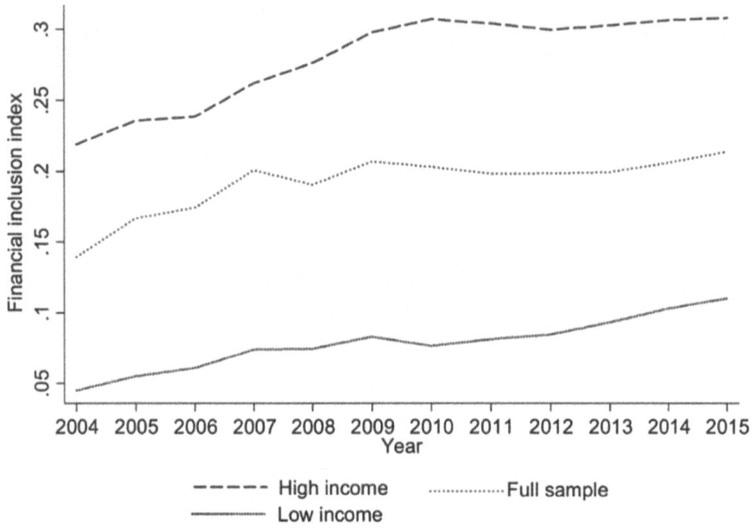

There are 17 SDGs and, although they do not explicitly mention financial inclusion, studies conducted by the World Bank have determined that such inclusion is a factor that facilitates 7 out of those 17 goals (GBM 2018). Klapper et al. (2016) found solid evidence of the link between financial inclusion and development. They specifically claim that access to financial services may directly support the achievement of the SDGs related to poverty reduction, zero hunger, well-being, quality education, gender equality, and economic growth. In the SDGs, financial inclusion is not an end but the means to increase household income; improve nutrition and access to healthcare provision and education; and empower people, especially women (Demirguc-Kunt et al. 2018).

According to the Alliance for Financial Inclusion (2013), in order to align financial inclusion with the SDGs, in addition to an adequate service, financial inclusion demands specific conditions that are necessary to take the idea to productive levels and a social scope based on four fundamental points: political commitment, a regulatory framework, open and inclusive networks, and equitable conditions.

4.2 Political Commitment

According to Sarma (2016), all the countries in Latin America and Asia, as well as some in Africa and Europe, have included in their government agendas passages related to financial inclusion as part of their strategies in government policies. These countries are members of the Global Policy Forum (GPF), which seeks to establish a global agenda, develop policies, and formulate the guidelines that are necessary to achieve an actual implementation of such agenda and sustainable growth in said nations. Among these policies, there are multiple actions aimed at micro, small, and medium-sized enterprises that involve efforts to mitigate climate change and can be related to green finance and Inclusive Green Finance (IGF). The objective of this is to generate financial services focused on climate action, thus adapting the financial structure to create investment products related to sustainable assets (Institute for Global Environmental Strategies 2019). The same institute has suggested that governments should act based on financial inclusion in order to develop policies that provide (help to ensure the availability of financial services for themselves, private enterprises, and households), promote (motivate the private sector to offer quality and inclusive financial services), prevent (reduce the effect of undesired outcomes due to macroeconomic situations), and protect (reduce financial risk with insurance and social payments) (Institute for Global Environmental Strategies 2019).

4.3 Regulatory Framework

According to the Bank for International Settlements (BIS) (2020), only 16.5% of the countries reported information on the implementation of regulations for inclusion initiatives as national policies; as a result, they are created by start-ups or financial institutions that seek to diversify their portfolios and extend their reach. These regulatory frameworks should first clarify the concept of financial inclusion considering the possibilities of each country and subsequently establish possible agencies, approaches, scopes, and responsible parties for said regulation (Yoshino and Morgan 2016).

For that purpose, Emara and Ayah (2019), in their study about the role of governments in the Middle East and the North of Africa in financial inclusion and growth, established four fundamental points that should be prioritized to develop regulations that can deal with problems and turn them into opportunities for stability. These are (1) management and development of financial markets with more inclusive products, (2) expansion of the market agents currently based only on financial institutions, (3) sufficient security assurance for users, and (4) low or nonexistent costs of transactions, ideally unlimited. Governments should normalize and legalize the financial activities, mainly, of their small and medium-sized enterprises, basing public-private relationships on opportunities for access to bank financing and structuring the system as a network of services focused on collaboration.

4.4 Open, Inclusive Networks

One of the best modern allies of financial inclusion is the development of technologies and applications of the Industry 4.0 and the so-called fintech era, whose goal is to boost finance based on information and communications technologies. According to Asli Demirgüç-Kunt et al. (2017), digital financial services can also help people manage financial risk because they enable them to collect money from friends and distant relatives in difficult times. In African countries such as Kenya and Zimbabwe, stable users of mobile money manage their household spending better when their income is reduced than those who do not use financial systems (whose number of transactions decreased between 7 and 10%). There are 469 million people using Web services in the region and an estimated 181 million of them are active users of financial services and mobile money. In the continent, 23.8 billion transactions were recorded in 2019 for more than 456 billion dollars (Velluet 2020).

Connectivity, defined as a technological infrastructure and the equipment that enables the articulation with a global information network, is the basis not only of financial inclusion but also of digital inclusion toward the ultimate goal of overcoming, changing, and reducing the current panorama of territorial disparities caused by the digital gap we are subjected to. Efforts should be made by all the actors (government, private companies, universities, and society) based on communicability, which is the free use of ICTs by communities without restrictions, in order to eliminate barriers for cooperation and expand the social coverage of accessibility and communication based on geography (Duarte and Pires 2011).

4.5 Equitable Conditions

Roa (2013) referred to financial inclusion as a multidimensional element because it should be examined from the standpoints of access, use, and, more recently, quality and nature of the said access and use. Those who do not have access to financial services (except for the self-excluded) are people and/or communities that cannot find financial instruments or services adapted to their needs and have no place in the formal system.

Financial systems, in addition to being generalized, should incorporate strategies that simplify their operation and represent low costs, with less information asymmetry and more peace of mind for users who are not interested in risky investments but rather in controlling their resources. The financial system, just as all the other economic and social actors that generate capital flows, should focus on developing transformative innovation policies that include an active vision of society as an actor that not only is a receiver but also proposes solutions (Villa et al. 2020). The generalization of financial resource management, an evaluative assessment of its use, and good levels of financial trust enable people to make financial decisions that improve their integration to the formal financial system and the individual and collective well-being.

4.6 The Pillars of Financial Inclusion

According to the GPFI (2016, p. 3), financial inclusion is measured in three dimensions: (1) access to financial services based on the provision of convenient and affordable services; (2) use of financial services based on the diversity of possibilities to enter the world of savings, credit, and investment; and (3) quality of the products and the service provision from the standpoints of transparency, security, and effectiveness. The following are some of the concepts included in these dimensions; they are fundamental pillars to achieve financial inclusion.

4.6.1 Microfinance

Two mechanisms of financial inclusion that have shown positive results are microcredit and microfinance, which are defined as small financial projects with an individual, family, or community scope that generate self-employment and sustainability (Crépon et al. 2011). Buera et al. (2013), adopting the approach of the general and partial equilibrium models, obtained results that indicate that (1) microfinance is a redistributive policy in favor of the poor; (2) the typical program of "small credits for small enterprises," when widely available in an economy, can have important added and distributive impacts; and (3) the effects of the general equilibrium, through salaries and interest rates, are quantitatively significant. Multiple experiments in which capital has been made available for small productive processes of villages and agricultural and stock-raising communities have resulted in empowerment (especially of women) and children education as business ideas were financed and poverty was reduced (Beaman et al. 2014; Desai et al. 2013).

From a different perspective, microfinance has also been misused by institutions to exceed high-interest-rate policies and high transaction costs by taking advantage of the segments, justified on the high probability of loan losses, once this type of loan has a high rate of being unproductive and generating low expected profit margins (Campion et al. 2010). This behavior has shown that the financial system still has a long way to go to get into the same dynamic of financial innovation, inclusive finance, and microfinance from its conception, because this capitalist behavior which is based only on financial risk differs from the proposals noticeable in financial inclusion (Sandberg 2015). The implementation of programs for households in vulnerable communities has had effects on their spending and their need to use loan sharks with exaggerated charges and high fees (Attanasio et al. 2013). Microfinance is therefore becoming a stronger tool for social intervention and consolidation of the role of society in transformative innovation policies and a measure to face recent economic decelerations; in fact, society and governments are currently producing tangible results, especially in the rural sector (Jha 2019).

4.6.2 Financial Education

Initiatives for financial inclusion and consumer protection are complementary to financial education to ensure safe access to adequate products for everyone, maximizing opportunities' success and promoting entrepreneurship and inclusive growth. Lack of financial education may result in behaviors that produce increased spending due to a lack of information about basic topics such as the terms of loans, mortgages, interests, or rates that are not paid in loans (A Lusardi 2011).

Different studies have tried to measure the relationship between financial education and financial behavior. For example, a report on the financial education of adults in G20 countries presented important findings that highlighted worrisome low levels of financial education. It found that less than half of the adults (48%) could correctly answer 70% of a set of questions on basic financial knowledge, and only 27% of the participants could calculate a simple interest. Additionally, on average, two-thirds of the respondents in G20 countries had a specific payment product, as compared to less than 10% of the participants in Brazil (8%) and Indonesia (2%). However, in the latter, most respondents had a savings product, which suggests that it could be used as a substitute. In all the G20 countries, 19% of the respondents contacted relatives or friends to ask for informal financial information (OECD & INFE 2017).

Financial education is a fundamental tool to project individuals' lives and plan the future in the short, medium, and long term. According to Lusardi and Mitchell (2017), consumers with lower financial literacy think less about retiring, do not make plans for it, and make less assertive decisions regarding daily purchases. Those who do not know how to correctly calculate an interest rate or a future value apply for more loans, save less, prefer shorter terms for loans and assets, and are, on average, more vulnerable, single, less educated, and unemployed or with low income (De Beckker et al. 2019).

4.6.3 Fintech Products

Great advancements in communications and mobile technologies, as well as payment systems, have created new opportunities to connect people with affordable financial systems. This has been called "digital financial inclusion," and it can be defined as the digital access to and use of formal financial services by population groups formerly excluded and neglected (Lauer and Lyman 2015). In high-income economies, 51% of adults reported having made at least one financial transaction using their cell phone or over the Internet in 2016. Technological innovations, as well as the high mobile penetration (reaching around 1.1 billion or two-thirds of all the unbanked adults around the world), represent a great opportunity to reduce financial exclusion using different fintech ventures (Demirgüç-Kunt et al. 2018).

The International Development Bank (IDB) and Finnovista (2018), in their annual report, identified a substantial increase in the number of young fintech enterprises in Latin America, which grew from 703 ventures in 2017 to 1166 in 2018. Their most representative sectors are payments, remittances, loans, and corporate finance management. Furthermore, financial technology companies presented a historical growth: their income rose from US$18 billion in 2017 to US$39.57 billion in 2018 (CB Insights 2019, p. 9). Ensuring that people benefit from the digitization of financial services requires developments in payment systems, infrastructure, regulations, and consumer protection, in addition to an implementation divided into stages that include financial literacy and numerical skills. This is because nearly 50% of the fintech ventures in Latin America reported orienting their solutions to sub- or unbanked consumers and small and medium-sized enterprises.

5 Conclusions

The definition of financial inclusion has certainly evolved over time. Currently, it is focused on the equitable availability, affordability, and usability of financial services provided by a wide variety of institutions in a financial system permeated by technology, virtual environments, product customization, access to microcredit, and knowledge about the workings of products and services, which are part of a responsible use aimed at development.

Financial inclusion, although not directly defined as a strategy in the SDGs, is an implicit factor that enables development; however, it is not an end. As shown in empirical studies conducted by different analysts, it is a means to orient efforts for the economy, nutrition, education, and the environment that should be complemented with political commitment, development of the associated regulation, and extension of coverage and connectivity as strategies for an effective and continuous implementation.

This literature review about the expansion of financial inclusion and its impact on development reveals many opportunities to include households that are still outside the financial system and new ways to make the most of products and technologies in order to promote their use among those who are in the system but do not have adequate practices due to their lack of financial literacy. Several studies indicate that resource digitization produces benefits beyond the expansion of the system, i.e., improvements in its efficiency.

Furthermore, the role of women in the process of financial inclusion is also essential because their empowerment has resulted in outstanding behavior when they are included in the system.

Nevertheless, financial education also poses challenges for governments, companies, and society (all of them fundamental actors in an inclusive innovation system) and requires shared willingness to expand technologies; create open, equitable networks; enforce government policies; develop easy-to-access, affordable financial products; and carry out teaching-learning processes for financial education.

A recommendation is made here that financial inclusion be considered as an 18th SDG which is an enabler and critical factor for giving life to the other 17 insofar as their implementation requires capital.

References

Achugamonu, U. B., Adetiloye, K. A., Adegbite, E. O., Babajide, A. A., Akintola, F. A. (2020). Financial exclusion of bankable adults: implication on financial inclusive growth among twenty-seven SSA countries. *Cogent Social Sciences*.

Aldasoro, B. (2020). *Mejores prácticas sobre iniciativas para promover la inclusión financiera de receptores de remesas en el ámbito rural*. 87. Mexico.

Allen, F., Demirguc-Kunt, A., Klapper, L., & Martinez Peria, M. S. (2016). The foundations of financial inclusion: Understanding ownership and use of formal accounts. *Journal of Financial Intermediation, 27*, 1–30. https://doi.org/10.1016/j.jfi.2015.12.003

Alliance for Financial Inclusion. (2013). Measuring financial inclusion core set of financial inclusion indicators. *Alliance for Financial Inclusion*.

Attanasio, O., Augsburg, B., De Haas, R., Fitzsimons, E., & Harmgart, H. (2013). Group Lending or Individual Lending? Evidence from a Randomized Field Experiment in Rural Mongolia. *SSRN Electronic Journal*, (136). https://doi.org/10.2139/ssrn.2365432

Banco Central do Brasil. (2015). Relatório de Inclusão Financeira. *Relatório de Inclusão Financeira*, 1–167.

Banco Interamericano de Desarrollo (BID) y Finnovista. (2018). *Informe Fintech en América Latina 2018: crecimiento y consolidación*. https://doi.org/10.18235/0001377

Beaman, L., Karlan, D., Thuysbaert, B., & Udry, C. (2014). *Selection into Credit Markets: Evidence from Agriculture in Mali*. (May).

Buera, F., Kaboski, J. P., & Shin, Y. (2013). The Macroeconomics of Microfinance. *SSRN Electronic Journal*. https://doi.org/10.2139/ssrn.2350287

Cámara, N., & David, T. (2015). Factors that matter for financial inclusion: Evidence from Peru. *AESTIMATIO*. https://doi.org/10.5605/ieb.10.1

Campion, A., Ekka, R. K., & Wenner, M. (2010). Interest Rates and Implications for Microfinance in Latin America and the Caribbean. *IDB Working Paper*. https://doi.org/10.1080/15320383.2014.896860

CB Ivghts. (2019). *Fintech Trends To Watch in 2019*. Retrieved from http://www.ghbook.ir/index.php?name=فرهنگ و رسانه های نوین&option=com_dbook&task=readonline&book_id=13650&page=73&chkhashk=ED9C9491B4&Itemid=218&lang=fa&tmpl=component

CEPAL. (2019). Tendencias recientes de la Población de América Latina y el Caribe. *Día Mundial de La Población 2019*.

CGAP. (2011). *Advancing financial access for the world's poor. Annual Report 2011*. Washington, DC.

CNIF. (2016). *Política Nacional de Inclusión Financiera*. https://doi.org/10.1017/CBO9781107415324.004

Crépon, B., Devoto, F., Duflo, E., & Parienté, W. (2011). Impact of microcredit in rural areas of Morocco: Evidence from a Randomized Evaluation. *IPA Working Paper*, (March).

De Beckker, K., De Witte, K., & Van Campenhout, G. (2019). Identifying financially illiterate groups: An international comparison. *International Journal of Consumer Studies*, 1–20. https://doi.org/10.1111/ijcs.12534

Demirgüç-Kunt, Asli, Leora Klapper, Dorothe Singer, Ansar, S., & Hess, J. (2018). La base de datos Global Findex 2017: Medición de la inclusión financiera y la revolución de la tecnología financiera. Cuadernillo de reseña. In *Banco Mundial*. Washington, DC.

Demirguc-Kunt, A., Klapper, L., Singer, D., Ansar, S., & Hess, J. (2018). The Global Findex Database 2017: Measuring Financial Inclusion and the Fintech Revolution. In *The Global Findex Database 2017: Measuring Financial Inclusion and the Fintech Revolution*. https://doi.org/10.1596/978-1-4648-1259-0

Demirgüç-Kunt, Asli, Klapper, L., Singer, D., Ansar, S., & Hess, J. (2017). The Global Findex Database 2017. In *REWAS'04 - Global Symposium on Recycling, Waste Treatment and Clean Technology*.

Demirgüç-Kunt, Aslı, Beck, T., & Honohan, P. (2008). Policies and Pitfalls in Expanding Access. In *Finance*. https://doi.org/10.1596/978-0-8213-7291-3

Desai, J., Johnson, K., & Tarozzi, A. (2013). On the Impact of Microcredit: Evidence from a Randomized Intervention Rural Ethiopia. *Demography, 48*(2). https://doi.org/10.1007/s13524-011-0029-0

Duarte, F., & Pires, H. F. (2011). Inclusión digital, tres conceptos clave: conectividad, accesibilidad, comunicabilidad. *Ar@cne*.

Emara, N., & Ayah, E. S. (2019). Financial Inclusion and Economic Growth: The Role of Governance in Selected MENA Countries. *Munich Personal RePEc Archive*. Retrieved from https://mpra.ub.uni-muenchen.de/99257/

Fressoli, M. (2015). Movimientos de base y desarrollo sustentable: la construcción de caminos alternativos. *CIENCIA E INVESTIGACIÓN*, *65*, 56–68.

Gabor, D., & Brooks, S. (2017). The digital revolution in financial inclusion: international development in the fintech era. *New Political Economy*. https://doi.org/10.1080/13563467.2017.1259298

GBM. (2018). Inclusión financiera. Retrieved from https://www.bancomundial.org/es/topic/financialinclusion/overview

The Global Findex Database. (2018). Mundial Bank. https://globalfindex.worldbank.org/

GPFI. (2016). *G20 FINANCIAL INCLUSION INDICATORS*. 1–6.

Institute for Global Environmental Strategies. (2019). *Inclusive Green Finance: A Survey of the policy Landscape*. Retrieved from https://www.afi-global.org/sites/default/files/publications/2019-06/AFI_IGF_report_AW_digital.pdf

Jha, S. K. (2019). Role of microfinance institutions in social development. *Apeejay School of Management--Centre for Public Policy & Governance Discussion Paper*, *3*.

King, R. G., & Levine, R. (1993). Finance and Growth: Schumpeter Might Be Right. *The Quarterly Journal of Economics*, *108*(3), 717–737. https://doi.org/10.2307/2118406

Klapper, L., El-Zoghbi, M., & Hess, J. (2016). *Achieving the Sustainable Development Goals: The Role of Financial Inclusion*. Retrieved from www.cgap.org

Lauer, K., & Lyman, T. (2015). *Digital financial inclusion*. Retrieved from http://www.cgap.org/publications/digital-financial-inclusion

Lusardi, A. (2011). *Americans' Financial Capability*. Retrieved from http://www.nber.org/papers/w17103

Lusardi, Annamaria. (2019). Financial literacy and the need for financial education: evidence and implications. *Swiss Journal of Economics and Statistics*. https://doi.org/10.1186/s41937-019-0027-5

Lusardi, Annamaria, & Mitchell, O. S. (2017). How Ordinary Consumers Make Complex Economic Decisions: Financial Literacy and Retirement Readiness. *Quarterly Journal of Finance*, *7*(3). https://doi.org/10.1142/S2010139217500082

Nasir, M. A., Ahmad, M., Ahmad, F., & Wu, J. (2015). Financial and economic stability as 'two sides of a coin': Non-crisis regime evidence from the UK based on VECM. *Journal of Financial Economic Policy*. https://doi.org/10.1108/JFEP-01-2015-0006

OECD/INFE. (2013). Financial Literacy and Inclusion. *Financial Literacy & Education*, (June), 1–175. 10.1787/9789264254855-en

OECD & INFE. (2017). *G20/OECD INFE report on adult financial literacy in G20 countries*.

Oz-Yalaman, G. (2019). Financial inclusion and tax revenue. *Central Bank Review*. https://doi.org/10.1016/j.cbrev.2019.08.004

Padoa-Schioppa, T. (2003). Central Banks and Financial Stability: Exploring a Land in Between. *The Transformation of the European Financial System*. https://doi.org/10.1093/0199270562.003.0008

Roa, M. J. (2013). Inclusión Financiera en América Latina y el Caribe: acceso, uso y calidad. *Boletín Del CEMLA*.

Rueda, L., García, N., Esguerra-Umaña, M. del P., Cano-Sánz, C. G., & Velasco, A. M. (2015). Inclusión financiera en Colombia. In *Política monetaria y estabilidad financiera en economías pequeñas y abiertas. Capítulo 6. Inclusión financiera en Colombia. Pág.:143-211.*

Ruiz Ramírez Héctor. (2011). Conceptos sobre educación financiera. *Observatorio de La Economía Latinoamericana*.

Sandberg, J. (2015). What's wrong with exorbitant interest rates on microloans? In *Microfinance, Rights and Global Justice*. https://doi.org/10.1017/CBO9781316275634.008

Sancho, R. (1990). Indicadores bibliométricos utilizados en la evaluación de la ciencia y la tecnología. Revista Española de documentación científica.

Sarma, M. (2008). Index of Financial Inclusion. In *Indian Council for Research on International Economic Relations Working Paper No 215*.

Sarma, M. (2016). Measuring Financial Inclusion using Multidimensional Data. *World Economics*.

Sarma, M., & Pais, J. (2011). Financial inclusion and development. *Journal of International Development*, *23*(5), 613–628. https://doi.org/10.1002/jid.1698

Schot, J., & Steinmueller, W. E. (2018). Three frames for innovation policy: R&D, systems of innovation and transformative change. *Research Policy*. https://doi.org/10.1016/j.respol.2018.08.011

Sha'ban, M., Girardone, C., & Sarkisyan, A. (2020). Cross-country variation in financial inclusion: a global perspective. *European Journal of Finance*, *26*(4–5), 319–340. https://doi.org/10.1080/1351847X.2019.1686709

Sinclair, S. (2001). Financial exclusion: An Introductory Survey. In *CRSIS, Edinburgh College of Art/Heriot Watt University*.

Transformative Innovation Policy Consortium (TIPC). (2019). Tip criteria.

United Nations. (2015). *Millennium development goals*. 1–75. Retrieved from http://www.un.org/millennium-goals/2015_MDG_Report/pdf/MDG 2015 rev (July 1).pdf

United Nations Statistics Division. (2016). SDG Indicators. *UN Chronicle*. https://doi.org/10.18356/0aec43e1-en

Velluet, Q. (2020). Africa: over 500 million mobile-money users expected in 2020. *The Africa Report*. Retrieved from https://www.theafricareport.com/25846/africa-over-500-million-mobile-money-users-expected-in-2020/

Villa, E., Cardona Valencia, D., Valencia-Arias, A., Hormechea, K., & García, J. (2020). *Transformative Innovation Policy, SDGs, and the Colombian University*. https://doi.org/10.1007/978-3-030-26157-3_14

World Bank. (2018). Financial Inclusion. Retrieved from https://www.bancomundial.org/es/topic/financialinclusion/overview

Yoshino, N., & Morgan, P. (2016). Overview of Financial Inclusion, Regulation, and Education. *SSRN Electronic Journal*. https://doi.org/10.2139/ssrn.2838114

Part III

Leadership in Implementing SDGs

7 Mentoring Women in the Resources Sector: A Leadership Case Study on Mentorship Practices for Effective SDG 5 Implementation

Bjanka Korb and Dawieș (D.A.J.) Bornman

Abstract

The mining sector globally has been a male-dominated industry and in South Africa, women were prohibited from working underground, until the promulgation of the Mine Health and Safety Act of 1996. The Mining Charter of 2002 attempted to facilitate the increase of women into the mining sector by stipulating that at least 10% of employees in core mining operations needed to be women. Despite the legislation, women are still in the minority in the mining sector constituting about 12% of the 2018 total mining workforce. The scarcity of women in the mining sector spurred the creation of the International Women in Resources Mentoring Programme (IWRMP), which aims to develop and retain women in the mining sector. This case study of the IWRMP examines the mechanisms, leadership, and effectiveness of the programme. Mentoring was found as a valuable tool for the development of female leaders as it facilitates the formation and nurturing of relationships between female mentees and mentors, thereby addressing the fifth Sustainable Development Goal. Several leadership capabilities were responsible for the adoption and implementation of the IWRMP, which itself cultivates leadership capabilities in future leaders. Furthermore, a structured mentoring programme should form part of the career development pathway of all women entering the mining sector.

Keywords

Gender equality · Mentoring · Mining · Leadership · SDG 5 · Women

1 Introduction and Background

The mining sector globally has traditionally been a male-dominated industry. Historical contributing factors to this were the harsh physical conditions underground, physically demanding nature of manual mining work, international standards, and South African legislation. The International Labour Organization (ILO) Convention 45 (C045) of 1935 and the South African Minerals Act (Act 50 of 1991) prohibited the employment of women in underground mining work

B. Korb
SRK Consulting, Northlands, South Africa
e-mail: bjanka.korb@gmail.com

D. (D.A.J.) Bornman (✉)
University of Pretoria, Pretoria, South Africa
e-mail: dawie.bornman@up.ac.za

G. Nhamo et al. (eds.), *Sustainable Development Goals for Society Vol. 1*, Sustainable Development Goals Series, https://doi.org/10.1007/978-3-030-70948-8_7

(Mudimba 2017). This meant that women were only allowed to be employed on surface operations of mines which limited them to administrative or laboratory work with the exception of geology work, which is a technical role.

Globally, countries began to denounce ILO C045, and so began the gradual incorporation of women into the mining industry. In South Africa, the promulgation of the Mine Health and Safety Act (Act 29 of 1996) saw the legislative ban on women working underground being lifted. The first Mining Charter of 2002 stipulated that at least 10% of employees in core mining operations needed to be women (Mudimba 2017). These two legislative instruments led to an increase in the percentage of women employed in the South African mining industry; however, progress has been slow. In 2019 the Minerals Council South Africa (2019) indicated that women in the mining workforce fill 16% of top management positions, 17% of senior management positions, and 18% of skilled technical profession positions. It is clear from these numbers that, despite the enabling legislative environment, the mining industry remains largely male dominated throughout all levels of mining organisations. Gender inequality is also evident in the mining consulting services industry, although to a lesser extent, because much of the work is conducted in an office environment.

There are several initiatives that enable and encourage women to work in the mining sector; however, there are several factors that continue to make mining unappealing to women, including harassment, cultural perceptions, and facility design. Leaders of mining organisations and mining consulting organisations alike have a role to play in achieving gender equality in the mining sector.

The scarcity of women in the mining sector spurred the creation of women's associations (including the South African Women in Mining Association, and Women in Mining South Africa), women's development programmes, and bursaries for women in the mining sector (Mudimba 2017, p. 23). These initiatives attempted to make mining more appealing to women, and to address challenges faced by women. One such initiative, which will be examined in this chapter, is the International Women in Resources Mentoring Programme (IWRMP). The IWRMP is a mentoring programme for women working in the resources sector which aims to develop women to become leaders in the mining sector, and to retain women in the mining sector. The IWRMP was founded by the founder of two women's networks in the mining sector: Women in Mining UK and International Women in Mining (IWiM). The founder is a woman that has worked in the human resources field within the mining sector for many years, and as such, she understood the value of mentoring as a critical component in any career and decided to create a programme that makes mentoring accessible to women in the natural resources sector. In 2018, in collaboration with Metisphere, a company specialising in organisational psychology, the founder created the IWRMP. The organisational psychology partner provides the structured framework, the mentor and mentee matching service, and the training of mentees and mentors for the IWRMP.

The purpose of the case study was to determine if the IWRMP is successful at improving gender equality in the mining sector, and if its adoption by companies in the mining sector can be attributed to leadership capabilities within the organisation. "Improving gender equality" refers to contributing to the achievement of Sustainable Development Goal 5 ("gender equality") of the United Nations Sustainable Development Goals (SDGs). This specifically focuses on SDG Target 5.5, which states: "Ensure women's full and effective participation and equal opportunities for leadership at all levels of decision-making in political, economic and public life" (UN DSDG 2019). The specific indicator that is relevant to this study is SDG 5.5.2: "Proportion of women in managerial positions" (UN DSDG 2019).

According to the UN SDG Knowledge Platform on SDG 5, in 2018, women represented 39% of the world employment and only 27% of managerial positions (UN DSDG 2019). It is interesting to note that the percentage of women in management has increased (since 2000) in all regions except in the least developed countries

(UN DSDG 2019). The IWRMP aims to address the imbalance of men and women in the resources sector, by mentoring women to become leaders in their respective organisations. If the programme produces 50 women each year (their target) who are ready to face the leadership challenges that await them, this will incrementally increase the proportion of women in managerial positions.

Many mentoring programmes have been trialled by companies operating in the resources sector. What differentiates the IWRMP from these programmes is that it is international, and is not tailored to any specific stream in the resources sector, but rather targets the leadership capabilities across the sector that have produced successful leaders in the sector. Mentors and mentees are employed by different companies, and are matched using the services of Metisphere. Mentees are matched with high-profile leaders in the resources sector.

Mining companies from around the globe have adopted the IWRMP as a leadership development tool for women at their companies. SRK Consulting (SRK) became involved with the IWRMP when a senior female leader in the organisation volunteered as a mentor for the programme in its inaugural year in 2018. Later that year, she pitched the idea of the IWRMP at the global company meeting, and proposed that SRK becomes involved with the initiative as a sponsor. The global leaders of SRK unanimously accepted the proposal. Two senior employees applied to be mentors for the IWRMP, and several employees were nominated as mentees.

SRK took the decision to adopt the IWRMP as a mentoring programme to develop the leadership skills of their female employees. At SRK Consulting, mentoring takes the form of on-the-job training in the technical aspects related to their core business of mining consulting, and this programme complements their current mentoring strategy.

This chapter focuses on how women in the mining sector are carving a place for themselves, how the SDGs play an important part in assisting women to form part of the mining sector, and the importance of mentoring programmes in this regard. The proposition of the study was that SRK Consulting's participation in the programme was driven by leadership capabilities that aggregate into a contemporary leadership style (or a combination of styles) as described in the literature. The research question for the case study was: "How did the leadership capabilities of SRK Consulting's leaders influence the organisation's participation in the International Women in Resources Mentoring Programme, and hence contribute to the realisation of SDG 5.5.2?" A literature review will follow, which places focus on the role of leadership in adopting mentoring programmes, followed by a brief look at leadership theories, rise of female leadership, leadership in the mining sector, and then how mentoring programmes assist women to gain entry into the mining sector. This will be followed by a discussion on the study's research design, and the results and findings, and then conclusion with recommendations in terms of the study's key concepts and ideas.

2 Literature Review

2.1 A Brief Review of Leadership Theory

Although there is no single, all-encompassing definition of leadership, the majority of scholars agree that leadership is an influencing process between a leader and followers that is dependent on leader attributes, leader perceptions, and context of the leadership situation (Day and Antonakis 2012). For the purposes of this study, the researchers will adopt the Northouse (2016, p. 6) definition that states: "Leadership is a process whereby an individual influences a group of individuals to achieve a common goal".

The purpose of the review is to provide a summary of leadership theory, with a specific focus on female leadership and leadership in the mining sector. Concise summaries of the main leadership theories are presented by Day and Antonakis (2012) and Northouse (2016), as the introductory chapters to the textbooks The Nature of Leadership and Leadership: Theory and Practice, respectively. Both chapters consider scholarly literature about leadership over several decades, and both reference the Leadership Quarterly journal (LQ) as a contemporary source of leadership literature. Both textbooks contain a chapter on female leadership which, although not a stand-alone leadership theory, is an increasingly pertinent category in modern leadership, as numbers of female leaders continue to grow (Gardner et al. 2010). The theories examined are summarised in Table 7.1.

2.2 The Rise of Female Leadership

There is no specific theory on feminine leadership; however, certain female characteristics do add a different flair to some of the existing leadership theories. Both Day and Antonakis (2012) and Northouse (2016) contain a chapter on "Gender and Leadership". In addition, there are many articles in the LQ on female leadership, but mostly in the context of business environments, such as how female executives perform compared to their male counterparts. The current primary research questions in the scholarly domain regarding gender and leadership are these: "Do men and women lead differently?" "Are men more effective leaders than women?" (Northouse 2016, p. 398).

Despite the increase in women in the general workforce, and the increase in female university graduates, women are still underrepresented in the upper echelons of business and political organisations. The "glass ceiling" is a term that was coined in 1986 by two Wall Street Journal reporters (namely Carol Hymowitz and Timothy D. Schellhardt) to describe why the gender gap in leadership occurs. The concept of the glass ceiling is that men and women start out on equal footing in an organisation, but inevitably, women hit this glass ceiling that cannot be surpassed, and they cannot progress to higher levels in the organisation.

One theory of why women do not reach the upper echelons of leadership in organisations is that although there are enough women entering the pipeline, many of them "leak" out of the pipeline. There are several theories for why the pipeline leaks, but one of them is women choosing to start a family. As soon as women have children, they break the continuity of their career, and lose experience during this time. This is largely due to

Table 7.1 Leadership theories reviewed in the case study

Theory	Premise
Trait	The foundation of trait theory is that a person is a successful leader due to certain personality traits, with which they are born, e.g. intelligence (Northouse 2016, p. 19)
Skills	Skills theory comprises three components: "individual attributes, competencies, and leadership outcomes" (Northouse 2016, p. 48)
Behavioural	Behavioural theory focuses on what leaders do and how they act (Northouse 2016, p. 71). It is defined by two types of leadership behaviours: "initiating structure" and "consideration" (Northouse 2016, p. 71; Conger and Kanungo 1987)
Contingency	Day and Antonakis (2012, p. 9) summarise contingency effectively in a single sentence, stating "leadership effectiveness is contingent on leader–member relations, task structure, and the position power of the leader"
Situational	The premise of situational theory is that "different situations demand different kinds of leadership", and to be an effective leader, a person must "adapt his or her style to the demands of different situations" (Northouse 2016, p. 93)
Path-goal	The leader chooses a leadership style that is suited to the motivational needs of his/her followers (Northouse 2016, p. 116). In the application of this theory, leaders must be able to identify leadership behaviours that are missing in the work setting, and display the missing behaviours. They must also elucidate the path to the goal and remove obstacles on the path to the goal. Leaders must also provide rewards for followers that attain workplace goals

(continued)

Table 7.1 (continued)

Theory	Premise
Information processing	Categorisation theory is a basic information-processing theory. In the context of leadership, categorisation theory "deals with three distinct but interrelated areas: specifying the internal structure of leadership categories, showing how properties of categories can be used to facilitate other information-processing tasks … and explaining leadership perceptions in terms of categorisation" (Lord et al. 1984)
Relational (also known as leader-member exchange)	Dansereau et al. (1975) posited that leadership effectiveness is based on the quality of the relationships between leaders and followers. Relational theory was initially termed "vertical dyad linkage". Later, the name evolved into "leader-member exchange" (LMX) theory. All leadership theories up until the development of LMX and transformational leadership (discussed below) have relied on transactional (i.e. social exchange) obligations (Day and Antonakis 2012). Graen and Uhl-Bien (1995) positioned LMX within the transactional and transformational leadership frameworks. They concluded that while LMX may start out as a transactional social exchange, it ends up as a transformational social exchange (Graen and Uhl-Bien 1995; Ilies et al. 2007)
Transformational leadership	Transformational leadership comprises four defining factors, the "four Is": idealised influence (also known as charisma), inspirational motivation, intellectual stimulation, and individualised consideration (Northouse 2016). These factors indicate that transformational leadership transcends the prior transactional leadership theories, and like LMX, it focuses on followers, i.e. not only leaders (Hater and Bass 1988)
Sceptics-of-leadership	Rush et al. (1977) suggested that leaders are assigned attributes based on the leadership theories toward which researchers are biased. Meindl and Ehrlich (1987) proposed that leadership attributes are retrofitted onto a leader to assign causes to organisational outcomes. They went so far as to say that leadership does not exist and is not necessary to achieve organisational outcomes
Responsible leadership	Responsible leadership emerges at the junction between stakeholder theory and leadership theory (Frangieh and Yaacoub 2017). This is also seen as the juncture between ethics, leadership, and corporate social responsibility which is more applicable in a business context. It is a complex and constantly evolving theory. Responsible leaders are positioned at the centre of a network of stakeholders and are viewed as equals, in contrast to top-down leadership hierarchies where leaders are considered to be more important than their followers.
Authentic leadership	The accepted definition of authentic leadership is the one set forth in an article by Walumbwa et al. (2008, p. 94), who defined authentic leadership as "a pattern of leadership behaviour that draws upon and promotes both positive psychological capacities and a positive ethical climate, to foster greater self-awareness, an internalised moral perspective, balanced processing of information, and relational transparency on the part of leaders working with followers, fostering positive self-development"

(continued)

Table 7.1 (continued)

Theory	Premise
Servant leadership	In servant leadership, the leader initially has the desire to serve, which then develops into a desire to lead. Although this was already evident in the scripture of the Bible, in terms of servant leadership in the context of organisational theory, the primary aim is to serve the interests of their (the leader's) followers, ahead of serving their own interests (Northouse 2016)
Adaptive leadership	This is similar to path-goal theory, because both rely on understanding motivation and enabling action, to achieve a goal. Adaptive leaders prepare followers to adapt to change, and mobilise them. Adaptive leaders also have the ability to mobilise people to solve complex problems. Adaptive leaders challenge others to face difficult challenges, providing them with the space or opportunity they need, to learn new ways of dealing with the inevitable changes in assumptions, perceptions, beliefs, attitudes, and behaviours that they are likely to encounter in addressing real problems (Northouse 2016)

the disproportionate responsibility women assume for child-rearing and domestic duties (Northouse 2016). This is changing globally, and in many countries, fathers are being allocated significantly longer periods of optional (or even compulsory) paternity leave, in an effort to divide child-rearing responsibilities. Despite this, women continue to do the majority of the childcare responsibilities and household chores (Northouse 2016). Women with children face work-home conflicts, and those who take advantage of workplace flexibility programmes are often marginalised, and those who take time off from their careers often find re-entry difficult (Northouse 2016).

Northouse (2016) explains that mentorship is an important developmental experience, and effective mentorship relationships influence career success. This author adds that it is more difficult for women to establish informal mentor relationships than it is for men, and that women receive fewer developmental opportunities than men do. In terms of leadership, despite gender stereotypes, women do not lead more interpersonally than men do, but they do lead in a more democratic manner (Eagly and Karau 2002). Studies have shown that men devalue women who lead in a masculine manner, in a masculine role. Devaluation of female leaders by male subordinates extends to female transformational leaders (Northouse 2016, p. 402).

Another meta-analysis described by Northouse (2016:402-403) states that "women and men were more effective in leadership roles that were congruent with their gender". In role congruency theory, "agency" is a group of behaviours known as masculine behaviours (e.g. dominance), and "communion" is a group of behaviours known as feminine behaviours (e.g. nurturing). Where there is role incongruity, i.e. where the leader role conflicts with the gender role, women tend to be less effective as leaders. This is another reason for the leaky pipeline described above.

Two leaders in the one researcher's organisation commented on female leadership. Both leaders stated that women have an empathetic leadership style, and that as such, they were more willing to listen to their employees' concerns than their male counterparts. This ability to listen leads to better communication, and in many instances better team performance.

2.3 A History of Leadership in the Mining Sector

There is a lack of scholarly literature on leadership in the mining sector. A thorough literature search was conducted on multiple scholarly platforms, which yielded several theses and/or dissertations of leadership in contemporary mining. There are very few publications describing the notorious, but not widely publicised, autocratic leadership style that was prevalent in mining operations. The researchers therefore used historical accounts to provide a flavour of what a so-called mining boss was like. The researchers also spoke to several older employees in her organisation, to obtain first-hand accounts of mining leadership styles.

In this discussion, the researchers would like to draw attention to the distinction between leadership and management. This distinction is debated among scholars, but has shown itself in the one researcher's work experience. The distinction lies in how the leader came to be a leader: i.e. did the leader acquire followers using their own influence, or was the leader appointed into a hierarchically higher position, by a third party (i.e. not the followers)? A person that has been appointed into a position of power might not be a good leader, but might wield power and use that power to force action. People may also be appointed as managers in an organisation, due to their technical capabilities, and not their leadership capabilities, and as such, are not necessarily good leaders.

People who have worked in mining have referred to "mine bosses" or "mine chiefs" (colloquial terms for senior managers in mining) as autocrats. As mentioned in the paragraph above, there are no formal social studies showing this; however, there are historical accounts. One such example is taken from a history book on Welsh coal miners in America in the late 1800s (Lewis 2009, p. 198), where a miner recalls: "Above ground, Welsh [mine] bosses might have been respectable citizens who attended chapel regularly, but underground they became crude tyrants who ruled corrupt regimes". The same miner commented on the way the bosses "strutted about full of their own importance" and that they were surrounded by men who smiled in appreciation while poking fun at the bosses' behaviour in their hearts. He states: "I have had to listen to a boss in this way but there was no point in my saying anything derogatory, otherwise I would surely feel his displeasure in my work". Lewis (2009, p. 238) also quotes the Mine Workers Journal, who referred to Gomer Jones, a man infamous for his "tyrannical methods as a mine boss".

A further reference is taken out of another history book on mining at the Lake Superior copper mines in the early 1900s (Lankton 1991), where the author was fascinated by "autocratic mine boss" Alexander Agassiz. He writes about how Agassiz "did not let outsiders—either radicals or reformers—tell him how to run his company ... as a staunch autocrat, he was driven to control events and not let problems get out of hand". As Agassiz's labour force got bigger and "potentially more fractious [he and his managers] devised a hierarchy of tactics intended to keep their hegemony intact" (Lankton 1991, p. 200). Another mine boss, Big Jim McNaughton, is described as being "[an] autocrat seemingly barren of self-doubts and indecision ... he could get done whatever needed doing using finesse or force" (Lankton 1991, p. 201).

Two leaders in the one researcher's organisation provided valuable insights into mining leadership. One leader described mining bosses in the 1980s and 1990s as autocratic and militaristic. He stated that this leadership style was, in part, due to the extremely high health and safety risks associated with underground mining. He said that everyone involved in underground work is required to be extremely disciplined, and as such, the shift bosses and superintendents ran their crews like military crews.

The other leader worked as a miner when the ban on women working underground was lifted. He said that it was extremely difficult for the male miners to accept that they had to work with women, because they inherently believed that mining is a man's role. He added that the first women to work underground were exceptionally tough and resilient. The men were not comfortable with seeing women conduct manual labour. They were even less enthused by the idea of women becoming shift bosses, because in their culture, women were considered subordinate to men. Some men went as far as to offer women assistance with their manual tasks in exchange for sexual favours. Some men even raped women underground. However, as time passed, and more women started working in mining, the resistance to women working in mines decreased.

2.4 A Mentoring Approach to Enable Women's Entry into Mining

The IWRMP is a structured mentoring programme that provides mentoring to women in the mining sector. Mentors volunteer their time to the programme, and mentees apply for the programme or are put forward by their employers. Mentees are usually identified and sponsored by their organisations. Mentors and mentees are matched manually by the organisational psychologists at Metisphere (IWRMP 2019).

According to their website, the IWRMP "aims to empower and promote the career prospects of women working in the international resources industry by developing productive mentoring partnerships to assist in creating a more diverse and gender progressive industry" (IWRMP 2019). The objectives of the IWRMP website are to:

- "Enhance global retention and support for women in mining;
- Give programme participants global opportunities to connect and network;
- Provide superior role models to women working in mining in different cultures and global communities;
- Assist women with setting goals, career plans and achieving them;
- Increase participants' self-confidence and self-assurance;
- Create the pipeline and leaders of tomorrow;
- Reduce cross-cultural barriers;
- Make sure women from all continents are represented and that we have diversity and intersectionality within, strong focus on emerging markets;
- Engage globally recognised industry sponsors to contribute to a mentoring programme that aligns with their values in diversity and inclusion".

Now that the foundation of the literature review has been investigated and developed, this chapter will continue with a look into the research design of the study.

3 Research Design

An inductive research approach was employed for this case study (Yin 1994). The researchers first reviewed the various leadership theories, developed a semi-structured discussion guide, and then conducted semi-structured interviews (i.e. qualitative research) (Lune and Berg 2017). Thereafter, the researchers used thematic analysis to analyse the data and look for leadership capabilities. The researchers employed purposive, targeted sampling to identify key people to interview to gather data on the IWRMP and leadership capabilities in the context of the IWRMP. Four individuals from two organisations (i.e. SRK and IWiM) were interviewed as part of this research case study. In terms of their roles in the case study, the participants were the founder of the IWRMP, the implementer of the IWRMP in SRK globally (the primary female leader in the study), the adopter of the IWRMP into SRK South Africa (the male leader in the study), and a mentee of the programme. The researchers used the literature review on leadership to make sense of codes, compared, and contrasted findings with the research. This approach allowed the researcher to improve the reliability and validity of the study through data triangulation, thus probing the dependability of evidence across different data sources such as the semi-structured interviews, comparing data with secondary data, and contextualising it with the literature. The unit of analysis for this single case study was the perception of the leadership capabilities of SRK's leaders that enabled participation in the IWRMP. The study was limited to interview data, scholarly literature, historical literature, and technical information on mining, the IWRMP, and the SDGs.

Qualitative methods, as their name suggests, analyse "qualities". Quality refers to the what, how, when, where, and why of a thing, i.e. its essence and ambience. Using this concept of qualities, qualitative research can be defined as the "meanings, concepts, definitions, characteristics, metaphors, symbols, and descriptions of things" (Lune and Berg 2017, p. 12). Case studies are robust tools for the investigation of complex social phenomena, which can be used in the exploratory, descriptive, and explanatory phases of an investigation (Yin 1994, p. 4). Research case studies need to present empirical data rigorously and fairly (Yin 1994, p. 2).

There are two organisations from which research participants (interviewees) were selected: SRK Consulting and International Women in Mining (IWiM). SRK Consulting refers to the entire international organisation which encompasses SRK Consulting (South Africa) (Pty) Ltd (SRK SA) and SRK Consulting (UK) Ltd (SRK UK). SRK (Global) Limited (SRK Global) is the strategic management arm of SRK Consulting, which holds an annual practice leaders meeting, where practice leaders from all SRK's branches meet to discuss progress and develop company strategy.

The study includes four participants: Participant 1, founder and director of IWiM and the IWRMP; Participant 2, principal consultant at SRK UK; Participant 3, managing director of SRK SA; and Participant 4, environmental scientist at SRK SA.

The primary data that was collected is transcribed texts from semi-structured interviews with four key participants, and secondary data was collected from textbooks, websites, and articles pertaining to the categories identified in the conceptual framework. In addition, informal discussions were held with two senior leaders in the researcher's organisation, and a brief Skype call was held with one of the participants.

The data was analysed using thematic analysis, specifically, open coding. Lune and Berg (2017) defines open coding as the "intensive process by which data are rigorously analysed for embedded phenomena, patterns, concepts, and themes". Codes were assigned to main themes and sub-themes which enabled the researchers to link the findings to the literature reviewed. All necessary ethics clearance for this study was obtained from the SRK and from each of the interviewees. All interviewees agreed to having their names disclosed in this case study report.

4 Findings and Results

4.1 Finding 1: Mentoring Is a Valuable Tool for Achieving Gender Equality

The benefits and power of mentoring were mentioned repeatedly in the interviews. Mentoring can take different forms; but in the context of this case study, the IWRMP mentoring approach is what is discussed. In the IWRMP, mentoring serves the function of personal and career development for mentees, by guiding them through the social, technical, and emotional aspects of their jobs. The IWRMP also allows mentees to develop leadership skills, as the mentors are senior technical professionals who are also leaders in their respective organisations.

The founder of the programme alluded to her lack of a mentor, and how she felt that it would have been valuable to her career development to have someone to guide her and teach her non-technical skills that could grow her emotional intelligence. Another interviewee gave testament to the power of mentoring by stating that if it were not for his mentors, he would not be where he currently is in his career. The mentee that was interviewed also attested to the benefits of the IWRMP by stating that the skills and lessons that she learnt from the IWRMP are of great benefit to her career.

Relationships are another theme that emerged in the interviews. Historically, most mining workplaces were patriarchal, and male-male relationships were stronger than the relationships that men formed with women, because of the "old-boys' club" construct, where men had access to senior men in their organisations, simply by being men, but also because they could participate in social activities with male leaders.

In summary, mentoring contributes to achieving SDG 5.5.2 (increasing the "proportion of women in managerial positions") by developing women's skills and emotional intelligence, thereby giving them the confidence to excel at their jobs which ultimately results in the filling of the female leadership pipeline.

4.2 Finding 2: Traditional Gender Roles and Stereotypes Are Barriers to Gender Equality in the Mining Sector

Gender inequality was mentioned numerous times in the interviews, which is consistent with the one researcher's professional experience, as well as the Minerals Council South Africa's statistics, which state that only 16% of people in executive leadership in the mining sector are women. In the 23 years since the prohibition on women working underground was lifted, South Africa's mining sector has not meaningfully transformed, with women only forming 12% of the total mining workforce, despite the promulgation of subsequent legislation such as the Mining Charter.

What was interesting to note is that male leaders are not always cognisant of gender inequality, and the female leader in the study alluded to the fact that most men would disagree with this assertion. The male leader in the study made allowance for the fact that may be occurring without his being aware of it, which shows lack of emotional intelligence. There are several generally accepted facts about the mining sector. The first one is that historically, only men worked underground. The reasons for this are twofold: (a) it was heavy manual labour that required physical strength and stamina, and (b) women were expected to rear children instead of being formally employed. The first reason also resulted in legislation which prohibited women from working underground. Therefore, the only women who would have been employed on mines, while the prohibition was in force, are women in administrative roles or quality assurance roles. Some of the lack of progress on gender equality can be ascribed to this legacy of prohibition.

One of the participant's mother was an influential female figure in the South African mining sector in the mid-1980s, and she actively fought to have the legislation amended so that women could work underground. This participant was also one of the first women (possibly the first) to work underground in South Africa. With the lifting of the prohibition in 1996, more women in technical roles entered the mining sector, for example, geologists, geotechnical engineers, and mining engineers. These first women would have been in the vast minority. An informal discussion with a senior leader at SRK explains that men were extremely displeased with the entry of women into technical mining roles, because the leading perception at the time was that mining was "men's work". In the early days following the prohibition being lifted, men strongly resisted working alongside women, and even more strongly resisted being led by women. As time passed, women infiltrated the mining sector workforce at all levels, and everyone became accustomed to women on the mines. It was not possible, within the scope of the study, to infer how traditional autocratic mining leadership mixed with the more empathetic and democratic style of female leaders, but it can be surmised that female traits and leadership styles would have been incongruent with mining bosses' recruitment and advancement requirements. The female leader spoke about how it was, and still is, more difficult for women to advance in their careers in the mining industry than it is for men.

The international mining sector is still largely patriarchal, and male executives still exhibit misogyny, with an example provided by the female leader where the chairman of listed company stated that he would not have a woman on his board and joked about the woman bringing her children to board meetings. Some mining clients go so far as to blatantly state that they do not want female consultants to lead their projects, and even if they are the project lead, they are told that a man must present the results. Some clients shake the hand of the female leader's translator, but will not look her in the eye, and others ask her directly if they must work with a woman.

Female roles were also mentioned repeatedly in the interviews. The break that women take for maternity leave, coupled with their disproportionally large caretaker roles (when compared to their male counterparts), makes it difficult for women to become leaders. In SRK, some women with children work part-time, due to family obligations, and as such, they lose experience. Both the male and female leaders in the study mentioned that women are still the ones expected to "rush out to attend to kids" or "do the school run", which makes it clear that traditional female roles are still being enacted largely by women, in 2019.

The female leader hints at the complexity of traditional gender roles by stating that gender roles are intrenched in human history and our DNA, and as such, women are not seen as equal. She went on to say that subconsciously, women are not the image or the voice of authority. Women are physically weaker, their voices are softer, and they cry rather than bottle everything up, which are all seen as weaknesses. The mentee in the study also spoke about this perception of women being weak, and that she thinks it is due to women having to balance work life and household duties. The mentee also thinks that this is used against women.

SRK was historically a patriarchal company, due to its close ties to the mining industry (SRK was founded in 1974 as a geotechnical engineering consulting company, to service the growing mining industry). Over the past few years, SRK has become increasingly aware of gender inequality in its business, and as such, it has implemented a general recruitment strategy to employ more women. In addition, SRK has spoken out publicly against gender bias in the science, technology, engineering, and mathematics (STEM) fields. Prior to adopting the IWRMP, SRK was looking at ways to improve gender equality in the company, and by extension, the mining industry". SRK started out by sponsoring the IWRMP, and then nominated a mentee to join the programme in 2019.

In SRK, the partner group is still predominantly male, but in the past few years, the number of female partners (including those with young children) has increased. The male leader is a strong proponent of transformation along gender lines within SRK, and his decision for SRK to participate in the IWRMP was an obvious choice, and is a testament to SRK's willingness to achieve gender equality.

4.3 Finding 3: Several Leader-Centric Characteristics Enable Leadership

The primary female leader in the case study is uniquely formidable. She is a strong, confident, and resilient woman. She is technically competent and well respected in the international mining sector, both by men and women, and by senior people and people at lower levels in SRK. She has had to fight for respect, equality, and equal rights in the mining sector from the age of 16. Her experiences are unique, and when it came to women working underground, she was a pioneer.

Five prominent traits emerged from the interviews, as representative of the leaders in this study: goal orientation, generosity (wanting to give), resilience/inner strength, perseverance, and confidence.

One prominent skill emerged from the analysis: the ability to identify opportunities. One prominent behaviour emerged from the analysis, and that is "initiation", i.e. taking initiative or initiating action. This behaviour is responsible for the creation of the IWRMP, and SRK's participation in the IWRMP. Life and/or work experience emerged as significant, and was mentioned numerous times in the interviews. The life experiences mentioned by the female leader were the formative ones that shaped her character. She has had to fight continually, to be heard and respected as a female leader.

Achievement and/or reward as intrinsic motivation for leaders emerged as significant in the interviews. The female leader recalled that there were several instances where a client specifically requested that she work on their project, and stated that it made her feel like her hard work was rewarded.

There is a lack of women in the leadership pipeline, and the founder of the IWRMP reiterated that the IWRMP is about positive action to develop women to fill the pipeline. These findings corroborate and are in line with the literature findings of Day and Antonakis (2012) and Northouse (2016).

4.4 Finding 4: Mentoring Is a Form of Relational Leadership (Literature)

Mentoring is about relationships and personal development. In a mentoring relationship, mentors elucidate the path for mentees to achieve a goal. Mentors also impart emotional intelligence skills to mentees. This makes mentoring consistent with relational leadership theory.

4.5 Finding 5: Women Need Certain Leader-Centric Characteristics to Succeed as Leaders in the Mining Industry (Literature)

Although trait, skills, and behavioural theories are too simplistic on their own to describe the leadership capabilities that contributed to SRK's participation in the IWRMP, these "leader-centric characteristics" certainly played a role. This is evidenced by the more than 200 times traits, skills, behaviours, and experiences are mentioned in the interviews.

4.6 Finding 6: Authentic Leadership Contributed to SRK's Participation in the IWRMP (Literature)

The four components of the theoretical model are "self-awareness, internalized moral perspective, balanced processing, and relational transparency" (Walumbwa et al. 2008). The leaders in the study strive to improve the personal development journeys of women in mining, in order to give them the best chance of success. They genuinely want to see women succeed in the mining sector. The prominent codes related to authentic leadership that were mentioned in the interviews are diligent, admirable, optimistic, motivational/encouraging, humble, worried about perception of themselves, having conviction, polite, responsible, being the best version of yourself, and doing the right thing.

The intrapersonal approach to authentic leadership states that "authentic leaders exhibit genuine leadership" and "lead with conviction" (Northouse 2016, p. 196). The male leader's perception that the female leader's belief that women should be [in leadership positions] and deserve to be there and on merit shows authenticity. He also mentions conviction when he says that the female leader has a strong conviction to drive gender equality and get more women into the mining sector.

The female leader's background was filled with struggle to be respected and heard. This further talks to the intrapersonal approach to authentic leadership that states that "a leader's life experiences and the meaning he or she attaches to those experiences [are] critical to the development of the authentic leader" (Northouse 2016, p. 196). This authenticity emerges when the female leader explains why she wanted to participate in the IWRMP; she thought that she has something to give and wanted to give it, to make a positive change.

4.7 Finding 7: Servant Leadership Contributed to SRK's Participation in the IWRMP

The seven behaviours of servant leaders are "conceptualizing, emotional healing, putting followers first, helping followers grow and succeed, behaving ethically, empowering, and creating value for the community" (Northouse 2016, p. 238). The beneficiaries of the IWRMP are primarily the mentees, and as such, the desire to participate in the IWRMP and the actions taken to initiate participation in the IWRMP are actions of service to the mentees. This is in line with the behaviours of servant leadership, where the leaders help mentees to grow and succeed, instead of primarily seeking benefit for themselves.

The founder, female leader, and male leader mention "wanting to give" in their interviews. This code, along with several other service-related codes, is listed as want to give/want to give back, supportive, share experience(s), contributing, and want to help others.

The female leader in the study wants to benefit women in mining, even though she gets very little for it (she works on the initiative outside of work hours). The male leader is passionate about the transformation agenda, and he also clearly wants to give back, and to help women succeed. These behaviours resonate with the servant leader behaviours of "putting followers first", and "creating value for the community".

5 Conclusions and Recommendations

Mentoring is a valuable tool for developing female leaders. Mentoring facilitates the formation and nurturing of relationships between female mentees and their mentors. Mentoring also enables the transfer of crucial technical and non-technical skills onto mentees (including leadership skills).

The leadership capabilities that contributed to SRK's participation in the IWRMP are primarily capabilities from authentic leadership theory and servant leadership theory. In addition, some capabilities from the trait, skills, and behavioural theories contributed to SRK's participation in the IWRMP. Some traits, skills, and behaviours contributed as stand-alone capabilities, and some as components of the authentic or servant leadership theories. These capabilities contributed to the female leader's interest in gender equality, her decision to initiate action to participate in the IWRMP, and her success in gaining support for the initiative from the executive leaders in SRK. These capabilities also contributed to the male leader's interest in gender equality and his decision for the SRK South African office to participate in the IWRMP.

At the current rate of progress on gender equality in the South African mining sector, we will not be able to achieve SDG 5.5.2 by 2030. The mining sector needs to do more to actively and vocally break through gender inequality. Society and the workplace need to evolve to the point where men and women are not held back in their leadership journeys due to family responsibilities.

SRK is addressing gender inequality and female leadership pipeline. One successful intervention that was implemented to retain women in the South African organisation is flexible work practices. The IWRMP is a more recent initiative adopted to address gender inequality and to develop the leadership pipeline in the organisation globally. The success of the IWRMP is evaluated annually, but feedback from mentors and mentees on the first two years of the programme indicates that it is a resounding success.

In order to accelerate the progress on gender equality in the mining sector, the researchers recommend that all mining companies (corporate and consulting) formalise a mentoring work package as a component of a formal employment portfolio, so that the work is accommodated within working hours.

A structured mentorship programme, such as the IWRMP, should be considered for all female employees of mining companies. Flexible work practices are an enabler for the retention of women in the mining sector, and as such, more companies in the mining sector should adopt flexible work practices.

The researchers recommend that further studies be conducted on gender equality in the South African mining sector, to develop a better understanding of the reception of female leadership into a historically patriarchal sector. The specific aspects that require study are role congruency and perceptions about family responsibility.

Men in the mining need to become aware of gender inequality, so that they may contribute to solutions to gender inequality. Although this can be achieved by various communication and education programmes in the workplace and in public society, it is the researchers' opinion that there is more merit to initiating gender equality awareness at university level, so that future leaders begin engaging with these challenges before entering formal employment.

References

Conger JA & Kanungo RN (1987). Toward a behavioural theory of charismatic leadership in organizations. Academy of Management Review, 12, 637–647.

Dansereau F, Graen G & Haga WJ (1975). A vertical dyad linkage approach to leadership within formal organizations: A longitudinal investigation of the role making process. Organizational Behavior and Human Performance, 13, 46–78.

Day DV & Antonakis J (2012). *The Nature of Leadership*, Sage, CA.

Eagly A & Karau S (2002). Role congruity theory of prejudice toward female leaders. Psychological Review, 109(3), 573–598.

Gardner WL, Lowe KB, Moss TW, Mahoney KT & Cogliser CC (2010). Scholarly leadership of the study of leadership: A review of The Leadership Quarterly's second decade, 2000–2009. The Leadership Quarterly, 21, 922–958.

Graen GB & Uhl-Bien M (1995). Relationship-based approach to leadership: Development of leader-member exchange (LMX) theory of leadership over 25 years: Applying a multi-level multi-domain perspective. The Leadership Quarterly, 6, 219–247.

Frangieh C & Yaacoub H (2017). A systematic literature review of responsible leadership: Challenges, outcomes and practices. Journal of Global Responsibility, 8(2), 281–299.

Hater JJ & Bass BM (1988). Superiors' evaluations and subordinates' perceptions of transformational and transactional leadership. Journal of Applied Psychology, 73, 695–702.

Ilies R, Nahrgang JD & Morgeson FP (2007). Leader-member exchange and citizenship behaviors: A meta-analysis. Journal of Applied Psychology, 92, 269–277.

International Women in Resources Mentoring Programme. (2019). *IWRMP Programme,* viewed 4 April 2019, from https://www.iwrmp.com/iwrmp-programme/.

Lankton L (1991). Cradle to Grave: Life, Work, and Death at the Lake Superior Copper Mines, Oxford University Press, New York.

Lewis RL (2009). Welsh Americans: A History of Assimilation in the Coalfields, University of North Carolina Press, North Carolina.

Lord RG, Foti RJ & De Vader CL (1984). A test of leadership categorization theory: Internal structure, information processing, and leadership perceptions. Organizational Behavior and Human Performance, 34, 343-378.

Lune H & Berg BL (2017). Qualitative Research Methods for the Social Sciences. 9th ed., Global ed., Pearson Education, England.

Meindl JR & Ehrlich SB (1987). The romance of leadership and the evaluation of organizational performance. Academy of Management Journal, 30, 90–109.

Minerals Council South Africa. (2019). *Women in Mining in South Africa: Fact Sheet*, viewed 17 September 2019, from https://www.mineralscouncil.org.za/industry-news/publications/fact-sheets.

Mudimba PC (2017). 'Re-defining Gender Equality in the South African Mining Sector', Master's thesis, Dept. of Public Law, University of Pretoria.

Northouse PG (2016). Leadership: Theory and Practice, 7th edn., Sage, CA.

Rush MC, Thomas JC & Lord RG (1977). Implicit leadership theory: A potential threat to the internal validity of leader behavior questionnaires. Organizational Behavior and Human Performance, 20, 756–765.

UN DSDG. (2019). Sustainable Development Goals Knowledge Platform: Sustainable Development Goal 5, viewed 3 May 2019, from https://sustainabledevelopment.un.org/sdg5.

Walumbwa FO, Avolio BJ, Gardner WL, Wernsing TS & Peterson SJ (2008). Authentic leadership: Development and validation of a theory-based measure. Journal of Management, 34(1), 89–126.

Yin RK (1994). Case study research: design and methods. 2nd ed., Sage, CA.

8 Exploring Leadership Capabilities in a Multi-Sector Road Infrastructure and Innovation (SDG 9) Partnership (SDG 17) in South Africa

Samantha Castle and Dawies (D.A.J.) Bornman

Abstract

An efficient national road transport network is a critical enabler to economic growth and sustainable development, hence the demand of a partnership between the University of Pretoria (UP), the South African National Roads Agency SOC Ltd (SANRAL) and the Council for Scientific and Industrial Research (CSIR), institutions with different but complementary mandates contributing to the sustainability of the road transportation and innovation of the industry. Grounded in complexity leadership theory, this chapter focuses on leadership features, which enabled the successful establishment of a multi-sector partnership, which links to SDG 17 (i.e. partnerships for the goals) and SDG 9 (i.e. innovation and infrastructure). This chapter offers evidence gathered from semi-structured interviews with project leaders on key leadership capabilities which enabled the establishment of this partnership. The findings reveal that the parties gravitate towards the complexity leadership theory to solve problems. This included enabling and administrative leadership practices (i.e. formalising the partnership) and adaptive leadership practices which smoothed the negotiation of boundaries. A longitudinal study is recommended to understand the impact of the partnership over time.

Keywords

Leadership · SDGs · Partnership · Road infrastructure · South Africa

1 Introduction

The countdown is on! With a decade left to achieve the 2030 Agenda for Sustainable Development and its 17 interlinked Sustainable Development Goals (SGDs), the need for radical innovation, impactful partnerships and inventive leadership is critical for the domestication and realisation of the SDGs. There is no doubt that leadership is accepted as a critical feature and driver in solving complex social challenges (Mumford 2000). However, traditional leadership theories have come under scrutiny in more recent times. The traditional conceptualisation of leadership, which historically was located in a "top-down, bureaucratic paradigm", has been brought into question (Uhl-Bien et al. 2007) and this raises further questions, including whether traditional theories are suffice to help society

S. Castle (✉) · D. (D.A.J.) Bornman
University of Pretoria, Pretoria, South Africa
e-mail: samantha.castle@up.ac.za;
dawie.bornman@up.ac.za

G. Nhamo et al. (eds.), *Sustainable Development Goals for Society Vol. 1*, Sustainable Development Goals Series, https://doi.org/10.1007/978-3-030-70948-8_8

solve the multifaceted and intractable problems highlighted in the SDGs.

These concerns are timely, as the Fourth Industrial Revolution (4IR) has, as projected, changed the way we live, work and how society function whilst cultivating a more knowledge-oriented economy. This reality renders the traditional conceptualisation of leadership "suitable for an economy premised on physical production but not well-suited for a more knowledge-orientated economy" irrelevant in the current context (Uhl-Bien et al. 2007, p. 298). A thought-provoking question raised by researchers has been to what extent does traditional leadership support innovation by joining skills and knowledge of diverse individuals (Mäkinen 2018).

The current context has created momentum for greater consideration of leadership in the context of partnerships. Partnership is as a vehicle that can bring diverse stakeholders together. Particularly, multi-sector institutions can advance a common goal, with the objective of unlocking innovation, impact and transformation. However, partnerships are complex and if not managed effectively can fail dismally in delivering on the dire transformational change required.

With the help of complexity leadership theory (CLT), this study examines the leadership capabilities that led to the successful establishment of the multi-sector partnerships between the University of Pretoria (UP), the South African National Road Agency SOC Ltd (SANRAL) and the Council for Scientific Research and Industrial Research (CSIR). This partnership serves as a model for a knowledge-intensive project that requires dynamic, adaptable and enabling leadership. It was created to establish an Integrated Education, National Certification, National Reference and Research Laboratories Facility as part of Engineering 4.0 (Kannemeyer 2016).

The UP, SANRAL and CSIR partnership is an example of three stakeholders with different but complementary mandates working together to contribute to the sustainability of the road transportation and innovation of the industry, a critical catalyst for economic growth. However, what kind of leadership is needed to unlock and facilitate these types of partnerships? Despite much-cited examples of multi-sector partnerships, there is scant evidence in the literature on the role which leadership played in unlocking and forming successful multi-sector partnerships in South Africa. Moreover, there is very limited research on the role leadership plays in managing and advancing transdisciplinary knowledge organisations such as the one which the UP, SANRAL and CSIR partnership seeks to deliver through Engineering 4.0. This study is opportune, as the President of South Africa, Cyril Ramaphosa, announced in February 2019 a R100 billion infrastructure investment to stimulate the ailing South African economy.

The intractable nature and complexity of today's sustainable development challenges demand responsive solutions from diverse and multiple arenas. Responses ranging from knowledge, resources and services are often located with different partners (Xue et al. 2020). Scholars Xue et al. (2020) highlight the fact that multi-sector partnerships comprise institutions, individuals and agencies across multiple sectors, coming together and sharing resources in the fulfilment of a mutual objective of an explicit project. Multi-sector partnerships can play an important role in addressing the SDGs and scaling impact.

Why is the UP, SANRAL and CSIR partnership noteworthy in terms of the SDGs? Evidently, SDG 17 focuses on partnerships as a stand-alone goal and an enabler for the other 16 goals. SDG 17 stresses the importance of cooperation and collaboration and advocates for an integrated approach which is important for the advancement of other goals. Therefore, the partnership explored in this study is directly linked to Sustainable Development Goal 17.

In addition, this partnership is similarly linked to SDG 9 which looks at innovation and infrastructure. Goal 9 underscores that investment in transport and information technology is crucial to reaching sustainable development and empowering communities (UN 2019). The importance of investing in road infrastructure has long been linked to that of "growth in productivity and incomes, and improvements in health and education outcomes" (UN 2019).

Apart from contextualising the partnership as a construct in relation to the SDGs, the chapter also reviews scholarly literature associated with complexity leadership theory, which is used as a framework for this study. This is followed by an outline of the methodology employed as well as a background of the challenges faced in the road transport sector. In addition this chapter discusses the UP, SANRAL and CSIR partnership, the rationale for the partnership, the partners and the project leaders. Subsequent sections discuss the data analysis and present the research findings on the perceived leadership capabilities by the project managers. The chapter concludes with recommendations for future research based on the potential impact of the UP, SANRAL and CSIR.

2 Literature Review

2.1 Complexity Leadership Theory

The literature (Bryant 2003; Lakshman 2007; Politis 2002; Srivastava et al. 2006) underlines leadership as being imperative for knowledge creation. Yet knowledge creation, enabled through multi-sector partnerships, has received little empirical consideration (Mäkinen 2018). There is a need to examine the leadership roles and practices in complex partnerships when seeking to promote collaboration, and there has also been limited research, particularly related to the role of leadership in academic institutions that promotes transdisciplinary research (Mäkinen 2018). It thus illuminates the need to understand how transdisciplinary research collaborations, which are facilitated through multi-sector partnerships, are managed. Further, this type of scientific research requires a particular brand of academic leadership that is able to foster cooperation across various types of "experts and knowledge boundaries" (Mäkinen 2018, p. 134).

This study employs a framework of complexity leadership theory which functions as a lens to understand complex adaptive systems (Mäkinen 2018). This study is largely shaped by the work done by the scholar Elina Mäkinen (2018) in the field of "Complexity Leadership Theory and the leaders of transdisciplinary science" as well as the need for further exploration in this area of limited research.

Academics (Hazy 2007; Uhl-Bien et al. 2007) argue that complex adaptive systems are spaces for participants to participate in knowledge-intensive collaboration in a cooperative network that has a mutual objective (Mäkinen 2018). Holland (2014) describes complex adaptive system as a "model of a special case of a complex system: one that exhibits the capacity to change and learn from experience. This can be attained through learning and adaptability of dynamic interaction between agents" (Holland 2014, p. 8).

Multi-sector partnerships that seek to promote and establish transdisciplinary research organisations like UP, SANRAL and CSIR can be viewed as complex adaptive systems. This is due to the partnership unifying different stakeholders from diverse backgrounds whilst encouraging a sense of interdependence between them (Mäkinen 2018, p. 134) which ultimately creates an influential shared collective with joint objectives (Mäkinen 2018, p. 134). Therefore, having an appreciation for "leaders and their actions in complex adaptive systems in academic" as well as industry context is vital. They can be considered influencers, who impact "organisational learning through social interactions in countless ways" (Mäkinen 2018, p. 135).

Complexity leadership theory looks at leadership that oscillates between multiple spheres due to "administrative, enabling and adaptive dynamics" (Mäkinen 2018). Scholars (Marion and Uhl-Bien 2001; Schneider and Smoers 2006) argue that traditional bureaucratic organisations are becoming inadequate in dealing with knowledge-intensive projects operating in a fast-paced world (Mäkinen 2018).

Subsequently various scholars (Lichtenstein et al. 2006a) unlocked debates on leadership as practices spread throughout the organisation instead of actions of a minority of leaders at the top. Therefore, greater emphasis was placed on relational leadership with a focus on dynamic and collaborating social practices (Lichtenstein et al. 2006b). This study employs complexity leadership theory as a research framework and uses key constructs such as administrative leadership, enabling leadership and adaptive leadership to understand the leadership capabilities that led to the successful establishment of the multi-sector partnerships between UP, SANRAL and CSIR.

Scholars such as Uhl-Bien and Marion (2009), and Uhl-Bien et al. (2007), argue that enabling leadership manifests in response to the tensions between administrative and adaptive leadership. It considers manners in which leaders can create environments that are ideal for "problem-solving, adaptability, and new learning" (Mäkinen 2018, p. 137). Administrative leadership characterises the "traditional, bureaucratic, and hierarchical type of leadership" (Mäkinen 2018, p. 136). For example these activities are considered to be part of administrative leadership practices, developing a vision, delegating responsibilities and enforcing strategy (Mäkinen 2018). Adaptive leadership is fluid and not a role but rather a complex dynamic (Mäkinen 2018). It is characterised as an interactive type of leadership that underlies emergent change activities (Mäkinen 2018, p. 137).

2.2 Leadership and Partnership

The literature abounds with debates on leadership characteristics, traits and styles. For example, scholars such as Weiss et al. (2002), Jian and Fairhurst (2016), Antonakis and Day (2018), and Farrell (2017), discuss this in their publications. Despite the rich scholarly literature, much of this research pays attention to single organisations located in the private sector (Armistead et al. 2007). Little attention has been given to leadership across organisational boundaries (Armistead et al. 2007). This hiatus prompts an exploration of how inter-organisational collaborative functioning takes place in complex multi-sector partnerships. However scholars such as Crosby and Bryson (2004), Huxam and Vangen (2005), and Murrel (1997), offer some direct thoughts on the discussion of leadership and multi-sector partnership. According to Huxam and Vangen (2005), unlike in single organisations, the question and uncertainty about who leads and who follows come to play in multi-sector partnerships. Multi-sector partnerships by its very nature are complex and uncertain as the relationships among the organisations invested in the partnerships can fluctuate between horizontal and hierarchical (Armistead et al. 2007). This chapter offers reflections of a multi-sector partnership in a South African context which highlights key leadership attributes that are valuable in forming such a partnership and simultaneously expanding on the literature of leadership and multi-sector partnerships.

3 Research Methodology

This study employed semi-structured interviews, as well as secondary data to explore the perceived leadership capabilities of the project leaders in the UP, SANRAL and CSIR multi-sector partnership. Moreover, the study used purposeful sampling. In doing so it assisted the researchers to choose the contributors who held valuable insights on the topic under study (Leedy and Ormrod 2010). This was intentional as the study was specifically aimed at only soliciting the perceived leadership capabilities from the project managers responsible for driving the partnership. The study was also shaped by the secondary data collected in particular the business case (Kannemeyer 2016) which serves as the founding document for the partnership.

The main benefit in collecting data in this manner is that it affords the researchers an opportunity to gain detailed information together with broad insights of the project leaders' perceived leadership capabilities. In order for this method to produce significant results, it required one-on-one interaction with the researcher and participant (Dei 2014).

The data collected through the semi-structured interviews was transcribed and analysed by summarising and developing codes that reflected the discussions. The transcribed interviews were coded by categories according to themes that were identified not only through interview findings but also linking it to the literature review. The relentless practice of inquiring and relating served to limit the researchers' subjectivity (Price 2012). The researcher developed analytical codes in line with the complexity leadership framework to establish how leaders in the multi-sector collaboration contributed to and managed the establishment of this partnership. The data was coded according to three core focus areas (anchored in the complex leadership theory) and analysed accordingly. These codes included administrative leadership, enabling leadership and adaptive leadership. Further, thematic codes such as organisational mandates, collaboration, governance, trust, bureaucracy, consultation and boundaries were developed to illustrate the significant commonalities. This data assisted in understanding the shared practices and leadership styles that participants believe were important for the formation of a partnership.

The researchers focused on categorical content analysis, which assisted with structuring the above-mentioned categories of the relevant topics related to the study. This allowed for deducing themes by describing recurring patterns and unpacking those themes in an organised manner. Furthermore, the thematic analysis assisted in connecting and linking the relationship between categories, allowing for a richer interpretation of the study. Lastly, the researchers used the literature review on leadership to make sense of codes, compared and contrasted research with the complexity leadership theory. This approach allowed the researchers to improve the reliability and validity of the study through data triangulation, thus probing the dependability of evidence across different data sources such as the semi-structured interviews, comparing data with secondary data and contextualising it with the literature.

3.1 Background to the Road Transport Sector

South Africa faces a number of challenges within the road transport industry, which includes the lack of quality material testers on construction sites, the lack of quality testing in laboratories and the skills required to do the material testing. These difficulties signalled a serious gap in the infrastructure that has the potential to hamper the delivery of quality roads, thus leading to the "poor state of the roads, vehicle overloading and financial constraints" (Chikagwa 2014, p. 49). Integral to this problem is the educational training system. It is evident that South Africa lags behind the global standard for engineers. In addition, scholar Chikagwa (2014) argues that the "number of engineers is insufficient and is dominated by ageing whites" (Chikagwa 2014, p. 49).

There is a strong correlation between the critical skill constraints and the severe challenges faced in transportation engineering in the country (Kannemeyer 2016). The need for a sound national road and transport system is a critical contributor for economic growth and sustainable development. Part of growing an economy requires good road infrastructure to move goods and services (Trimbath 2010). South Africa is recognised as one of the most well-organised transport systems in Africa, and has a responsibility to maintain and lead innovation in this area (Doke 2015). If the country wants to have a competitive edge among its "African and Asian rivals" in the global economy, the transport industry requires serious attention (Doke 2015).

3.2 The Partnership: UP, SANRAL and CSIR

With the aforementioned in mind, the UP, SANRAL and CSIR invested collaborative efforts to establish a formal partnership. The aim of the partnership is to work together to transform the road transportation industry by producing the laboratory facilities and human capacities needed to address the above challenges and enhance the skill gap (Kannemeyer 2016). The partners aspire to create an internationally renowned platform for academic and vocational training support in transportation infrastructure material testing, a national transportation material reference testing platform and high-quality research facilities and skilled staff. For example, autonomous or driverless cars will be tested in laboratories (Kupe 2019a).

Simultaneously, contributing to the quality and quantity of production whilst circumventing expensive duplication of laboratories, this initiative will form part of the "newly developed Engineering 4.0 facility at the University of Pretoria, and will consist of the SANRAL National Road Materials Reference Laboratory, SANRAL Training laboratory, an Accelerated Pavement Testing facility, an Active traffic track and an upgraded Concrete research laboratory" (Kannemeyer 2016). It also allows for the "evaluation of models in the APT and live traffic lanes by monitoring pavement responses under both accelerated and live traffic" (Steyn 2018, p. 31). Combining skills in road engineering, data science and sensor technology will help with intelligent transportation systems and cost-effective solutions (Kupe 2019b).

3.3 Rationale for the Partnership

The UP, SANRAL and CSIR partnership is indicative of creatively modernising models for business and government. In addressing the lack of civil engineers and related skills in transportation engineering in South Africa, development and maintenance skills in this area became imperative (Kannemeyer 2016, p. 1). Through smart optimisation and usage of "current facilities and staff available at UP, SANRAL and CSIR, these objectives can be achieved" (Kannemeyer 2016, p. 1). A key characteristic which makes this partnership successful is understanding the different institutional mandates and how they converge. In doing so, it was vital to differentiate "between the research conducted at Tertiary Education Institutes (UP in this instance), Research Councils (CSIR in this instance) and industry/public sector (SANRAL in this instance) to clarify the focus and mandates of each of the partners" (Kannemeyer 2016, p. 2), which could be likened to the parameters of the Frascati Manual as indicated in Fig. 8.1 (Kannemeyer 2016, p. 2).

The perceived success of the partnership is based on interrelationships between institutions, in order to produce different forms of research (Kannemeyer 2016:2). Moreover, the partnership will also create a channel for "stable research funding and alignment of bursary student programmes at UP, SANRAL and CSIR [including SANRAL Academy Trainees]" (Kannemeyer 2016, p. 2). This will assist in contributing to the critical mass from which the much-needed research skills can be recreated (Kannemeyer 2016, p. 2).

Besides formal training, students will also be mentored by UP, SANRAL and CSIR staff (Kannemeyer 2016, p. 2) and in addition, they will also have access to a "group of professionals that have expertise of various aspects of transportation engineering to complement academic supervision" (Kannemeyer 2016, p. 2). The outcome suggests that there will be more skilled professionals in transportation engineering which is of great benefit for South Africa's transport industry and economy.

3.4 About the Project Leaders

For a partnership to be successful, it requires the willingness of project leaders to work together in order to achieve a common goal. Partnerships require leaders who are enabling and adaptive and who are willing to engage in shared leadership (Clark 2008). The three key project leaders of the UP, SANRAL and CSIR partnership, responsible for driving this partnership and the achievement of its objectives, epitomised shared leadership. They were responsible for the groundwork and spearheading the collaboration on behalf of their respective institutions. Whilst the project leaders share commonalities, none of them are the heads of their institutions.

In spite of not being the head of their institutions, they were all sufficiently senior to convince management of the value of this project and partnership. More importantly, they are technical experts in their fields and have a deep appreciation for the technical challenges, nuances, com-

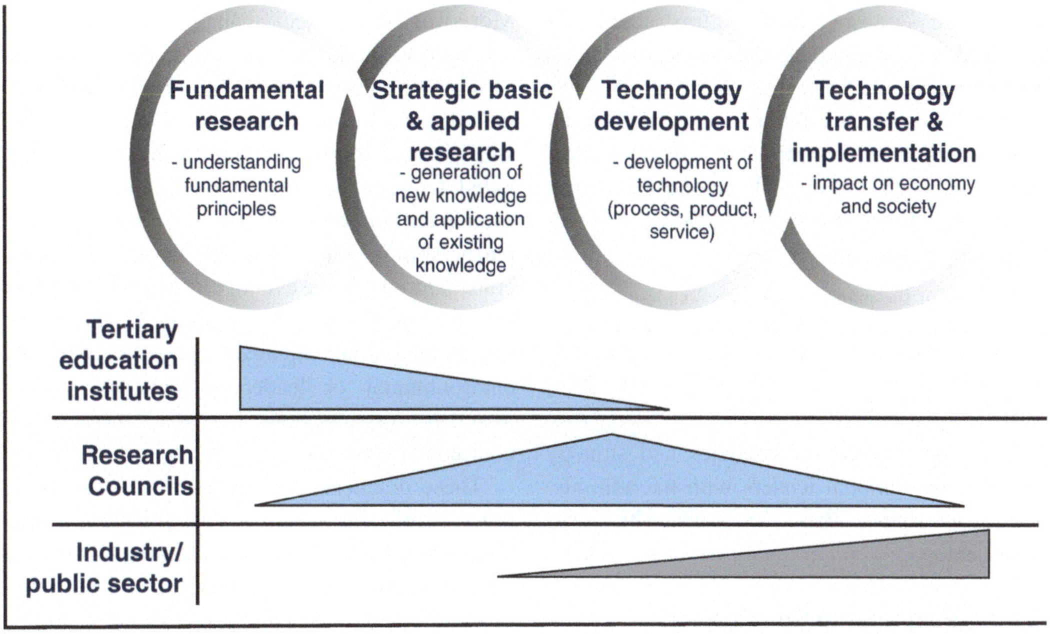

Fig. 8.1 Research and development classification based on OECD Frascati Manual and strategic positioning of the three partners. Source: Kannemeyer (2016)

plexities and risks that the transport industry face. The project leaders co-created solutions from the bottom up. These solutions were born from their experiences, challenges in the industry and a vision that dared to reimagine the transport industry.

The project leaders held numerous meetings, consultations and workshops to convince executives and boards of why this partnership matters. In spite of several rejections and reiterations of the business case, the project leaders did not give up. Nor were they discouraged by the protracted bureaucratic delays, countless presentations and resistance of sceptics. The project leaders found solace in their long-standing relationship with each other, which served as the glue that bonded them.

4 Data Analysis and Findings

The high-level findings on the project managers' perceptions of their leadership capabilities in the partnership, discussed below, include administrative, enabling and adaptive leadership practices.

4.1 Administrative Leadership Practices

Administrative leadership practices seek to create a stable environment to stimulate innovation and creativity. As Mäkinen (2018, p. 136) points out, "the goal is not to spin out of control … but that complexity leadership theory recognizses their importance for creating managed chaos". Administrative leadership practices hold the partnership together in a traditional and legal manner and can assist in preventing, circumventing and managing potential misunderstandings arising in the partnership. It further serves to limit some of the risks involved in "jeopardising the creation of adaptive dynamics" (Mäkinen 2018, p. 137).

4.1.1 Convergence of Mandates

UP, SANRAL and CSIR are publicly funded institutions, each having a defined and distinctive legal mandate, and are required to have duly approved strategic plans stating among other their institutional vision and strategic goals. Whilst the responsibility for the state of the country's road transportation infrastructure is SANRAL's core mandate, the partnership could be grounded in the convergence of the mandates of the respective institutions, as well as their visions and strategic goals. The project leaders understood each institutional mandate and had an appreciation for each institution's priorities, challenges and constraints.

> In terms of the mandates for sure, looking at the three organizations UP is supposed to train people and do research. SANRAL is supposed to look after the national road network. CSIR is supposed to do research [ID:A].

This convergence of mandates and strategy provided institutional leaders with the administrative assurance that the partnership was compliant.

4.1.2 Partnership Governance

In order for a collaboration to progress to an effective partnership, project leaders recognised the importance of formalising the partnership as well as appreciating that this would be non-negotiable for their respective institutions. Thus, the partnership was established as a legal entity with founding documents such as the Business Case for the Proposed Integrated Laboratory Facilities and a Memorandum of Agreement (MOA), which were critical for the governance and operation of the partnership. To this end, the MOA sets out clear roles and responsibilities for the parties involved, as well as details how the partnership will operate, what each party contributes, how financial reporting will occur and how disputes will be managed.

> So there's a really detailed MOA, which sets out the basic things like, what the steering committee looks like. What should the steering committee do? What should the broad financial model look like … we had a lot of discussions about that, because without that agreement, it doesn't help to say thank you for the money, we put this up, and suddenly, I thought you're going to pay for the electricity, and you think I'm going to pay ... It makes it easier to manage some things [ID.A]

A Steering Committee is the partnership governance structure and the MOA stipulates that all three partner organisations must be represented at Steering Committee meetings for it to quorate. Moreover, the project managers also had the foresight to think about the partnership in terms of long-term sustainability. Therefore, the MOA includes a clause that stipulates that the partnership is valid for 30 years and cannot be terminated without the consent of all three parties. This is an important element to be noted, as this partnership is bound by a time frame and shows consideration for the eventuality beyond the project managers' own involvement. It also safeguards the partnership against potential political, environmental or leadership changes at the respective organisations that may threaten its continued existence.

These practices are associated with administrative leadership which is a key element of the complexity leadership theory (Mäkinen 2018). In this instance these administrative leadership practices created an enabling traditional governance framework for managing the partnership.

4.1.3 Navigating Bureaucracy

Bureaucracy is a key pillar of administrative leadership as it, among other, enables separation of roles and responsibilities as well as accountability. As neither of the project managers were the heads of their respective institutions, it called on their leadership to strategise and skilfully navigate the corridors of power in their respective institutions, and beyond.

> To get a business case to be approved by the various, if I can call it governing bodies of the three institutions. So here our board, CSIR board, and the University of Pretoria because that was in the end, the biggest challenge was getting the approval at these higher levels of the initiative. So at a technical level, we could all see the benefits. And the importance for the country. But it was convincing the relative boards and the chancellor, in parting with money for this initiative. Because in the end, different people are putting in different amounts of money, or resources or people into the project for you to work. [ID:B]

As National Treasury funds SANRAL's non-toll road operations, obtaining National Treasury's buy-in and approval was an unavoidable administrative barrier which had to be overcome. National Treasury insisted that the partnership should be compliant with the national regulatory Supply Chain Management (SCM) policy and enforced competitive bidding. Ensuring the attainment of the partnership objectives required significant groundwork, information gathering and other interventions to convince National Treasury to provide a special deviation from the SCM prescripts to proceed with this collaboration. These negotiations to obtain National Treasury approval and unlocking the funds partnership caused major delays.

> … applying to Treasury, getting the approval to deviate, going through the tender process, adjudicating them, took me nearly two and a half years. In that two and a half years, I could not issue or continue with any research work [ID:B].

At SANRAL, the project leader first had to secure the support of the CEO, whereafter they could together approach the SANRAL Board.

> At the time, in terms of I was the one writing all the memos. So It took several board meetings to get approved [ID:B]. It wasn't approved at the first two attempts. Not at all [ID:B].

SANRAL overcame this administrative or bureaucratic challenge by holding multiple meetings with its Board and a special 3-h Board workshop. This workshop served as an opportunity to present a detailed project business case of the project and created a platform to persuade Board members to approve the partnership. Similarly, at the University of Pretoria, several stakeholders needed to be consulted to get buy-in, before this project could proceed, particularly because the laboratories were being built as part of the Engineering 4.0 facilities on UP land and this partnership required a substantial financial investment by the University. The University Council eventually approved enabling the partnership to advance.

> And I think similarly, at the University of Pretoria at first attempt, It wasn't necessarily plain sailing. We had to go and meet and explain and meet again. And explain again to the relevant stakeholders [ID:B].

Due to all the delays, the establishment of the partnership took 3 years since the initial discussions and conceptualisation. As mentioned above, the project leaders were in fact selling a dream that required the buy-in from stakeholders outside their immediate circle of trust. In some cases, project leaders also went along to each other's institutional meetings to field questions and assist with convincing interested parties.

4.2 Enabling Leadership Practices

Uhl-Bien and Arena (2018), point out that "enabling leaders are individuals who adopt behaviours for enhancing interactive and adaptive dynamics" (Mäkinen 2018, p. 137). They can serve as brokers who mobilise people, concepts and resources and facilitate the sharing of information (Mäkinen 2018, p. 137).

4.2.1 Appreciating Institutional Collaboration

The partnership had its genesis in informal discussions between the project leaders where the initial plan for collaboration was crafted after several meetings. The project leaders understood each institution's mandate and had an appreciation for each other's challenges and constraints. For example, below one of the participants provided a case in point of his understanding of the challenges, the need and how they arrived at a solution:

> It started with a need that needed to be addressed with regard to the quality of the material testers on the construction site. And also the quality of the testing within the laboratories on our construction, because in the end, the quality control is what really ensures the product is delivered as per the standard …. So when we looked at it,…, we identified a gap with regard to the training of material testers in the country … We're not really geared to training people. We are there to operate and maintain the network.
> … the CSIR,… had a dire need in terms of in the long-term in having access to better manpower, more people and a more steady flow in terms of income.
> … the problem with the University is they have no people to do these lectures and things because that's not really the area and the level they're focusing at with regard to material testers. The CSIR however ever had the skilled laboratory personnel that could assist with the training. Okay, the problem was that the CSIR's laboratory facilities are setup for research orientated work, and not accom-

> modating large numbers of students in a kind of a training environment in terms of that. So that was the problem there. So that's how in the end we then saw that the existing facilities at the University of Pretoria will not cater because they cannot accommodate for these large number of students, as it is. The CSIR's one where there are facilities can either not cater for accommodating these large numbers of students. Nor for running a proficiency testing. So it became apparent that we do need a separate facility [ID:B].

By looking beyond the individual institutional challenges, they connected across academia, industry and research. The project leaders' modelled behaviour enabled the grounds not only for a multi-sector partnership but also for the potential of establishing a transdisciplinary research facility. The partnership started to come to life much earlier than the MOA was signed between the different institutions. Below, one of the participants indicates how they felt in the initial phase:

> And then a business plan had to be written. And I was quite intricately involved in the writing of the business plan, with ID:A and ID:B obviously giving inputs into it. It was kind of like a joint thing when we started to physically writing things down [ID:C].

Like any other project or partnership, the above section draws attention to the importance of the ideation phase as it enables the project leaders to approach the next steps sequentially.

4.2.2 Sector Consultation

In a deliberate effort to address the challenges in the road transport sector, the project leaders created an enabling environment for transdisciplinary collaboration. It started with discussions at industry forums where serious concerns were raised. As one participant indicates:

> … in terms of our industry, we meet twice annually at a thing called the roads pavement forum. So the Roads Pavement Forum is an interaction of everybody in the industry, contractors, consultants, research client bodies etc. that gets together to discuss problems in the industry. So that helped a lot in putting everyone on the same page, in terms of what really is the problem … How are we going to solve and address this need that is aligned with what is identified by the industry as a problem. What is identified by client bodies as a problem, what is identified by research houses as a problem [ID:B].

Through this industry consultation, project leaders validated their initial hypothesis and amplified the initial business case with the sector's perspective of the problems and the solutions. Emboldened by a "mandate" to build an environment for problem-solving, the project leaders at UP, SANRAL and CSIR came together to deepen the discussions on how they can work together to develop solutions.

> "It's about synergy … For us, it's about a need. It started with a need to be addressed …" We then started having discussions to say can we not as a group work together on this … [ID:B].

4.2.3 Grit

No partnership is possible without leaders who are willing to work together and to champion a collective vision. In the interviews, the researcher asked "What in your opinion were the critical leadership features which served as a catalyst to the formation of this partnership?" The project leaders believed that it was a combination of things.

Firstly, the technical understanding of the real problem and the ability to turn the problem into a real technical solution:

> I think there's a few things … it's an understanding at a technical level, what the real problem is, and to be able to put the problem into a solution [ID:B].

Secondly, the passion for wanting to make a difference and believing that cooperative efforts could produce solutions and impact far beyond the individuals:

> . . .this passion, this commitment are crucial, because it's difficult things to make happen. And if you don't put that amount of energy and passion and commitment in, you usually come short. So it's a very rare animal that's prepared to do that. And that's the real leadership, but then the passion and the commitment, and the perseverance comes with it [ID:C].

Thirdly and more importantly was the issue of diligence and perseverance. The project leaders could have abandoned the idea of establishing this partnership several times after each rejection and disappointment. However, their tenacity and grit are what led to the successful establishment of this partnership:

> But I'd say the other issue is persistence. Because I mean, as I said, we could have given up after the first board, No [ID:B]. And the third thing would be diligence and perseverance ... [ID:C].

Among themselves they understood the need for this partnership and its potential, but it was not necessarily supported by everyone at first. It required a healthy dose of the four Ps, namely, passion, persuasion, persistence and perseverance, to get the buy-in from the relevant stakeholders.

4.2.4 Circle of Trust

Project leaders worked with each other earlier in their careers and appeared to have stayed in touch with each other's work at different institutions. In addition to the understanding of and respect for each other's professional expertise and knowledge, the long-standing relationships among them generated an enabling circle of trust.

There is an inherent reliance and trust in each other's words and promises:

> There is a major trust element in terms of I think, if I discussed this with you that there's a potential that you should be able to get it done. Yeah. And from our side for trust in the people of SANRAL for instance, that if they tell us, we're going to support you financially, that it's really going to happen. And so yeah, trust is the major issue [ID:A].

Notably, the project leaders are all white males. Given our fractured past as a nation, the inherent distrust of "different others" and the implicit comfort with "sameness" in terms of appearance, language and history, the element of "sameness" among the project leaders is a significant contributory factor which enabled collaborative leadership in this partnership. It can be assumed that this "sameness" made it easier to negotiate complexities and to reach consensus.

4.2.5 Proximity

The spatial proximity of the three institutions also played an important role in the decision to collaborate. Even though SANRAL owns enough land to construct laboratories, the project managers saw the value in constructing new facilities at the University of Pretoria. This meant that the facilities not only would be used to meet SANRALs needs but will be used on an ongoing basis for training. These facilities are an essential part of Engineering 4.0 which is in the process of being constructed at the University of Pretoria's Hillcrest Campus. This means it is 7.8 km from SANRAL and 9.3 km from the CSIR. This area has been coined the "most innovative square mile in Africa", reinforcing the idea that within a 10 km radius, this significant partnership has a valuable role to play in improving the transport industry of South Africa (Kupe 2019a).

4.3 Adaptive Leadership Practices

Adaptive leadership is not an allotted role but it is viewed as a complex dynamic (Mäkinen 2018, p. 137). Mäkinen (2018) describes it as "an interactive type of leadership that underlies emergent change activities" (Mäkinen 2018, p. 137).

The project leaders illustrated elements of adaptive leadership and flexibility. This became apparent when the partnership was negotiated. Bringing together three complex institutions each with its own vested interest is bound to present some degree of push-pull dynamics. Therefore interaction, flexibility and adaptation were important for the partnership to succeed. Below a participant gives some practical insights into the process of negotiation and adaptation:

> To a large extent one had to take the needs of all three partners, into consideration. And you had to see how one achieves each partner's needs through a process ... So we agreed upfront that we are not going to be duplicating test equipment ... So we're not going to be trying to the one taking over the others current area of knowledge or specialization. Yeah, there will be this joint cooperation. There will be a process of how easy access is ensured from the CSIR to the facility and vice versa. So those are like the kind of things we needed to clearly say, as part of this agreement, this is being reserved for the CSIR, the concrete part, because of the special knowledge is the university we retain that for your part. So CSIR is not going to set up a concrete laboratory ... [ID:B].

Setting boundaries among project leaders and for their institutions was important in order for the partnership to come into fruition. This is often where a partnership can fall apart, especially when there is an unwillingness to be open to an alternative perspective, or having the ability to review and adapt your position in the mutual interests of the partnership. Below is another example of how a participant described his understanding of adapting:

> … if we look at what the initial plan was, it will be a laboratory ABC and the way that it developed and the way that it changed. Although the major ideas are still the same, you have to have flexibility … So what's the stuff that we can be open minded and flexible about. And what's the stuff that sorry, non-negotiable, this needs to be there. This is important to have that understanding, because it makes it much easier, also, that I don't try to convince them of something that I know is a non-negotiable … [ID:A].

5 Discussion

Complexity leadership theory offers a useful framework for exploring how project managers of multisectoral institutions try to cultivate and bring together collaboration across sectoral borders. In the past complexity leadership theory has been critiqued for failing to show how complexity ideas influenced leadership behaviour in organisations (Rosenhead et al. 2019). However complexity leadership theory holds a

> … valuable metaphor for leadership practice and research. This metaphor holds the possibility of conveying the intricacies and tensions generated in milieu of radical indeterminacy in which, nevertheless, organisations need to take action (Rosenhead et al. 2019).

Through recognising various leadership roles and practises, this study was able to provide insights from the perspective of project managers that demonstrated some of those metaphoric leadership aspects. Based on the project managers' experience, it also illuminated what they perceived as the leadership capabilities that made this partnership work.

The findings reveal that when the parties gravitated towards each other to solve a complex problem, they relied on a number of leadership roles and practices, highlighted in the complexity leadership theory (Mäkinen 2018). For example at the beginning of the collaboration the project managers drew on enabling leadership. They were able to capture varied viewpoints and angles to the complex problems and needs by establishing a project team that worked together. Project leaders modelled collaborative behaviour and moved outside of their circle of trust to broker support from management structures and National Treasury. They were able to win over the sceptics and sell the potential impact and importance of the partnership. Drawing on the data collected, this partnership demanded certain personal leadership traits. According to the project managers, these traits included perseverance, commitment and passion for the project. These traits enabled them to move from ideation to implementation of the partnership.

As activities started to progress, leadership practices moved towards administrative leadership, formalising the partnership by developing a MOA. This allowed the different institutions to thrash out how they envisage this partnership operating for the next 30 years, therefore allowing project managers to influence and direct the partnership. As literature shows, immense bureaucracy, often associated with administrative leadership practices, poses an inherent risk to creativity and innovation; it does however offer a useful guide for managing certain interactions and ensuring long-term sustainability.

Adaptive leadership practices were drawn upon especially as it pertains to negotiating boundaries in establishing the new integrated laboratories. However, the researchers believe that more research should be done, in particular, on the complex adaptive dynamics as this partnership progresses. This will allow for better insights for understanding the impact of leadership as the partnership evolves.

Although this study categorised some of the leadership practices within the complexity leadership theory framework, the theory by its very nature is complex and argues that leadership is entangled, interdependent and interrelated. Thus, upon reflection, an interplay between enabling, administrative and adaptive leadership is evident in the UP, SANRAL and CSIR partnership. As literature confirms "complexity leadership theory does not ignore or deny the influence of traditional and bureaucratic tendencies" (Mäkinen 2018, p. 136). In connecting the threads, it appears that based on insights provided by the research, this study leans more towards convergence with the complexity leadership literature.

6 Conclusion

This chapter highlights the valuable role that partnerships can play in addressing complex leadership issues when integrated with the SDGs. The UP, SANRAL and CSIR partnership was born out of the need to find solutions for problems in the road transport sector. The project managers came together and exhibited enabling leadership practices by creating solutions in a collaborative manner. This resulted in the construction and the launch of the Engineering 4.0 facility at the University of Pretoria's Hillcrest Campus in 2020 5 years after the partnership discussion was initiated.

This partnership was not devoid of challenges; however the project leaders were able to navigate these through perseverance and drawing on each other for support. The findings presented in this chapter show that leadership plays an important role in a complex collaboration. The complexity leadership theory was a very useful framework for understanding the interplay and interdependencies of these leadership practices. Through this study, it was evident that project managers utilised their leadership capabilities in the partnership by means of administrative, enabling and adaptive leadership practices. In addition, this study affirms the important role of SDG 17 (partnerships) in contributing to SDG 9 (innovation and infrastructure), in this case, the transport industry. The UP, SANRAL and CSIR multi-sector partnership has the potential to have both a national and international impact through providing quality road infrastructure that is a key enabler for global markets and economic growth. Although this partnership is at its early stages, this study demonstrated the importance of strategically planning for the future through multi-sector partnerships and enabling implementation to benefit in the long term. The study recommends that a longitudinal study be done on the UP, SANRAL and CSIR partnership to understand the impact of the partnership over time and "how the actions of leaders shape knowledge creation activities in complex adaptive systems" (Mäkinen 2018, p. 136).

References

Armistead C, Pettigrew P, & Aves S (2007) Exploring Leadership in Multi-sectoral Partnerships. Sage Publications 3(2): 211-230.

Antonakis J & Day D. V (2018) Leadership: Past, present, and future.The nature of leadership 1(1): 3–26.

Bryant SE (2003) The role of transformational and transactional leadership in creating, sharing and exploiting organizational knowledge. Journal of Leadership and Organizational Studies 9: 32-44. https://doi.org/10.1177/107179190300900403

Clark J (2008) Developing Collaborative Leadership: A study of organizational change toward greater collaboration and shared leadership.

Chikagwa E (2014) A policy analysis of the Gautrain Public–Private Partnership in South Africa.http://researchspace.ukzn.ac.za/xmlui/bitstream/handle/10413/12156/Chikagwa_Edith_Wakondiye_2014.pdf?sequence=1

Crosby B & Bryson J (2004) Leadership for the Common Good: Tackling Public Problems in a Shared-power World. San Francisco, CA:Jossey-Bass.

Dei F (2014) A new evaluation of school feeding program: Case study of Magog Primary School. University of South Africa, Pretoria.

Doke L (2015) Investment still needed in transport infrastructure. https://mg.co.za/article/2015-07-17-investment-still-needed-in-transport-infrastructure.

Farrell M (2017) Leadership Reflections: Extrovert and Introvert Leaders. Journal of Library Administration 57(4): 436-443.

Hazy JK (2007) Computer models of leadership: Foundation for a new discipline or meaningless diversion?. Leadership Quarterly 18(4): 391-410. https://doi.org/10.1016/j.leaqua.2007.04.007

Holland J. H (2014) Complexity: A very short introduction. Oxford: Oxford University Press.

Huxam C & Vangen S (2005) Managing to collaborate: The Theory and Practices of Collaborative Advantage. London: Routledge.

Jian G & Fairhurst G.T. (2016) Leadership in Organizations. In: The International Encyclopedia of Organizational Communication.

Kannemeyer L (2016) Business Case: Integrated Laboratory Facilities. South Africa.
Kupe T (2019a) Inauguration Speech as Vice-Chancellor and Principal at the University of Pretoria. Pretoria.
Lakshman C (2007) Organizational knowledge leadership: A grounded theory approach. Leadership & Organization Development Journal 28: 51-75. https://doi.org/10.1108/01437730710718245
Leedy P & Ormrod J (2010) Practical Research: Planning and Design. 9th Edition.
Lichtenstein B, Uhl-Bien M, Marion R, Seers A, Orton JD & Schreiber C (2006a) Complexity Leadership Theory: An Interactive Perspective on Leading in Complex Adaptive Systems. Emergence 8: 2-12.
Lichtenstein B, Uhl-Bien M, Marion R, Seers A, Orton J & Schreiber C (2006b) Complexity leadership theory: An interactive perspective on leading in complex adaptive systems. Emergence: Complexity and Organization. Edition 1. 10.emerg/10.17357.289e129e99e80aa333abfde5fb792d3d.
Mäkinen E (2018) Complexity Leadership Theory and the Leaders of Transdisciplinary Science. Informing Science: The International Journal of an Emerging Transdiscipline 21:133-155.
Marion R & Uhl-Bien M (2001). Leadership in complex organizations. The Leadership Quarterly 12:389–418.
Mumford M (2000) Leadership Skills for a Changing World: Solving Complex Social Problems. https://www.researchgate.net/publication/222072620_Leadership_Skills_for_a_Changing_World_Solving_Complex_Social_Problems.
Murrel K (1997) Relational Models of Leadership for the Next Century Not-for-profit Organization Manager. Organization Development Journal 15(3):35-42.
Politis J D (2002) Transformational and transactional leadership enabling (disabling) knowledge acquisition of self-managed teams: The consequences for performance. Leadership & Organization Development Journal 23: 186-197. https://doi.org/10.1108/01437730210429052
Price J (2012) Coding: Open Coding. Encyclopedia of Case Study Research. Sage Publishing House.
Rosenhead J, Fr3anco L, Grint K & Friedland B (2019) Complexity theory and leadership practice: A review, a critique, and some recommendations. The Leadership Quarterly 30(5).
Schneider M, Somers M (2006) Organizations as complex adaptive systems: Implications of complexity theory for leadership research. Leadership Quarterly 17(4): 351-365. https://doi.org/10.1016/j.leaqua.2006.04.006
Steyn W (2018) Intelligent and smart transportation systems. Innovate 13: 30-31.
Srivastava A, Bartol K, & Locke EA (2006) Empowering leadership in management teams: Effects on knowledge sharing, efficacy, and performance. Academy of Management Journal 49: 1239-1251. https://doi.org/10.5465/AMJ.2006.23478718
Trimbath S (2010) The Economic Importance of Transportation Infrastructure.
UNDP (2019) Goal 9: Industry, innovation and infrastructure https://www.undp.org/content/undp/en/home/sustainable-development-goals/goal-9-industry-innovation-and-infrastructure.html on 19-03-2019.
Uhl-Bien M, Marion R, & McKelvey B (2007) Complexity leadership theory: Shifting leadership from the industrial age to the knowledge era. Leadership Quarterly 18(4): 298-318. https://doi.org/10.1016/j.leaqua.2007.04.002
Uhl-Bien M & Marion R (2009) Complexity leadership in bureaucratic forms of organizing: A meso model. The Leadership Quarterly, 20:631–650.
Uhl-Bien M & Arena M (2018) Leadership for organizational adaptability: A theoretical synthesis and integrative framework. The Leadership Quarterly, 29:89–104.
Weiss E.S, Anderson R.M & Lasker R.D (2002) Making the Most of Collaboration: Exploring the Relationship between Partnership Synergy and Partnership Functioning. Health Education and Behaviour 29(6): 683-98.
Xue Y, Salaj A, Engebø, A & Lohne J (2020) Multi-sector partnerships in the urban development context: A scoping review. Journal of Cleaner Production.

The Role of Leadership Capacities in the Response of the South African National Statistics System to the Sustainable Development Goals

9

Joe de Beer and Willem Fourie

Abstract

The Sustainable Development Goals (SDGs) require the National Statistics System to reconsider the ways in which it organises the collection of statistics. In South Africa the SDGs are perceived to have posed a disruptive challenge to Statistics South Africa, the key institution in South Africa's National Statistics System (SANSS). This study identifies the leadership capacities that enabled a successful response to the SDGs by the SANSS. It uses a single case study approach to identify these capacities, focusing on the trait and the neoclassical schools of leadership. The study examines the range of perceptions of the leadership capacities displayed between 2015 and 2017 and between 2017 and 2019 by two different leaders. These perceptions were collected through discussions with the two leaders themselves, a peer as well as three followers, and then analysed through a structured coding system. The frequency of occurrences of perceived capacities was subsequently linked with the two leadership theories. The study found that different leadership capacities were prominent during the two periods. The capacities associated with the trait leadership school were successfully used during the implementation period of the response to the SDGs, whereas the second maintenance period demonstrated capacities associated with the neoclassical school. The study concludes that different capacities are required at different stages of a successful response.

Keywords

Leadership · Trait · Transformation · Neoclassical · SDGs

1 Introduction

Statistics South Africa (Stats SA), South Africa's current national statistical office and the key institution in its National Statistics System (NSS), was established in its current form in 1999. However, the keeping of official statistics has a long history in South Africa. The first reference to the establishment of the statistical authority in the country can be found more than a century ago. The Statistics Act (38/1914) passed by the Union of South Africa mandates a body of government to "provide for the collection of statistics relating to agricultural and to industrial, commercial, shipping, fishing and other business

J. de Beer (✉) · W. Fourie
University of Pretoria, Pretoria, South Africa
e-mail: joe@debeer.za.org; willem.fourie@up.ac.za

G. Nhamo et al. (eds.), *Sustainable Development Goals for Society Vol. 1*, Sustainable Development Goals Series, https://doi.org/10.1007/978-3-030-70948-8_9

undertakings and other matters in the Union" (South Africa 1914).

Official statistics have always been collected by following a well-structured approach to meet the needs of the user community. This was articulated by the first Fundamental Principle of Official Statistics (United Nations 2015), which prescribes that the statistics must meet the needs of its users and potential users. The statistics system followed a programme of statistical surveys that has remained fundamentally unchanged, but has slowly evolved in response to the changing needs of stakeholders. These needs revolved around measuring the economy as well as the changing demographics of the population.

However, over the last 115 years of official statistics, there have been occasions which required a far more rapid response to changing user needs. These rare events completely disrupted the status quo of the statistical landscape and required significant adjustments. Examples include the onset of the two world wars as well as the establishment of the Republic of South Africa.

The most recent significant disruption occurred in the period between 1994 and 1996. With the introduction of democracy in South Africa, a single unified government was formed. This required the integration of statistical offices from the Republic of South Africa and the six statistical offices from the so-called self-governing territories (Gazankulu, Kangwane, Kwandebele, KwaZulu, Lebowa and Qwa-Qwa), as well as the four so-called independent states (Transkei, Bophuthatswana, Venda, Ciskei). In addition to the amalgamation of the 11 statistical offices with their own work programmes, staff and priorities, the remit of the Central Statistical Services was changed to allow it to measure the socio-economic conditions of all citizens of the Republic of South Africa as opposed to focusing on the traditionally White residential areas and businesses.

In 1999 the Statistics Act (South Africa: 6/1999) was amended to allow for the establishment of Stats SA as an autonomous national government department, headed by the Statistician-General (SG). This position entails two roles, namely to be the head of Stats SA and to be the SG of South Africa. In the former role, the office is responsible for delivery on the work programme of Stats SA, and in the latter for the coordination of all official statistics within the country as well as their certification as official.

Prior to this, although Stats SA's activities were seen to be independent of political influence, the head of the organisation reported to the Directors-General of other national departments such as Home Affairs or National Treasury. An often overlooked aspect of the 1999 Act is that the SG is now responsible for all official statistics, and must play a coordinating as well as a quality certification role for statistics released by other departments.

In addition to disruptions to the statistical landscape as a result of changes in the external environment, different frameworks being introduced on an international, continental or national level can also act to disrupt the balance of supply and demand of statistics in the NSS. The introduction of SDGs as a comprehensive, data-intensive organising framework for statistics is the most recent example of a disruption to the system of official statistics. Significant changes were required in the way national statistical agencies and national statistics systems went about implementing their mandate. The challenges included the need to produce official statistics to meet new data demands, the certification of the quality of unofficial data sets within a resource-constrained environment and the establishment of partnerships with other data owners, government departments, civil society, academia and interested parties. Previous disruptions from a statistical perspective occurred when the System of National Accounts, which provides the framework for economic statistics, changed from the 1968 version to the 1993 version and then again to the 2008 version. All these changes were made in response to a global economy that was changing and required that the measurement of these changes adapt to the same extent.

It could be argued that certain specific leadership qualities were required to ensure that the SANSS response to this disruption was effective. Research theory suggests that there is a link between leadership and successful management of change, but there are still instances where change within an organisation fails despite leadership. Lawrence (2015, p. 3) argues that such failures could be the result of leaders following sequential linear steps in addressing change rather than implementing a centrally articulated strategy.

The purpose of this case study is to examine demonstrated leadership capacities to respond to these challenges to the NSS. The central research question of the study is "Which leadership capacities from the trait and neoclassical theories enabled the South African National Statistics System's response to the SDGs?" The leadership capacities of the two SGs during the periods from 2015 to 2017 and 2017 to 2019, respectively, delimit the case study. The chapter provides a concise review of the literature on the two leadership schools before describing the research design. It concludes with a presentation and discussion of the findings.

2 Literature Review

There are numerous leadership theories, each highlighting different attributes, styles or capabilities. They are generally organised into leadership schools, given that "leadership is a complex and diverse topic" (Day and Antonakis 2012, p. 3), because of the multitude of changes that leadership research has had to contend with in history. As societies, cultures, technologies, economies and political landscapes change, so does the behaviour that is associated with appropriate leadership under the changing conditions.

Being personally familiar with the two leaders who form the focus of my case study assisted in shaping my thoughts about which leadership theories best describe their capacities. It is therefore possible to use an autoethnographic approach to prioritise which leadership theories we think will be most relevant, given our experiences and own observations. As Ellis et al. (2010, p. 4) note, "Autoethnographers must not only use their methodological tools and research literature to analyze experience, but also must consider ways others may experience similar epiphanies; they must use personal experience to illustrate facets of cultural experience, and, in so doing, make characteristics of a culture familiar for insiders and outsiders".

This study uses two of the most prominent leadership theories as theoretical points of reference: the trait leadership theory and the neoclassical theories, with the overall focus being on transformational leadership theory.

As the trait school of leadership theory is the oldest, we can trace its origins back to one of the earliest definitions of leadership. Thomas Carlyle uses the phrase "great man" 43 times in his series of lectures "On Heroes, Hero-Worship, and the Heroic in History", and he associates the exemplary man with being a hero of divine proportions, a prophet, a poet, a priest, a man of letters and a king (Carlyle 1840). In its purest form, the leader is therefore taken as a great man who can adopt any of these roles and possess the necessary characteristics appropriate for the role.

Galton (1869) identified the notion of genius as an example of a leadership trait that is passed down to generations in a hereditary manner. In his studies he aims to measure this phenomenon quantitatively in order to derive numerical results.

This school suggests that the characteristics of an individual determine whether he/she will be a leader or a follower. The trait leadership school evolved from the idea that a person was a born leader and that these inherent traits were passed down through the generations into theories that recognise that there are different sets of characteristics and that the importance of understanding the linkages between them and eventual effective leadership cannot be overemphasised. Some of these mechanisms are not necessarily inherent in a leader but can be developed over time.

Judge and Long (2012) identify three paradoxes in relation to this leadership theory, namely that the benefits of certain traits could be disadvantageous in different situations; traits rarely have complete and universal advantages in a particular situation; and individual traits have exponential advantages as they relate to leadership compared with other traits.

There is no clear consensus amongst scholars regarding the results of the studies that try to correlate the characteristics of leadership with actual leadership in practice. Various frameworks have been proposed to improve the understanding of the role that traits play in leadership behaviour. One of those is a five-factor model which was generally accepted as a summary of relevant characteristics of trait-based leadership. These are summarised by Judge and Long (2012) as the traits of extraversion, agreeableness, conscientiousness, emotional stability and openness. Each of these traits can have both a positive and negative impact on leadership. They also add to this model the traits of core self-evaluation and intelligence.

The idea is that traits have a direct impact on leadership style and quality. The proximal elements are viewed to be a direct influence on leadership and are "state-like" in nature. A quantitative study was conducted to determine whether the traditional traits or proximal traits had the most significant impact on effective leadership. The analysis included 25 different individual characteristics and was moderated for the level of leadership as well as type of organisation (Hoffman et al. 2011). The outcome of the study was that "on average, state-like and trait-like individual differences have a similar relationship with effective leadership" (Hoffman et al. 2011, p. 365).

Further insights into the trait leadership theory are provided by Derue et al. (2011, p. 9), who state that their research "reviews and integrates the literature on leader traits and behaviours, and takes a first step toward an integrative theory of how leader traits and behaviours influence leadership effectiveness". The area of interest in this model is therefore those behaviours (task oriented, relational oriented or change oriented) which will serve as the most appropriate mechanism to link leadership traits to leadership effectiveness. Their study concludes with two primary findings. Firstly it showed the importance of leadership being taken per se, finding that even poor leadership is better than the complete absence of leadership. Actively embracing the leadership role is vital. Secondly, all these leadership behaviours are required for effective leadership. A focus on only one of the dimensions cannot transform all the inherent traits into effective leadership (Derue et al. 2011).

Despite the changing perceptions of the requirement of a specific set of inherent traits for leadership, as opposed to the view that it is teachable, there are certain more specific perspectives that must be kept in mind when thinking about this leadership school. They are summarised by Judge and Long (2012) as follows: individual differences matter; demonstrated styles will differ depending on the leader's disposition; and context matters because leaders function in varying and diverse groups and environments.

The second school of leadership is called the neoclassical school; it adopts a charismatic approach to leadership, and focuses on how a leader influences the followers by appealing to their emotions, morals and intellect. The combination of *pathos, ethos* and *logos* can bring about social change and transform society, not just lead members of a society in a particular direction. The followers, therefore, are not simply led to meet their obligations to fulfil the needs of the leader and the followers, but they are transformed and view their participation as being for the greater good of the society.

The link between leadership and "charisma" is generally believed to originate in Weber (in Antonakis 2012, p. 258), who described the relationship between followers and a "charismatic leader". Bass (1990, p. 184) noted that "if successful, charismatic leaders bring about radical transformations in their groups, organisations and societies".

The success of a charismatic leader hinges on two related aspects. Firstly, the leader must possess the leadership capacities that are inherent in charismatic leaders. Secondly, the followers must have a strong urge to follow the leader and share a strong sense of belief. The characteristics of charismatic leaders can be defined in many ways, but would include aspects such as the requisite abilities and personality characteristics to function in the environment, expressive behaviour, self-confidence, self-determination, insight, freedom from internal conflict, eloquence as well as high activity and energy levels. Similarly, the followers must be open to a charismatic leader and be able to identify with the leader. Typically there will be a mystique around the charismatic leader, and the leader will be able to transform the followers, most likely in a time of crisis when the leader will seem to be able to provide a solution. Often there are also cultural and organisational expectations that demand of leaders to be, or to be seen to be, charismatic and transformational (Bass 1990).

Bass (1990, p. 32) identified "three transformational factors – charismatic leadership (including inspirational leadership), individual consideration, and intellectual stimulation" to transform followers. With the progression of the understanding of leadership qualities, these were later reformulated as "the four Is", by renaming "charismatic leadership" as "idealised influence" separately from inspirational leadership (Bono and Judge 2004, p. 901). These four components of transformational leadership form the basis of a number of studies on transformational leadership.

Leaders all have different personalities, which influence the leadership capacities that they embody. These capacities may be analysed through a framework that distinguishes five personality traits. This "Big Five Framework" includes extraversion, neuroticism, openness to experience, agreeableness and conscientiousness and "provides a meaningful framework for formulating and testing hypotheses relating to individual differences in personality" (Barrick and Mount 1991, p. 23).

Although the categorisation of leadership capabilities that form part of the neoclassical and transformational leadership theories is not always precise, scholars have reached a broad consensus about them. Identifying the leadership behaviours that stem from the personalities of the leaders assists in understanding their approach to leadership.

At the core of the neoclassical transformational leadership theory is the notion of a leader that has near-heroic attributes. This is not the same as Carlyle's "great man" described earlier in the trait leadership school, but is instead focused on the leader who is able to transform the followers into believing and doing everything that is required. However, Fourie and Hohne (2019, p. 54) point out that "the transformational leader is also in need of transformation". This is to make allowance for the leader to fail and for this to be included in the leadership theory. This brings an aspect of humanity to the leader, rather than being a mighty, infallible hero at all times.

3 Research Design

3.1 The Case

The NSS response to the SDG disruption started in 2015, with the first phase ending in 2017 with the publication of the South African Indicator Baseline Report. This coincided with the appointment of a new SG, who served as the formal leader of the project from 2017 to 2019 (phase 2 of the project). This is graphically indicated in Fig. 9.1.

Although the Millennium Development Goals (MDGs) that preceded the SDG initiative required Stats SA to collect and publish various indicators and to interact with a range of stakeholders, the scope of the project was not as comprehensive and sophisticated as that for the SDGs. The reporting on MDGs did, however, provide some experience for Stats SA officials that would set the scene for successfully participating in the SDGs.

In essence, the SG had to perform a number of duties to guide the NSS's response to the SDGs. On a global level, he was a member of the interagency expert group that developed the SDGs and their indicators. The group met frequently and was responsible for inter alia the formulation of the indicators as well as the finalisation of the methodology and required definitions.

On a continental level, the SG supported the United Nations Economic Commission for Africa, the African Development Bank and the African Union in formulating joint positions on matters relating to the SDGs as well as mobilising African countries to participate.

Within the country the SG is responsible for the collection, coordination and certification of the statistics required to populate the SDG and its

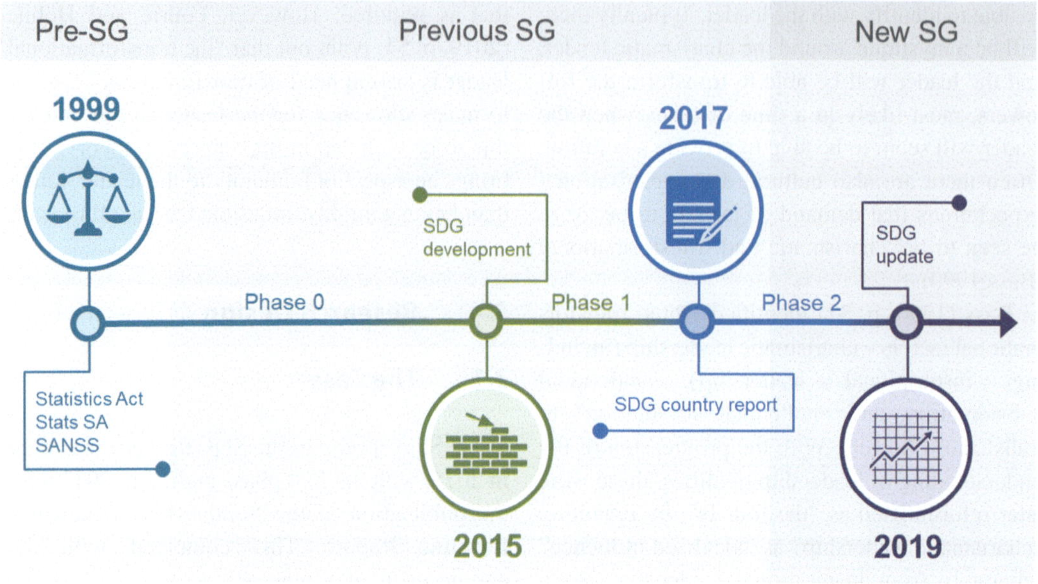

Fig. 9.1 Graphical representation of the phases of the case

indicators. Initially, Stats SA was the only driver of the SDG agenda, and relationships with other government departments such as the Department of International Relations and Cooperation (DIRCO) and the Department of Monitoring and Evaluation (DPME) needed to be established. This was strained at times as Stats SA had to retain the mandate to provide the official statistics for use in the SDG reports. It was during this era that Stats SA produced the Country Baseline Report (CBR), which was published in September 2017. The purpose of the report is to establish the benchmark against which progress will be measured in future.

After the publication of the CBR, work continued on the international and continental front to refine the SDG indicators and increase African participation. Nationally, there were ongoing efforts to coordinate the production and collection of statistics from various sectoral national departments as well as to domesticate the indicators in order to align them more closely with South Africa's own developmental needs and national strategies. Particular efforts were made to broaden the scope of stakeholders that are involved in the SDG. This was done through consultative processes with other organs of state, academia and civil society.

South Africa decided to participate in the Voluntary National Review, a project led by DPME. This review relies on statistics from the CBR, but the aim is to describe the preparedness of the policy environment to implement the SDGs. The report was presented on 17 July 2019 at the United Nations High-Level Political Forum (United Nations 2019).

Subsequently, the SG led the SANSS to develop the first Country Report on SDG. The basis of the report remains the CBR, although some indicators could be enhanced with updated data sources. Writing the report added complexity to the typical Stats SA reports as they were written by authors who are independent of government.

This is important in two respects, firstly, because the SDGs are not a government project, but require the entire country to participate and focus on sustainable development, and secondly, it is essential that the interpretation of the data is done objectively and cannot be, or even perceived to be, tainted by views that are subjective and favour government.

Table 9.1 Interview questions and purpose

Question theme	Purpose
Question 1: Was there leadership in the project? Yes/No	Was the project a success, and what contributed to the success?
Question 2: What leadership styles can be used in general?	General comments on perceptions about leadership in the project
Question 3: What leadership capacities did you observe in the project? This was done in two parts for each phase of the project: • XG (first Statistician-General) • SG (second Statistician-General)	The perceptions of the formal leadership style of Statistician-General in phase 1 The perceptions of the formal leadership style of Statistician-General in phase 2
Question 4: Were there elements of informal leadership in the project?	The perceptions of informal leadership in the project
Question 5: What leadership style would you use yourself in this project?	Perceptions about leadership if the interviewee had to lead the project

3.2 Data Collection

A questionnaire with five questions was administered to six respondents to collect information on their perceptions of leadership capacities in this case study. All respondents worked in Statistics South Africa. Five people were interviewed, and one leader responded to the questions in writing. The data from the transcripts were processed according to the practice of qualitative data analysis (QDA). Chenail (2012, p. 1) notes that QDA involves "collecting quality talk, observations, and/or documents", adding that QDA can be understood "as both the analysis of the data and the analysis of the analysis of the data". The responses to the five questions cf. (Table 9.1) in the interview were divided into seven data sets.

3.3 Data Analysis

The analysis had two underlying phases. Firstly, a thematic approach to data analysis was used. The analysis aimed to identify themes or sub-themes from the data in the transcripts, and then relate them through a coding frame to a standard list of codes. These codes were analysed to identify those codes that best described the leadership capacities that emerged during the project.

Secondly, the extent to which the clusters of codes aligned with leadership capacities foregrounded in the reference leadership theories was investigated. In this sense we followed the approach already proposed by Glaser and Strauss (1967), according to which data can be used both to generate new theory and to verify existing theory.

This was an iterative process because it was possible to make adjustments to the interview guide for subsequent interviews or change the direction of the initial analysis as other areas of interest came to the fore. Once data have been collected, qualitative analysis requires the coding of the text into meaningful categories that will form the basis for interpretation and analysis. Shahedul (n.d., p. 2) notes that "concepts can be classified differently; it depends on the different properties of data the researcher is focusing on and how he/she is translating them".

Once the interviews were conducted, transcribed and quality controlled, the processing took place in a number of steps. The purpose was to transform the transcripts into information that can be analysed.

The responses to the open-ended questions were analysed and coded. The information was anonymised to ensure that the confidentiality of the respondents was protected before the final tabulations were prepared. The responses were combined within each stratum in order to develop a response from the strata in totality. Emphasis was placed on identifying patterns in the responses to the questions. Both similarities and differences in the responses could lead to interpretations of whether the leadership theories adequately address the issues raised in the problem statement.

This was done as follows:

Step 1

The transcripts were carefully scrutinised for any references to the success (or not) of the project as well as references to leadership behaviours, styles and capacities. These sentence fragments were all marked for further coding. Subsequently, highlighted sections were interpreted in the context of the interview and provided with a first-level code which highlighted the key phrase, with a memo if necessary. A sixth set was developed that included all negative comments that were made in any of the five questions that were asked.

Step 2

The highlighted text and first-level codes were summarised in a spreadsheet, including the page number, to make it possible to revert back to the original text. This was often done during data analysis as new information was discovered and then previous codes or highlighted text had to be revisited and revisions made.

Step 3

All seven (the third question was separated into data sets for the XG and SG separately) data sets were constructed for each of the three strata in the study, yielding a data cube with potentially 21 combinations, as shown in Table 9.2.

This took the form of tabulations to which a second-level code was added for each observation. The purpose of the second-level code was to start aligning the first-level code with the nomenclature typically used in the leadership theories, as well as to introduce some standardisation to the codes. This code can be seen as the first identification of perceived leadership capabilities. In some instances, the notes made as memos to the highlighted text were used to make a judgement on which second-level code would be most appropriate. The addition of the second-level code once again allowed for a different perspective on the original highlighted text from the transcript, and with the data trail described thus far it was possible to return to the actual audio interview, transcript, memo and first-level code to decide on the most appropriate interpretation. In some cases, first-level codes had to be combined or new codes added to the analysis.

Table 9.2 Potential data cube

Data set	Followers	Peer	Leader
Yes/No	21 permutations		
Leadership			
XG			
SG			
Informal			
Self			
Negative			

Step 4

The analysis of the data resulted in 312 data points. In order to aid analysis and interpretation, frequency tables were developed based on the intersections of the variables. The first option is a cross-classification of the 3 strata, 7 data sets and 24 second-level code variables. The resulting matrix is shown in Table 9.3. The same information can be tabulated by changing the order of the stratum and data set. As shown in Table 9.4, this makes it easier to identify capabilities that were mentioned for the two formal leaders (XG in phase 1 and SG in phase 2), which is the crux of the research question.

A limitation of the second-level coding structure that was identified during the initial analysis is that the negative perceptions, as well as the notion of maturity, could have been dealt with differently. The current approach that relies on using them as part of the field notes or memos to inform the findings and conclusions has the inherent weakness that it is not as instructive as the capabilities that are explicit in the descriptive analysis in Tables 9.3 and 9.4. Maturity is treated as a capacity for the purpose of further analysis.

Additional analysis was therefore required relating to the negative data set. From Tables 9.3 and 9.4 it is evident that 51 negative comments were made, mostly (40) by the followers. All the negative extractions were reassessed to identify those that relate specifically to the SG and XG data sets.

Table 9.5 shows the count of the negative comments relating to phase 1 (XG) and phase 2 (SG) for each of the second-level codes.

Although there were 51 negative comments (Table 9.3), 8 of these did not relate to the SG or XG. There were 43 negative comments relating to the perceived leadership capabilities in the combined two phases, evenly distributed

Table 9.3 Frequency table of stratum, data set and second-level code

	Stratum and data set																								
	Follower								Leader								Peer								
Second-level code	Informal	Leadership	Negative	Self	SG	XG	Yes	Subtotal	Informal	Leadership	Negative	Self	SG	XG	Yes	Subtotal	Informal	Leadership	Negative	Self	SG	XG	Yes	Subtotal	Grand total
Adaptability		4	1		2			7				1		1		2		1	5			1		7	16
Assertiveness			6			9		15			3		1			4						2	21		
Authority		1						1																	1
Charisma					2			2				1		3		4									6
Collaborative	12	6	7	1	7	10	4	47	7	2	1		12	3	3	28	3	1		1	1			6	81
Competence	2	6		3		1	2	14	1							1				3	2	3		8	23
Continuity																					1			1	1
Decisiveness		3	1			1		5							1	1									6
Emotional			3					3																	3
Experience		1					2	3						2		2						1	2	3	8
Extraversion			1					1																	1
Inspirational	1	2	2		2	5	2	14						2	2	4		1						1	19
Leadership							5	5																	5
Maturing																		12						12	
Openness												1	1	1		3					2	1		3	6
Partial success			1					1			1					1									2
People							2	2							2	2		1	1					2	6
Perseverance	2		4		1	4		11				1	2	2		5		1						1	17
Processes	1						1	2															1	1	3
Responsibility		3						3	2	2		1	1			6									9
Systems		2			4	1	2	9							1	1					1		1	2	12
Visible			5			14		19																	19
Visionary			9		1	9		19						3		3						2		2	24
Yes							4	4							5	5							2	2	11
Grand total	18	28	40	4	19	54	24	187	10	4	5	5	17	17	14	72	3	17	6	4	7	10	6	53	312

Table 9.4 Frequency table of data set, stratum and second-level code

	Data set and stratum																												
	Informal				Leadership				Negative				Self				SG				XG				Yes				
Second-level code	Follower	Leader	Peer	Subtotal	Follower	Leader	Peer	Subtotal	Follower	Leader	Peer	Subtotal	Follower	Leader	Peer	Subtotal	Follower	Leader	Peer	Subtotal	Follower	Leader	Peer	Subtotal	Follower	Leader	Peer	Subtotal	Grand total
Adaptability					4		1	5	1		5	6		1		1	2			2		1	1	2					16
Assertiveness									6	3		9						1		1	9		2	11					21
Authority					1			1																					1
Charisma														1		1	2			2		3		3					6
Collaborative	12	7	3	22	6	2	1	9	7	1		8	1		1	2	7	12	1	20	10	3		13	4	3		7	81
Competence	2	1		3	6			6					3		3	6			2	2	1		3	4	2			2	23
Continuity																			1	1									1
Decisiveness					3			3	1			1									1			1		1		1	6
Emotional									3			3																	3
Experience					1			1														2	1	3	2		2	4	8
Extraversion									1			1																	1
Inspirational	1			1	2		1	3	2			2					2			5	2		7	2	2		4	19	
Leadership																									5			5	5
Maturing							12	12																					12
Openness														1		1		1	2	3		1	1	2					6
Partial success									1	1		2																	2
People							1	1			1	1													2	2		4	6
Perseverance	2			2			1	1	4			4		1		1	1	2		3	4	2		6					17
Processes	1			1																					1		1	2	3
Responsibility		2		2	3	2		5						1		1		1		1									9
Systems					2			2									4		1	5	1			1	2	1	1	4	12
Visible									5			5									14			14					19
Visionary									9			9					1			1	9	3	2	14					24
Yes																									4	5	2	11	11
Grand total	18	10	3	31	28	4	17	49	40	5	6	51	4	5	4	13	19	17	7	43	54	17	10	81	24	14	6	44	312

Table 9.5 Negative comments relating to phase 1 and phase 2

Second-level code	SG	XG	Grand total
Adaptability	1	5	6
Assertiveness	1	5	6
Collaborative	1	6	7
Emotional	1	2	3
Extraversion	1		1
Inspirational	1	1	2
Perseverance	2	2	4
Visible	5		5
Visionary	8	1	9
Grand total	21	22	43

between the two formal leaders in the case. In the case of phase 1, there were negative connotations attached to the leadership capabilities (second-level codes) of collaborative (27%), adaptability (23%) and assertiveness (23%). In the case of phase 2, there were negative connotations attached to the leadership capabilities (second-level codes) of visionary (38%) and visible (24%).

In the case of the notion of maturity, it was mentioned only by the peer stratum. This was done in the context of the leadership data set, although additional scrutiny of the transcripts and coding shows that there was an allusion to the first phase (XG). This will guide the finding and analysis.

Another limitation is that the coding concepts that were used are not always clearly distinct from each other. I used the context of the question, responses to other questions and stratum from which the response originated to interpret what the interviewees meant during their response to the question. This level of interpretation is not viewed as detrimental to the reliability of the results. I have an advantage in having myself observed the leadership behaviour of the two formal leaders during the two phases, although I was not directly involved in the project. This experience allowed me to have a better understanding of the points that the interviewees were trying to convey, and I could link their responses to leadership issues without having to ask them explicit or technical questions on it.

4 Presentation and Discussion of Findings

In order to address the problem statement, i.e. which leadership capacities from the trait and neoclassical theories enabled the South African National Statistics System's response to the SDGs, the data set was used to compare the perceptions of leadership with the leadership theories identified in the literature review. The aim is for the evidence to show to what extent the identified leadership theories were able to explain the perceived leadership capacities displayed by XS and GS during the two phases of the workplace initiative and whether these capacities were the primary reasons for the success of the project.

The actions of leaders will be perceived differently by followers. Bradley et al. (2006, p. 9) note that this is because of "factors such as leader behaviour, leader-follower relationship, and leadership prototypes". Pronin (2007, p. 37) notes that "people's perceptions can be biased by their beliefs, expectations and context, as well as by their needs, motives and desires".

Perceptions are the basis for the data in this study and it was therefore expected that the interviewees would perceive the same leadership behaviours differently. This does not invalidate any views that differ from the general consensus of opinion. However, there was sufficient conformity of views to allow for meaningful analysis.

The methodology allowed for analysis to be done in three different ways. Firstly, an analysis can be made of specific leadership capacities. This would include whether some capacities were more prevalent than others as well as which data set they were dominant in. This will correspond to the second-level codes in Tables 9.3 and 9.4.

Secondly, the analysis can be conducted on the basis of which leadership capacities were mentioned most often by the three different strata in the sample. This will correspond to the information in Table 9.3.

Thirdly, the analysis can be aligned to the original intention and design of the study, which is to analyse perceptions about the leadership capacities of the two formal leaders. In a similar fashion, a complementary analysis can be done for the other data sets that were developed during analysis, such as "informal leadership" and "self-leadership". This corresponds to the analysis in Table 9.4.

Analyses of a data set in different ways add to the robustness of the findings. Developing different scenarios of findings through the analysis of competing hypotheses enhances the value of the findings that can be extracted. Heuer (1999, p. 31) observes that analysts "should think about how they make judgments and reach conclusions, not just about the judgments and conclusions themselves". It is therefore not sufficient to follow only a single approach to the collection of information, coding and analysis in order to answer the primary question of the case, but alternatives should be explored as well. The subjectivity that is introduced during coding and interpretation could influence the findings if it is not explicitly recognised.

The primary aim of this study was to identify and analyse the leadership capabilities that were exhibited during the NSS's response to the challenges posed by the SDGs. We used arguably the two most prominent leadership theories as reference theories, even though our methodology allowed for the identification of leadership capacities not present in either of these theories.

This analysis is the nexus of the study. In order to understand which capacity was dominant for each of the formal leaders, it is necessary to draw findings from interpreting Table 9.3 in conjunction with the negative findings in Table 9.5. This allows for an adjusted frequency of the perceived leadership capacities, as negative observations can be contrasted with the initial observations. This provides the effect of a "nett value" of perceived leadership capabilities. These are shown in Table 9.6. Initially, there were 43 data points relating to leadership and the SG, but with 21 also made in a negative context, the nett effect covers only 22. Similarly, the data points reduce from 81 to 59 in the case of the XG (22%). The third most often cited capacity is collaborative (12%), but it declines by nearly half after incorporating the number of negative comments. The negative comments show strongly in collaborative (−27%) as well as adaptability and assertiveness (−23% each). The capabilities of adaptability and emotional have more negative than positive connotations and were therefore wholly absent. The perceived leadership capabilities tie in closely with the trait theory of leadership, based

Table 9.6 Nett value of perceived leadership capabilities

	SG			XG		
2nd level code	Initial	Negative	Nett value	Initial	Negative	Nett value
Adaptability	2	1	1	2	5	−3
Assertiveness	1	1	0	11	5	6
Charisma	2		2	3		3
Collaborative	20	1	19	13	6	7
Competence	2		2	4		4
Continuity	1		1			0
Decisiveness			0	1		1
Emotional		1	−1		2	−2
Experience			0	3		3
Extraversion		1	−1			0
Inspirational	2	1	1	7	1	6
Openness	3		3	2		2
Perseverance	3	2	1	6	2	4
Responsibility	1		1			0
Systems	5		5	1		1
Visible		5	−5	14		14
Visionary	1	8	−7	14	1	13
Grand Total	43	21	22	81	22	59

on the formal leader in phase 1 being highly visible and a great visionary. He led from the front and did not try to use emotions to transform his followers.

In the discussions phrases such as "leading from the front", "Pali's vision", "it was a very visionary type" and "you have to be known" were made.

Although the collaborative capacity was mentioned often, the frequent occurrence in a negative light suggests that this leadership style cannot be aligned to another leadership theory such as the neoclassical or the transformative. This indicates strongly that the followers expected him to play a more collaborative role, but he did not.

Perceptions included "he would have collaboration only if it suits the vision that he had" and "he displayed them as long as they ended up with him being in charge".

His lack of adaptability stems from his inability to change his way of doing things; this characteristic of stubbornness is another example of a leadership capability associated with the trait leadership theory. This can be seen in the context of the comments made by the peer stratum regarding mature leadership. The idea that was discussed revolved around how leadership changes over time as the leader becomes more experienced or mature. Initially, the leader was collaborative, but over time he became more autonomous and less collaborative. Prior to the project in this case study, the leader had a completely different leadership style compared to the style that was perceived during phase 1 of the project. He was perceived to have moved from a charismatic style to a trait-based leadership style as he began to think that he is the visionary, the person in total control, and could take charge of the project without the need for others.

This notion of leadership changing as the leader becomes more mature is described by Anderson and Sun (2017, pp. 90–92) as a particular set of attributes that a leader possesses which forms a self-identity. Depending on the situation, the leader will display a different self-identity, which develops over time as the ego of the leader.

Comments substantiating this interpretation in support of this change included "many of them many leaders when they start they talk collective, collective as we. And if you start to listen to the language, with time, it becomes to be I", "so in the early years, we learned, in the early years we coerce, in the early years you want people to hear him clearly, and you take everybody with", "but with time, you start to see leaders that more and more start to take decisions; this is the way it has to be done. And then that's when you start to see some rifts that you would have had a leader that was in this new leader that does not listen anymore", "when people have a lot of knowledge. So we, we, we get to a point where you could even believe you know more than the country itself" and "so as you get older, maybe you have that stronger character of not listening to others …".

In the case of phase 2, the formal leader's nett value is dominated by the collaborative capacity (86%). The only other capacity, as derived from the second-level code, of significance is systems (24%). The negative perceptions are dominated by the visionary (−38%) and visible (−24%) capacities.

Comments from the discussions included "The current SG seems to be much more inclusive", "The current SG is more reconciliatory in his approach" and "This one is more let's work together", sentiments that indicate that the leader supports a collaborative approach to leadership.

The negative perceptions included comments such as "where now we don't feel his presence in their leadership", "but I don't he doesn't portray that vision when doing certain things" and "I think because of the collective leadership approach, it's difficult to really say what his vision is for the SDG".

These perceived leadership capacities tie in closely with the principles of the neoclassical transformative leadership school. The leader acknowledges the need for teamwork, cooperation and coordination as well as makes use of existing systems that are in place and functional. There is no desire to try and convince the followers about a vision that must be followed, and there is no need for the leader to be visible. There are rather opportunities to be open to new ideas and to persevere.

In an analysis of leadership styles and success in projects, Raziq et al. (2018, p. 317) conclude that a "transformational leadership style is positively associated with goal clarity". This aligns with the findings of this case study, as the SG actively collaborated with key stakeholders during the project to ensure that there is a common goal and purpose.

5 Conclusion

The case study aimed to answer to the question of which leadership capacities from the trait and neoclassical schools of leadership enabled the response of the South African National Statistics System to the SDGs. It was established that the project was a success based on the achievement of two milestones, which also defined two phases in the project.

People who were interviewed included followers, a peer, as well as the two leaders, who all confirmed that some form of explicit leadership was necessary to ensure success. However, the style of leadership was perceived to be very different in the two phases. There is a need for both leadership styles to deal with a disruptive influence such as the SDGs on the SANSS. Initially, a new project requires a vision and a dominant leader aligned with the trait leadership school to force change, take risks and establish how things will be done. Once in motion, a neoclassical leadership style is necessary to ensure that productive partnerships are in play and can build on the initial successes. This is described by Bass (1990, p. 563) as meaning that the "situational demands and the personal attributes of the leader must both be considered in trying to understand the likely effectiveness of the leader".

It is not clear from the literature that was reviewed whether there is a strong link between the maturity of a project or its life cycle and the leadership style that is required, nor whether the maturity of the leader himself or herself leads to a change in leadership style. The study suggests that it does, based on the leadership capacity of "maturity" that was identified as a second-level code. This was raised in the context of the XG changing behaviour as he or she became more experienced in his or her leadership position, as well as their emerging views that he or she is leaving his or her followers behind as he or she develops and responds to the demands of his or her leadership.

The case study was able to eliminate the explanatory power of rival leadership styles by aligning the findings with the trait and neoclassical leadership theories that were identified in the literature review. The data were used to show that the evidence does not point to another conclusion and that the deductions and inferences made are sound.

The case study stimulates further questions: For example, can these findings be generalised to other statistical agencies within the same context, or other disruptions to the NSS? A challenge to an appointed leader is whether he or she knows different styles of leadership, is capable of using these leadership capabilities and knows when to use which. It is essential for a leader to recognise the preferred style of leadership and if it is applicable to the situation. If not, the leader should either be able to change their leadership approach or be aware of the limitations of the style, or actively work on improvements.

Future work on this topic must include similar studies that will be able to add richness to the understanding of leadership styles, especially on how to deal with disruptive elements in the public sector environment. An evaluation of whether these two leadership styles were effective will also be of interest for further research. This can be done through an integrated model of leadership traits and behaviours (Derue et al. 2011, pp. 10–11).

References

Antonakis, J. (2012). Transformational and Charismatic Leadership. In D.V. Day & J. Antonakis (eds.), *The nature of leadership* (pp. 256-288). London: SAGE Publications Inc.

Anderson, M.H., & Sun, P.T. (2017). Reviewing leadership styles: overlaps and the need for a new full-range theory. *International journal of management reviews* (19), 76-96.

Barrick, M.R., & Mount, M.K. (1991). The big five personality dimensions and job performance: A meta-analysis. *Personnel Psychology 44(1)*, 1-26

Bass, B.M. (1990). *Bass & Stogdill's Handbook of leadership*. New York: The Free Press.

Bradley, T.P., Allen, J.M., Hamilton, S., & Filgo, S.K. (2006). Leadership Perception: Analysis of 360-Degree Feedback. *Performance Improvement Quarterly 19(1)*, 7-23.

Bono, J.E., & Judge, T.A. (2004). Personality and transformational and transactional leadership: A meta-analysis. *Journal of Applied Psychology 89(5)*, 901-910.

Carlyle, T. (1840). On heroes, hero-worship, and the heroic in history. http://www.gutenberg.org/files/1091/1091-h/1091-h.htm. Accessed 5 August 2019.

Chenail, R.J. (2012). Conducting Qualitative Data Analysis: Qualitative Data Analysis as a Metaphoric Process. *The Qualitative Report 17(1)*, 248-253.

Day, D.V., & Antonakis, J. (2012). Leadership: Past, Present, and Future. In D.V. Day & J. Antonakis (eds.), *The nature of leadership* (pp. 256-288). London: SAGE Publications Inc.

Derue, D.S., Nahrgang, J. D., Wellman, N., & Humphrey, S.E. (2011). Trait and behavioral theories of leadership: An integration and meta-analytic test of their relative validity. *Personnel Psychology 4(1)*, 7-52.

Ellis, C., Adams, T.E., & Bochner, A. P. (2010). Autoethnography: An Overview. *Forum: Qualitative Social Research 12(1)*, 1-18.

Fourie, W., & Hohne, F. (2019). Thou shalt not fail? Using theological impulses to critique the heroic bias in transformational leadership theory. *Leadership, 15*, 44-57.

Galton, F.R.S. (1869). *Hereditary genius: an inquiry into its laws and consequences (27)*. London: R. Clay, Sons and Taylor printers.

Glaser, B., & Strauss, A. (1967). *The Discovery of Grounded Theory*. New Brunswick: Aldine transaction.

Heuer, R.J. (1999). Psychology of Intelligence Analysis. http://www.odci.gov/csi. Accessed 5 August 2019.

Hoffman, B.J., Woehr, D.J., Maldagen-Youngjohn, R., & Lyons, B.D. (2011). Great man or great myth? A quantitative review of the relationship between individual differences and leader effectiveness. *Journal of Occupational and Organizational Psychology 84(2)*, 347-381.

Judge, T.A., & Long, D.M. (2012). Individual differences in leadership. In D.V. Day & J. Antonakis (eds.), *The nature of leadership* (pp. 256-288). London: SAGE Publications Inc.

Lawrence, P. (2015). *Leading change: How successful leaders approach change management*. London: Kogan Page Ltd.

Pronin, E. (2007). Perception and misperception of bias in human judgement. *TRENDS in Cognitive Sciences 11*(1), 37-43.

Raziq, M.M., Borini, F.M., Malik, O.F., Ahmad, M., & Shabaz, M. (2018). Leadership styles, goal clarity, and project success: Evidence from project-based organisations in Pakistan. *Leadership & Organization Development Journal 39(2)*, 309-323.

Shahedul, H. (n.d.) Open Coding. http://pages.cpsc.ucalgary.ca/~saul/wiki/uploads/CPSC681/open-coding.pdf. Accessed 11 August 2019.

South Africa, 1914, *Statistics Act*, No. 38 of 1914, Pretoria.

South Africa, 1999, *Statistics Act*, No. 6 of 1999, Pretoria.

United Nations. (2015). United Nations Fundamental Principles of Official Statistics. https://unstats.un.org/unsd/dnss/gp/Implementation_Guidelines_FINAL_without_edit.pdf. Accessed 31 October 2019.

United Nations. (2019). RSA Voluntary National Review Report. https://sustainabledevelopment.un.org/content/documents/23402RSA_Voluntary_National_Review_Report__9_July_2019.pdf. Accessed 31 October 2019.

Role of Libraries in Promoting the SDGs: A Focus on the University of South Africa

10

Godwell Nhamo and Melanie Malan

Abstract

The International Federation of Library Associations and Institutions is on record indicating that libraries can drive progress across all 17 Sustainable Development Goals (SDGs) leading to the attainment of the 2030 Agenda. Yet, a 2018 study in Osun State, Nigeria, revealed that none of the libraries surveyed had immediate and/or long-term plans for scaling up awareness raising of the SDGs. This research took note of the foregoing, and emerges from a partnership between the SDGs for Society (SDGs4S) Research Stream and the University of South Africa (Unisa) Libraries. The SDGs4S Research Stream provides lenses through which trans-, inter- and multidisciplinary research takes place to address challenges focusing on people and the planet. Data was generated through participatory action research and a survey that sampled 72 respondents to reflect their experience of using the SDGs4S Library Guide, a quasi-open-access online portal. The Unisa libraries see this partnership as an opportunity to support Unisa's mission of advancing development and ensuring global sustainability. The key finding is that the SDGs4S Library Guide portal was established as a one-stop shop for resources on the SDGs, which will enable researchers to locate key information easily. The portal has, among other contents, key internet sites, open-access reports and statistics, open education resources, books, journal articles, theses and dissertations and newspaper links. Although relatively new, the portal has been attracting traffic, thereby promoting the Unisa libraries as agents of change in the SDG space, especially given that Unisa is host to over 350,000 students across the world, many of whom are from Africa and countries with inadequate library facilities.

Keywords

SDGs · Libraries · University of South Africa · ICT · IFLA · Stakeholders

G. Nhamo (✉)
Exxaro Chair in Business and Climate Change, Institute for Corporate Citizenship, University of South Africa, Pretoria, South Africa
e-mail: nhamog@unisa.ac.za

M. Malan
University of South Africa Library, Pretoria, South Africa
e-mail: malanmm@unisa.ac.za

1 Introduction

Soon after the 2030 Agenda for Sustainable Development (AfSD) and its 17 aligned Sustainable Development Goals (SDGs) were

G. Nhamo et al. (eds.), *Sustainable Development Goals for Society Vol. 1*, Sustainable Development Goals Series, https://doi.org/10.1007/978-3-030-70948-8_10

ratified in 2015, the International Federation of Library Associations and Institutions (IFLA) made it clear that libraries the world over had to harness the 2030 AfSD in its entirety (IFLA 2015a). This was to be done by, among other measures, promoting universal literacy, minimising gaps in access to information, advancing digital inclusion through access to information and communications technology (ICT) and playing a central role in serving the research and academic community. This call is even more pressing for public libraries (Edwards 2018) that provide traditional access to information and also embark on services and programmes that are community oriented. Libraries are, by their nature, information hubs (Igbinovia 2016) and are part of the DNA of public and other universities (O'Keeffe 2016). So important is the role of libraries in the 2030 AfSD that the IFLA put in place the International Advocacy Programme to promote and support the role of libraries in the attainment of the SDG agenda (Kear 2018). Earlier, and in line with the role of public libraries in the attainment of the Millennium Development Goals (MDGs), Forsyth (2005, p. 315) hinted that "[l] ibraries contribute to social capital and social inclusion, making communities with libraries stronger than communities without them".

Reference to the research and academic community takes us right into the university and other institutions of higher education space. This is the space where the University of South Africa (Unisa), a public open-distance and e-learning (ODeL) institution, with an annual student population averaging 350,000 mainly across Africa comes in (Nhamo 2019). As a matter of fact, Vyas-Doorgapersad (2011, p. 51) sees ODeL as "a significant medium for sustainable development in the information age of advanced learning". Drawing lessons from India's Indira Gandhi National Open University, the author indicates that ODeL provides a transformative form of capacity building, creating new opportunities for job seekers, as identified in SDG 8 which focuses on decent jobs (United Nations 2015). The observations align with Sedlacek's (2013) writing, which regards the library as central to the roles of universities in fostering sustainable development across different spatial levels. Hence, universities remain as change agents for sustainability (Peer and Stoeglehner 2013) and their role in providing open educational resources to increase access and equity in education (Abeywardena 2017; Banji 2017) cannot be overemphasised.

Drawing from the experience of African libraries in their commitment to the MDGs, Mulindwa (2015) calls upon all national libraries to take the experiences gained and apply them to the attainment of the SDG agenda. African libraries, assisted by investments in ICT, have positioned themselves strategically to align with the MDGs. Libraries have the added advantage of their neutrality and their position as safe havens. Among some of the strategies libraries used to embrace the MDGs, of which many remain relevant for the SDGs, the following may be highlighted: being people and community centric, building lasting partnerships, embracing ICT, and strategic positioning and re-positioning. Wand (2016) sees libraries gathering more data related to SDGs, thereby embracing the Big Data issue. Hence, online resources linking global databases like the World Development Indicators (WDI) and the United Nations SDG Indicators Global Database should be made available (DeLuca 2016). However, Igbinovia (2016) cautions libraries in developing countries not to be too ambitious in wanting to take on board the entire SDG spectrum, as resources will always be limited. The author advised libraries in developing continents to adopt SDGs that can be well managed through consolidated efforts like information literacy services.

A number of library initiatives aimed at propelling the 2030 AfSD have been instituted in several countries. For example, the National Library of Uganda (Jain and Jibril 2018) has an ICT training programme for female farmers (SDGs 2 and 5) that provides access in local languages to weather forecasts and crop prices and helps in setting up online markets (IFLA 2015a). In Finland, the Open Science Lab hosted by the National Library offers everyone access to scholarly work and data, thereby providing an innovative way (SDG 9) of doing things. Similar work is being done in South Korea (Garcia-Febo et al.

2017) where university libraries have embraced Industry 4.0 (Fourth Industrial Revolution) that emphasises digital transformation as part of SDG 9. In addition, Libraries Without Borders service refugee camps through the Ideas Box which allows access to information via satellite internet connections, thereby addressing SDG 10 requirements on reducing inequality (IFLA 2015a). The inaugural Beijing subway library, the "M Subway Library", was opened at the National Library of China Station in 2015, focusing on the provisions of SDG 11 on inclusive and sustainable cities. Lastly, the Biodiversity Heritage Library in the USA remains an ongoing open-access digital library for biodiversity literature, while the National Library Board Singapore has worked with partners to build a Children's Green Library hosting special collections on environmental conservation and climate change, some of which focus on children. These initiatives address SDGs 12–15 which embrace sustainable consumption and production, climate action, ocean economy and biodiversity (IFLA 2015a). However, similar initiatives are still lagging behind in many developing countries like Zambia (Chewe and Imasiklu 2018) where libraries are still to become acquainted with and acquire SDG-related information.

This chapter documents how Unisa libraries are engaging the SDGs, including the processes undertaken in establishing a Sustainable Development Goals for Society (SDG4S) Library Guide portal in partnership with the SDG4S Research Stream. The Library Guide is open to both Unisa researchers and those from other institutions. The objective is to determine the extent to which the Unisa libraries have localised the SDGs and how the SDG4S Library Guide portal is being utilised by users since its establishment in April 2019. The portal is accessible by any user, anywhere in the world, although there are some limitations in accessing some resource packs if one is not a registered Unisa student or staff member.

2 Literature Survey

The above quote is an inspiration in that it touches on the fundamentals of the 2030 AfSD, including the "let no one be left behind" motto (Nhamo et al. 2018). It also places SDG 4, with its focus on quality education, at the centre. The role of libraries in partnering for the attainment of the 2030 AfSD cannot be overemphasised and they remain an integral component of the SDG agenda. Meschede and Henkel (2019) alert us to the growing body of literature on libraries and sustainable development. In addition, matters pertaining to how (public) libraries should become involved in global development agendas are not new (Forsyth 2005). Njobvu (2007) argued that Zambian libraries could fight poverty through the provision of information, thereby assisting the country to attain the MDGs, an aspect revisited by Mulindwa (2015) at the African continental level. From Njobvu's (2007) perspective, information has the power to drive (sustainable) development. Hence, communities are likely to thrive if they have access to information and embrace its power. Reference is made to how the key libraries in the United States, namely, the Library of Congress, the National Library of Agriculture, the National Library of Medicine and the National Library of Education, go all out to gather information across the globe, an element attributed to the success of the US economy. This is an aspect supported by Bawack (2018, p. 63) who indicates that "the success of any society, country or institution greatly depends on the adequacy of its library collections, its infrastructure, and its dissemination policy".

Njobvu (2007) goes further to identify four library services that could have assisted in attaining the MDGs (and in the context of this chapter, the SDGs), including targeted reference services, ICT, transformation of crucial documents into local languages and organising outreach programmes. The author concludes by identifying several challenges that result in libraries failing to provide the identified services and hence to contribute significantly to national development agendas. Among such challenges are a lack of funding, inadequate training facilities aligning to global development agendas, underdevelopment of ICT infrastructure and lack of political will to develop new and/or update existing information policies. Drawing from the Nigerian experience, Abata-Ebire et al. (2018) add challenges of inadequate staff, low self-esteem among library and information professionals, poor lobbying and SDG advocacy-related knowledge and skills among librarians and limited partnerships among librarians. The lack of funding is also raised in line with the eight Cameroonian public universities that do not enjoy financial autonomy, thereby affecting the way in which their libraries are resourced (Bawack 2018) and deliver on the SDGs (Okoye et al. 2019). Nevertheless, the Consortium of Cameroon Universities and Research Libraries have agreed to proactive engagement to assist the government in achieving the SDGs as well as promoting SDG-related information. In this regard, a series of workshops and conferences on SDGs have been held, with a special focus on how libraries could be involved to achieve the 2030 AfSD and the 17 SDGs.

Recent work by Chewe and Imasiklu (2018) found no evidence of specific programming for SDGs by some libraries. Yet, in 2014, the IFLA came up with the landmark "Lyon Declaration on Access to Information and Development" (Wand 2016). This call would not have come at a better time because the MDG commitment period was coming to the end, with the world in full negotiations for the post-2015 development agenda (Edwards 2018). The Lyon Declaration highlighted that if the world were to move quicker in attaining sustainable development, the United Nations member states would have to commit to and ensure that every citizen had access to and was able to use and share information (Garcia-Febo et al. 2017). The call was heard; the Global Connect Initiative from the US State Department presented to the United Nations General Assembly in 2015 with the aim of bringing 1.5 billion users lacking internet worldwide access online by 2020 (Wand 2016).

The IFLA advocates that libraries take their place across all the 17 SDGs. To this end, the organisation has come up with a matrix on how this could be done (Table 10.1). The organisation believes that there is a need to increase access to SDG information and knowledge in the entire world through ICT (IFLA 2015a). The SDG 16.10 target aims to ensure public access to information and protect fundamental freedoms. This is to be done in line with national legislation and available international agreements. Several ICT-related targets (4.4, 4.a, 5.b, 9.c, 17.6 and 17.8) and indicators (4.4.1, 4.a.1, 5.b.1, 9.c.1, 17.6.2 and 17.8.1) also exist across the SDGs (Nhamo et al. 2019a). To this end, libraries are well set to harness this advantage and propel access to information through ICT. The major point to note is that libraries should work with other national key stakeholders to drive this agenda because the SDGs are implemented at the national and other subnational levels (Nhamo et al. 2018, 2019b).

In October 2015, the IFLA published the first edition of a toolkit focusing on libraries, development and the 2030 AfSD, which was revised in 2017. In its view, libraries had to show that they could drive progress towards the attainment of the SDGs (IFLA 2017). As governments developed their SDG localisation plans, the libraries had to take an active role in national consultation forums, presenting scenarios on how libraries could assist, as well as raising awareness (Kear 2018). In fact, the IFLA became an active stakeholder in the United Nations system in 2012 during the Rio+20 Summit (Garcia-Febo et al. 2017). Since then, the IFLA has been active, leading to the finalisation of the SDGs. Soon after the ratification of the SDGs, which were enshrined in the 2030 AfSD in September 2015, libraries had to organise meetings with policymakers following a six-step process (Fig. 10.1). In this process, Step 5 is critical in making sure that no one is left behind, which is the main driver of the 2030 AfSD (United Nations 2015).

In earlier work, Igbinovia (2016) focused on Nigerian libraries as a vehicle for attaining the 2030 AfSD. In response to the proclamation of the global development goals in 2015, the Registrar of the Librarian's Registration Council of Nigeria

Table 10.1 Libraries and the SDGs: an IFLA perspective

The SDG (abridged)	Libraries to provide
Goal 1: No poverty	• Public access to information and resources that give people opportunities to improve their lives • Training in new skills needed for education and employment • Information to support decision-making by governments
Goal 2: Zero hunger	• Agricultural research and data on how to make crops more productive and sustainable • Public access for farmers to online resources like local market prices, weather reports and new equipment
Goal 3: Good health and well-being	• Research available in medical and hospital libraries that supports education and improves medical practice for healthcare providers • Public access to health and wellness
Goal 4: Quality education	• Dedicated staff who support early literacy and lifelong learning • Access to information and research for students everywhere • Inclusive spaces where cost is not a barrier to new knowledge
Gaol 5: Gender equality	• Safe and welcoming meeting spaces • Programmes and services designed to meet the needs of women and girls, including rights and health • Access to information and ICT that helps women build business skills
Goal 6: Clean water and sanitation; goal 7: affordable and clean energy	• Access to quality information and good practices that support local water management and sanitation projects • Free and reliable access to electricity and light to read, study and work
Goal 8: Decent work and economic growth	• Access to information and skill training that people need to find, apply for and succeed in better jobs
Goal 9: Industry, innovation and infrastructure	• Widespread existing infrastructure for public and research libraries and skilled library professionals • Welcoming and inclusive public spaces • Access to ICT like high-speed internet that may not be available anywhere else
Goal 10: Reduce inequality	• Neutral and welcoming spaces that make learning accessible to all, including marginalised groups like migrants, refugees, minorities, indigenous peoples and persons with disabilities • Equitable access to information that supports social, political and economic inclusion
Goal 11: Sustainable cities and communities	• Trusted institutions devoted to promoting cultural inclusion and understanding • Documentation and preservation of cultural heritage for future generations
Goal 12: Responsible consumption and production; goal 13: climate action; goal 14: life below water; and goal 15: life on land	• Sustainable system for sharing and circulating materials that reduces waste • Historical records about coastal change and land use • Research and data needed to inform climate change policy • Widespread access to information needed to guide decision-making by local and national governments on topics like hunting, fishing, land use and water management
Goal 16: Peace, justice and strong institutions	• Public access to information about government, civil society and other institutions • Training in the skills needed to understand and use this information
Goal 17: Partnerships for the goals	• Global network of community-based institutions, primed to support local libraries and stakeholders

Source: Authors, based on IFLA (2015b, p. 1)

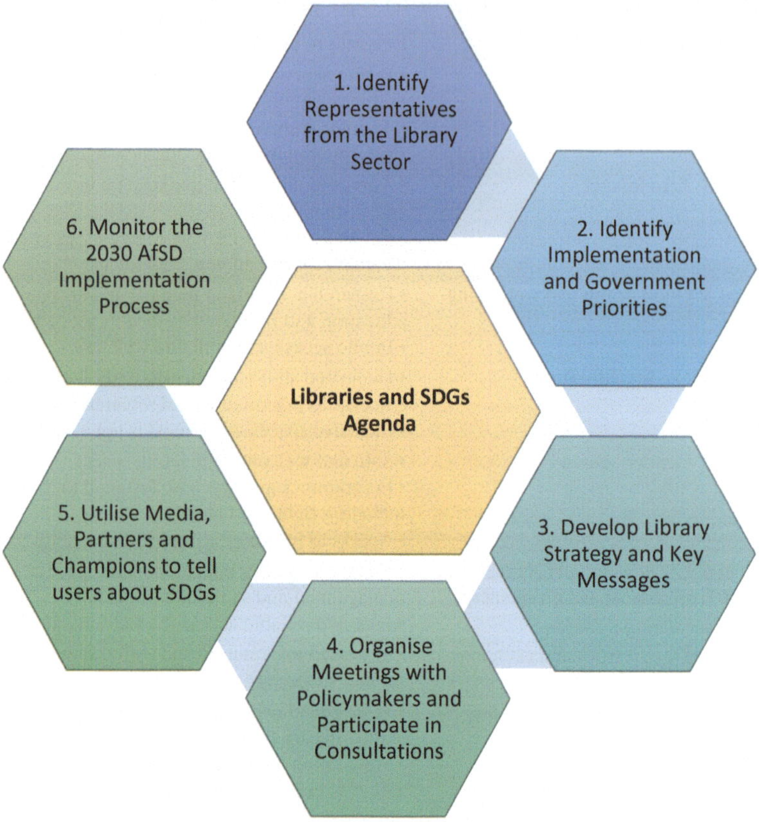

Fig. 10.1 Engagement between libraries and policymakers for SDG implementation. Source: Authors, based on IFLA (2017, p. 11)

declared that the profession would scale up its efforts to make the SDGs a reality. Thereafter, several initiatives were put in motion including the conclusion of the translation of the 2030 AfSD into Igbo (one of the local and indigenous languages) by the Anambra State Library Board and the creation of the Nigerian Librarians SDG Action Group from a Facebook platform (Jain and Jibril 2018). Other platforms to raise SDG awareness by the librarians and their partners included radio broadcasts, tweets (@SDGsNigeria Action, @SDGsNigeria, etc.) and posters (Igbinovia 2016). However, Okunlola et al. (2018, p. 28) found that many library and information personnel in Osun state were "aware of the sustainable development goals but not conversant with the details and modalities for its attainment". There were also no long-term plans on engagement with the 2030 AfSD from the 100 libraries sampled. To Abata-Ebire et al.'s (2018) way of thinking, Nigerian libraries had to provide relevant and sufficient SDG information to the public in an easily accessible way. The picture painted herein differs from libraries in Ogun state where ICT facilities and information personnel for the delivery of the SDG agenda are in place (Anasi 2018). The noticeable challenge has been inadequate power supply and lack of political accountability.

In Cameroon, the University of Yaounde I's main library started a project for the identification and collection of information resources about the SDGs from completed dissertations (Bawack 2018). The process involves reviewing all hard copies and classifying the topics according to related SDGs. This is done by library staff who have been trained on the 17 SDGs, with each librarian assigned 2 SDGs. Only dissertations that make significant contributions end up in the collections. However, owing to limited funding, the proposed open-access digital repository for the dissertations has not been finalised. Hence, the work is merely being preserved.

Public libraries in Botswana have initiatives in place addressing SDG 1 (no poverty), SDG 3 (good health), SDG 4 (quality education), SDG 5 (gender equality), SDG 11 (sustainable cities and communities) and SDG 17 (partnerships for the goals), according to Jain and Jibril (2018). The focus on SDG 3 is also emphasised by Popoola (2019). Botswana library projects for SDG 1 include basic ICT training programmes and the Business Café Service (Jain and Jibril 2018). As for SDG 3, there are health talks and partnerships with health-related NGOs for health activities. Among the initiatives to address SDG 4 are homework assistance, early childhood development clubs, indigenous language improvement programmes, a Braille training programme for the blind and a mobile library service for prisoners. Basket weaving, sewing and needlework, as well as capacity-building workshops, emerged as projects for SDG 5. Owing to limited space, details on other projects can be found from the cited publication. However, in addition to challenges raised earlier for the libraries, the authors also note the lack of mobility, a common platform for librarians, equipment such as computers and printers, space, local content for local needs and old collections (Jain and Jibril 2018).

Good progress has been made by Australian libraries regarding harnessing the energy from the proclamation of the SDGs in 2015. Much of the initiatives are from the Australian Library and Information Association (ALIA). Although initiatives exist across 10 of the 17 SDGs, only a few of these are profiled here. To address SDG 3, 328 health libraries and 1250 specialist library and information professionals provide expert support to medical practitioners and researchers (ALIA 2018). In New South Wales, public libraries provide drug and alcohol information, education and awareness programmes. In Tasmania, the Tasmanian Department of Education is working with libraries to remove existing barriers for some students and their parents to gaining access to library resources by issuing membership cards, thereby responding to SDG 4. In Victoria, libraries have programmes that encourage women to connect and interact socially, thereby enhancing new skills. For example, the Stepping Stones and Job Club are twin programmes offered in partnership with not-for-profit organisations that address the needs of migrant and refugee women (SDG 5). To bridge inequalities (SDG 10), there are 78 mobile libraries across Australia, most of which service rural and remote communities. For example, the Riverina Regional Library serves about 138,500 residents in a jurisdiction of 50,000 square kilometres. For over 25 years, public libraries in New South Wales have successfully delivered legal information services (SDG 16) to their communities. This is done in partnership with the Legal Information Access Centre hosted by the State Library of New South Wales.

Given the seriousness with which the Latin America and the Caribbean (LAC) region takes the role of libraries in the attainment of the SDGs, in April 2018, the LAC in partnership with the IFLA came up with a declaration, "The Santiago Declaration on Access to Information to Achieve Sustainable Development in Latin America and the Caribbean" (IFLA 2018). The Santiago Declaration reflects libraries' commitment to sustainable development in general and the SDGs in particular in the LAC. The LAC has called on governments to support the roles of libraries in their quest to be one of the key stakeholders in the attainment of the 2030 Agenda for Sustainable Development. The Santiago Declaration therefore remains a vital instrument for librarians, libraries, library associations and friends of libraries in LAC and other regions to lobby governments, the United Nations and other stakeholders for their cause (IFLA 2020). Further details on the Santiago Declaration are reflected in Box 10.1.

Of the 7% (eight in number) of library associations that have ratified the Santiago Declaration, focusing on "Access to Information to Achieve Sustainable Development in Latin America and The Caribbean" (IFLA 2020), half (four) are from Kenya (Fig. 10.2). The South–South and North–South collaboration among and between library associations is encouraged by Popoola (2019), who sees such collaborations as adding value through capacity development for health sciences library personnel, especially health sciences literacy.

Box 10.1: Commitments from the Santiago Declaration

- Reaffirm the importance of libraries as strategic allies in the fulfilment of the SDGs, with a particular focus on public access to information and knowledge for all citizens and all communities.
- Raise awareness among governments of the need to develop strategies and mechanisms, at the national and local levels, which will facilitate and reinforce the work of libraries in order to maintain and reinforce the work of libraries in Latin America and the Caribbean, and in particular which allow them to provide decent, egalitarian and trusted public spaces.
- Reinforce the role of libraries in the promotion and dissemination of information resources produced in Latin America and the Caribbean.
- Encourage governments to enter into agreements and pass laws, including on copyright, which facilitate public access to information and knowledge without barriers for all citizens of the region.
- Support the development and provision of the resources necessary for the involvement of libraries in Latin America and the Caribbean in national and regional projects which aim to implement one or more of the SDGs.
- Facilitate collaboration and exchange of knowledge, experience and good practices through inclusive national and regional initiatives in Latin America and the Caribbean.
- Facilitate discussion around library policies for access to information as part of a universal approach to human rights, as well as the right to knowledge.
- Call on the member states of the United Nations, in the context of discussions on SDG 16 at the 2019 High Level Political Forum, to include discussion on SDG 16.10 and relevance of public access to information and knowledge for all of the 2030 Agenda.

Source: IFLA (2018 online)

Samantaray (2017), however, observed that regardless of the paradigm shift in the field of library information science, many libraries remain passive repositories of SDG knowledge. This implies that the knowledge comes in, but does not filter out as it should. This scenario emerges from the fact that in many instances, librarians tend to view the users as potential threats, thereby reverting to their traditional role of information and knowledge preservation.

3 Materials and Methods

As indicated in Sect. 10.1, this work documents the way in which the Unisa libraries are engaging the SDGs, including the processes undertaken in establishing a SDG4S portal in partnership with the SDG4S Research Stream open to both Unisa students and postgraduate students from other institutions. The objective is to determine the extent to which the Unisa libraries have localised the SDGs and how the SDG4S portal is being utilised by users since its establishment in April 2019. The findings will be used to determine the extent of engagement with the entire SDG framework in line with the IFLA guidelines and also to enhance the use of the SDG4S portal in the lead up to 2030. In total, there are 14 Unisa libraries (13 in South Africa and 1 in Ethiopia). The spread of libraries in South Africa is shown in Fig. 10.3.[1]

The methodology draws from ongoing participatory action research (PAR) focusing on Unisa's sustainability journey that started in 2011 (Nhamo 2012, 2019). To incorporate the SDGs into the research agenda of Unisa, the SDG4S Research Stream was developed in December 2018 and became part of the ongoing Unisa Annual Interdisciplinary Academy and Summer School (Annual Research Academy). The Annual Research Academy runs along the lines of specific research streams led by senior academics, mainly professors. Activities are

[1]GIS map done by Hlengiwe Kunene under authors' supervision with data obtained by authors from Unisa main library

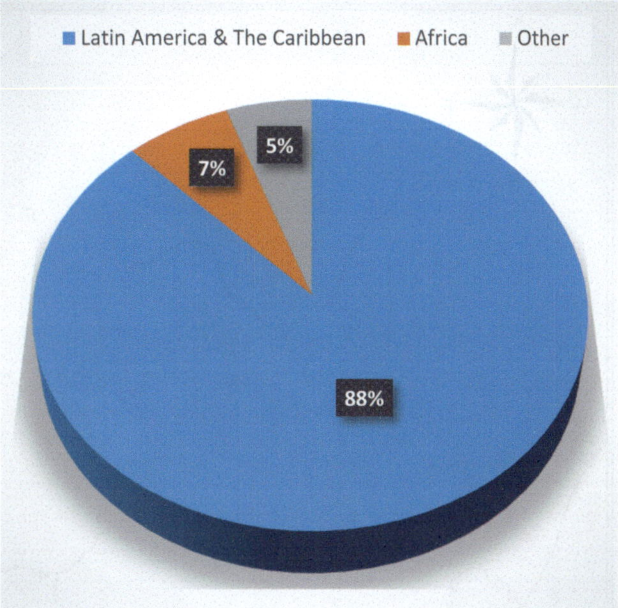

Fig. 10.2 Latin America and the Caribbean Declaration Signatories (n = 164). Source: Authors, data from IFLA (2020)

scheduled throughout the year starting with a 3-day introduction in April, several workshops and public lectures in between, leading to a week-long Summer School in November. Once the online and partially open-access SDG4S library portal (https://libguides.unisa.ac.za/SDGs4S) was established in April 2019, through a partnership with the Unisa library, a deliberate effort was made to promote it both internally and externally in faculty meetings, through regional and national public lectures, during international conferences and on any other relevant platforms. On a regular basis, the authors of this chapter took stock of SDG material that needed to be loaded on the portal, commonly referred to internally as the Library Guide. On a monthly basis, the authors of this work record a number of hits on the portal.

Apart from the PAR methods applied, a survey was administered during the Unisa SDGs Localisation in Institutions of Higher Education (IHE) Indaba[2] held on 29 November 2019 at the Unisa main campus in Pretoria. The Indaba was attended by 150 stakeholders from across IHE in South Africa, as well as students, embassies and other interested and affected parties. A total of 72 respondents completed the survey, giving a response rate of 48%. The demographics of respondents indicate 51% female, 47% male and 1% who wish not to disclose. Other details regarding affiliations are shown in Fig. 10.4.

Lastly, we also made use of the Unisa Library Federated Search Engine that searches across all the databases that the Unisa libraries subscribe to. The search engine "Sustainable Development Goals" was used and the results comprised extracted references containing the key words. The next section presents data and discusses the findings regarding the SDG4S portal and other SDG-related activities by the Unisa libraries in general.

4 Presentation of Data and Discussion of Findings

This section comes in two parts. These are (1) a general assessment on how the Unisa libraries and staff have been engaging with the 2030 AfSD and its 17 SDGs outside the SDG4S portal, and (2) documentation on the SDG4S website usage and an evaluation of the usefulness of the SDG4S portal.

[2]An Indaba may be translated as 'a gathering' in isiZulu.

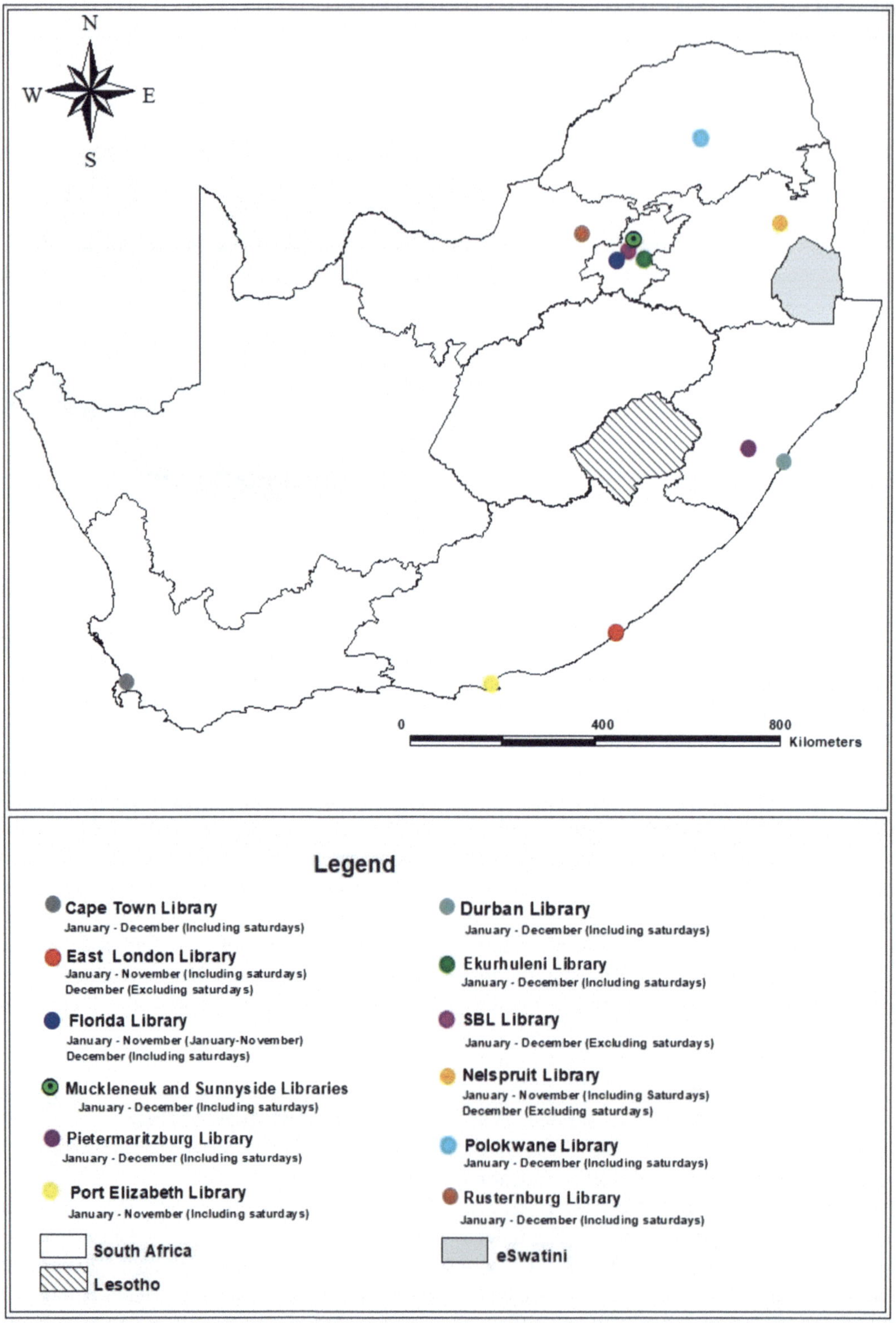

Fig. 10.3 Spread of Unisa libraries in South Africa and their opening times. Source: Authors

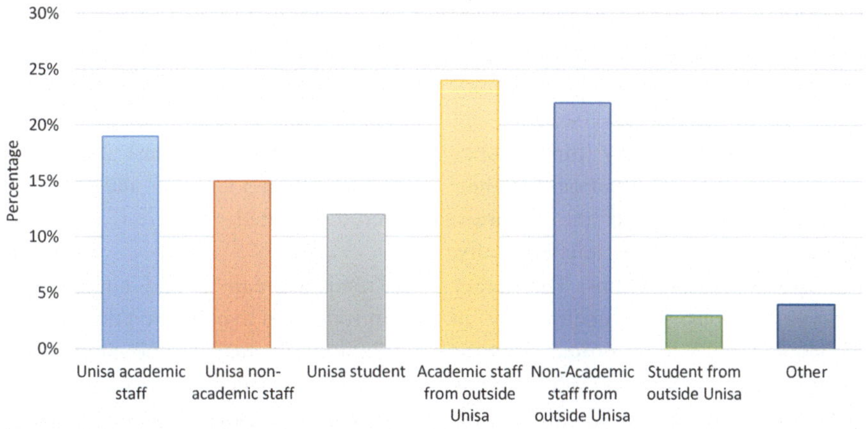

Fig. 10.4 SDGs Localisation Indaba participant affiliations. Source: Authors

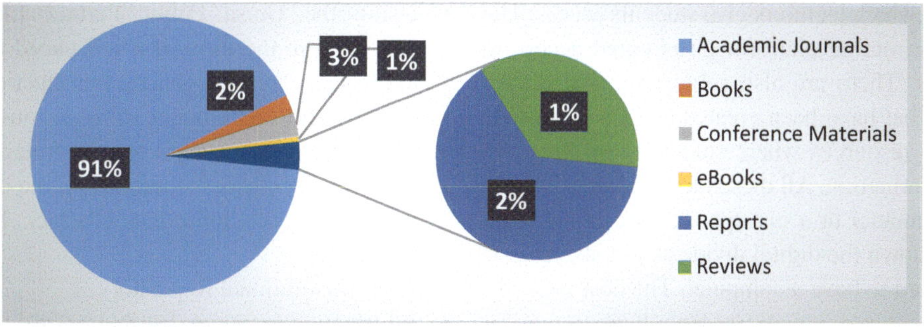

Fig. 10.5 SDG references from online Unisa-linked databases (n = 77,449). Source: Authors

4.1 General Engagements on the SDGs by Unisa Libraries and Staff

The entry point in terms of how the Unisa libraries engage with the SDGs was to check the availability of references across the many databases to which the libraries subscribe. As of 20 January 2020, 223,984 references had been extracted by the federated search engine "Sustainable Development Goals", with news databases contributing more than half of the references (124,184 hits) and magazines sitting at 12,706 hits. Further, 88 dissertations, 59 printed materials, 1 audio and 2 biographies were extracted. The share of the remaining 77,449 references from other databases is shown in Fig. 10.5. Hence, with over 77,500 online academic references directly mentioning the SDGs, the Unisa library has significant material in its system.

Following the IFLA (2015b) library engagement framework, an attempt was made to ascertain whether the Unisa libraries had done something on certain aspects. The findings were positive, although the SDGs had not received equal attention. This is not unusual as SDG localisation permits institutions to select specific SDGs they wish to run with. To permit public access to information and resources in line with SDG 1, the Unisa Libraries' Institutional Repository is an open-access resource providing access to theses and dissertations completed at Unisa and gives access to research conducted on all SDGs.

As for enabling access to information and research for students everywhere, and where cost is not a barrier to knowledge (SDG 4), the library has a number of initiatives. The students are trained on how to use the library; this includes training on how to find information for their research which is done at all the libraries. Personal librarians train students telephonically, in person or with Microsoft Teams. The library uses the Skype for business functionality in Microsoft Teams in order to do online group or one-on-one training, which gives all students equal access to training opportunities. Microsoft Teams provides an inclusive space for students. The libraries also develop training material and upload it on myUnisa proposal sites and other sites on which lecturers give students access. The myUnisa online platform gives equal access to students. There are also subject-specific library guides that have been created and these provide links to resources where students can find relevant information. All these facilities are available on a computer or a smartphone. Use of the latter breaks down the digital divide as not all students can afford to have a computer. The cost factor is, however, a concern, as students have to pay for data to be able to connect with us. Internet access is also a factor of concern. The library has been given responsibility for research data management. This open-access platform will provide reusable data sets and is available at https://unisa.figshare.com/.

One key drawback noticed pertains to the spatial distribution of brick-and-mortar libraries in the country. For example, of the 13 libraries, 5 are located in the smallest (but vastly populated) Gauteng province and in very close proximity is another library in North West province. Two provinces, the Northern Cape and Free State, do not have libraries, while there is one library apiece in the Western Cape and KwaZulu-Natal provinces which are on the coast. Such distribution patterns leave some students unserviced should they need to get into the library physically.

Regarding SDG 9, and specifically the call by the IFLA for skilled library professionals, all staff in the library have access to training funds that can be used to improve skills. The library also offered training to public librarians to give them the necessary skills to assist their clients. This programme was unfortunately stopped in 2018. The last 2-day programme was offered on 3 October and the key elements covered included:

- Trends in creativity and innovation
- Introduction to collection development
- Community analysis and interpretation
- Change a challenge into an opportunity
- Assisting Unisa students (myUnisa and Mylife)
- Twenty-first-century literacies (Partner Library LibGuide)
- Using the Unisa Public Partner Library's LibGuide in the form of a self-paced tutorial to create awareness (general introduction)
- Finding full-text content from various open-access resources using a search strategy (free e-books, open-access journals, online dictionaries and encyclopaedias (Partner Library LibGuide)
- Unisa Institutional Repository
- Information resource distribution/information search services
- Marketing your library

Still on SDG 9 and access to ICT like high-speed internet that may not be available anywhere else, all students who visit one of the Unisa libraries have access to free internet. Partnerships with public libraries also exist to ensure that students can make use of their facilities and internet. This is an element that also addresses SDG 17. Igbinovia (2017) considered the involvement of Nigeria's librarians in cross-disciplinary research and its implication for the SDGs. The key finding was that the librarians had high levels of knowledge in cross-disciplinary research, an element that could enhance the achievement of the SDGs. Given that there are 17 SDGs, 169 targets and over 250 indicators (United Nations 2015), this understanding is of great value as the libraries will be servicing a diverse range of researchers both within and outside their boundaries. However, for library personnel to contribute significantly to the SDGs, there was a need to ensure effective knowledge-sharing behaviour in regard to such information, an aspect that the management of libraries had to promote (Igbinovia and Osuchukwu 2018). Furthermore, library staff had to equip themselves with the necessary skills and knowledge in relation to the SDGs, which could stimulate higher rates of knowledge circulation.

On SDG 11, specifically the IFLA guideline on documenting and preserving cultural heritage for future generations, the Unisa libraries took on a project to digitise documents with historical value and which are open access. The Unisa library Digital Collection is available to all who need to use it at http://digilibrary.unisa.ac.za/digital/. Lastly, on SDG 16, with specific reference to the IFLA's call for training in the skills needed to understand and use information on peace, justice and strong institutions, the Unisa library has created a research skill library guide, which provides students with learning opportunities for finding and evaluating information. This link is available at https://libguides.unisa.ac.za/Research_Skills.

4.2 Documentation on the SDG4S Website

As indicated in the methodology section, the SDG4S portal emerged as a result of the SDG4S Research Stream that was part of the ongoing Unisa Annual Interdisciplinary Academy and Summer School. The SDG4S portal was developed following a generic template for Unisa library guides. It hosts a number of pages that include internet sites, open-access reports and statistics, open educational resources, books, articles, theses and dissertations, reference sources, newspapers, referencing, research support and apps for research. Figure 10.6 shows the SDG4S portal views from its inception in April 2019 to December 2019. Apart from the standard pages displayed, SDG-related academic calls and events are regularly posted. For example, by the time of completing this chapter there had been three such calls, two for book chapters for two books, one on Cyclones, Floods and SDGs in Southern Africa and the other on Sustainable Development Goals for Society. The third call was for Accelerating the Implementation of the Sustainable Development Goals in Africa International Symposium that was to be held in

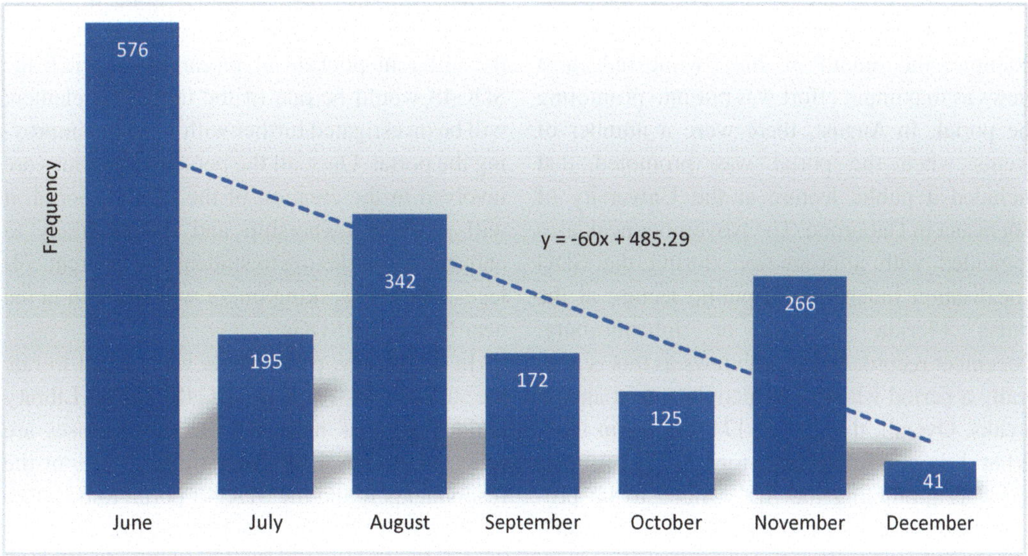

Fig. 10.6 SDG4S portal views, 2019. Source: Authors

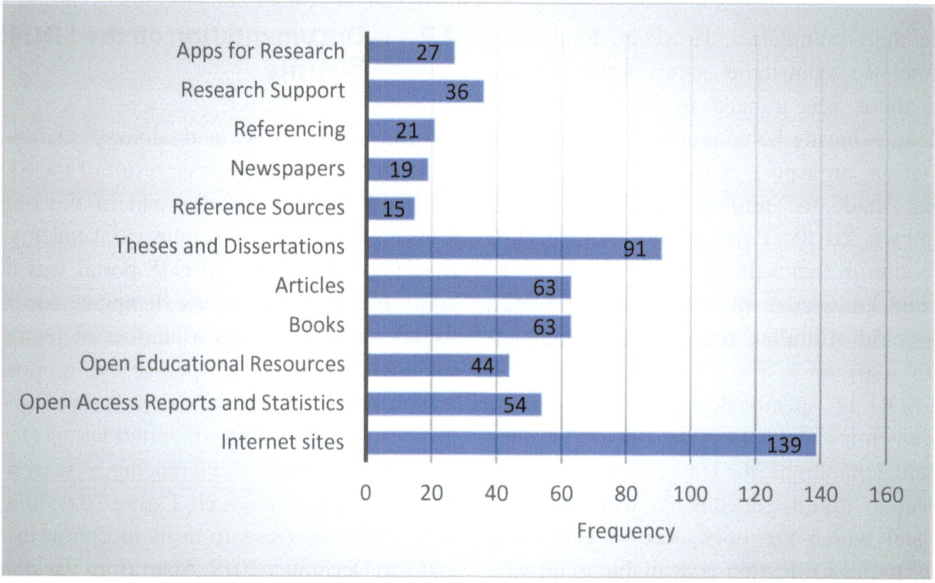

Fig. 10.7 Other website page views. Source: Authors

Pretoria, South Africa, from 5 to 6 March 2020. The call for the Sustainable Development Goals for Society book chapters resulted in two book volumes commissioned by Springer with expected full manuscript submissions on 31 October 2020.

Although there is a downward trend in views since the launch of the SDG4S portal, the peaks coincide with stronger promotion pushes. For example, the month of June witnessed most views as maximum effort was put into promoting the portal. In August, there were a number of events when the portal was promoted that included a public lecture at the University of Botswana in Gaborone. The November peak also coincided with a promotion during the SDG Localisation Indaba and a public lecture at the Unisa Florida Campus in Johannesburg. December recorded the least views as this is generally a period when many people are on annual breaks. Overall, there were 1717 views in total. What becomes clear therefore is that library products promoting the SDGs need to be promoted themselves on various platforms including social media. Details on statistics broken down into other page visits excluding the home page are presented in Fig. 10.7. Home page views numbered 1145 for the period under review. Internet sites recorded the highest hits, followed by theses and dissertations, with articles and books coming in shared third position. Open-access reports and statistics took the fourth position.

It is important to revisit the SDG4S portal to include all the SDG research done at Unisa. Perhaps the main SDGs "homepage" with tabs to the different pockets of research, of which the SDG4S would be one of the tabs. This element will be investigated further with a view to improving the portal. Once all the personal librarians are involved in the creation of the SDG4S portal, it will promote ownership and this is likely to enhance its marketing to students in different colleges who are working on, or will work on in the near future, the SDGs.

In the survey, participants were asked to rate the usefulness of the SDG4S Unisa Library Guide portal as marketed. Their responses are reflected in Fig. 10.8. An estimated 56% of the respondents found the SDG4S portal to be either very useful (39%) or somehow useful (17). Only 4% of the respondents indicated that the SDG4S portal was not useful, with 6% not sure and the remaining 34% having not had a chance to

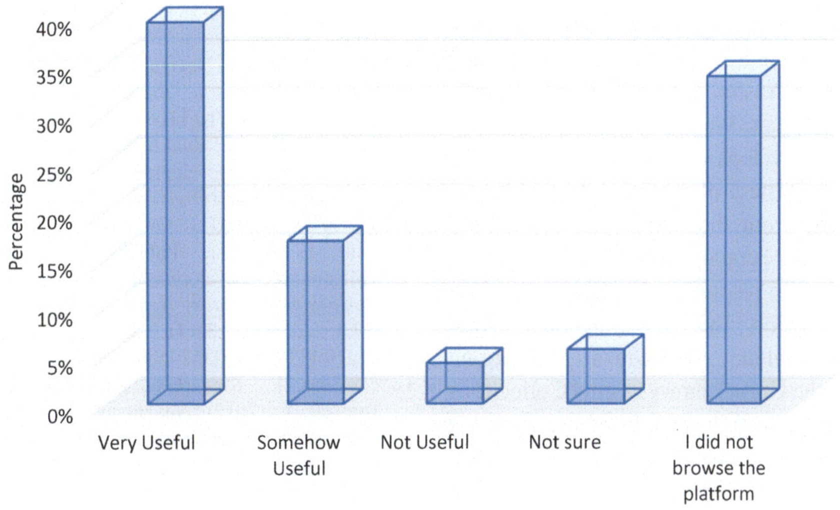

Fig. 10.8 Rating of the usefulness of the SDG4S portal ($n = 72$). Source: Authors

browse the platform. The fact that more than half of the respondents found the SDG4S portal generally useful is encouraging, especially for the staff working on the initiative.

From our assessment as authors, the importance of the SDG4S portal is the fact that it exposes the Unisa students and staff to the idea of finding literature to solve SDG-related issues in communities. They have access to subscription-based research material that they would otherwise not have access to. The library literacy training that they receive will ensure that they are set up for lifelong learning. Students will continue to use the portal after graduation, while staff have continuous access as the portal includes open-access resources. As observed by Forsyth (2005, pp. 315–316), "libraries have crucial roles in facilitating democracy and civil rights. They are places for people who read, or who are learning to read, or who do not read. There is a space for each member of a community in a library".

5 Conclusions

There is no doubt from the literature reviewed that libraries play and must play a central role in the attainment for the 2030 Agenda for Sustainable Development and its 17 inter-linked SDGs. This agenda has been at the epicentre of the International Federation of Library Associations and Institutions initiatives, leading to the ratification of the global development goals in 2015. Libraries remain hubs of information and data, working for, with and in communities. This mission is not going to be easy, especially for libraries in Africa and other developing regions that remain underfunded, undercapacitated and battling with an ad hoc electricity supply needed to maintain e-resources. Although resources may be constrained, Unisa has set the wheels in motion to become a change agent in line with the SDGs. Unisa libraries' SDG work is epitomised by the development and maintenance of the SDG4S, the quasi-open-access library guide portal. The portal was developed in partnership with senior academics running the SDG4S Research Stream which is integral to the broader Unisa Annual Transdisciplinary Academy and Summer School. However, the main challenge has been maintaining traffic volumes. Declines have been witnessed since the portal's inception in April 2019. We recommend that Unisa continues to promote the SDG4S portal and open up other avenues, programmes and activities, leading to the promotion of open access to information on the SDGs. This way, the

Unisa library can be part of proactive global libraries that are driving strong engagement with the 2030 Agenda for Sustainable Development like those from Australia, Europe, Latin America and the Caribbean. We further recommend that Unisa library develops a broad SDG library guide with tabs to clusters of research done that will include some from the current SDG4S portal. This way, more people will be likely to use it. Several current e-resources should be reoriented to better support the SDG agenda. These e-resources include the Unisa Institutional Repository, the Unisa Library Digital Collections, the Archival Collections and the Unisa Research Data Repository. Finally, the collection developers need to include SDGs in collection development policies and the Unisa library should harvest more open-access resources and be an advocate for open access.

References

Abata-Ebire, B.D., Adebowale, J.A., Ojokuku, B.Y. (2018). Achieving Sustainable Development Goals (SDGS) In Nigeria: The Roles of Libraries. *International Journal of Applied Technologies in Library and Information Management,* 4(2), 89-95.

Abeywardena, I.S. (2017). An Empirical Framework for Mainstreaming OER in an Academic Institution. *Asian Association of Open Universities Journal*, 12(2), 230-242. doi:https://doi.org/10.1108/AAOUJ-11-2017-0036

ALIA (The Australian Library and Information Association). (2018). *Australian Libraries Support the Sustainable Development Goals.* Canberra: The Australian Library and Information Association.

Anasi, S.N. (2018). University Libraries: Bridging Digital Gaps and Accelerating the Achievement of Sustainable Development Goals through Information and Communication Technologies. *World Journal of Science, Technology and Sustainable Development*, 15(1), 13-25. doi:https://doi.org/10.1108/WJSTSD-11-2016-0059

Banji, G.T. (2017). The Role of Libraries in Empowering Women Education through Open Access. *International Journal of Innovative Research and Advanced Studies (IJIRAS),* 4(7), 64-66.

Bawack, R. (2018). Academic Libraries in Cameroon: Achieving Agenda 2030 Goals. *International Information & Library Review,* 50(1), 63-66. doi:https://doi.org/10.1080/10572317.2018.1439148

Chewe, P., & Imasiklu, I. (2018). Centrality of Libraries in the Actualization of Sustainable Development Goals in Zambia. *European Journal of Social Sciences Studies,* 3(4), 253-265. doi:https://doi.org/10.5281/zenodo.2525432

DeLuca, L. (2016). Navigating United Nations Data Sources. *Collection Building*, 35(4):, 114–118. doi:https://doi.org/10.1108/CB-08-2016-0022

Edwards, J.B. (2018). "Vital Assets": Libraries as Partners in Community Development. *Advances in Library Administration and Organization,* 39, 1-14. doi:https://doi.org/10.1108/S0732-067120180000039001

Forsyth, E. (2005). Public Libraries and the Millennium Development Goals. *International Federation of Library Associations & Institutions – IFLA Journal*, 31(4), 315–323. doi:https://doi.org/10.1177/0340035205061389

Garcia-Febo, L., Kim, J.K., Lallaisangzuli, Jain, V., Tugwell, Y. (2017). Advancing the United Nations Sustainable Development Goals: Experiences of International Academic and Research Libraries. *C&RL News*, 516-520.

IFLA (International Federation of Library Associations and Institutions). (2015a). *Access and Opportunity for All: How libraries contribute to the United Nations 2030 Agenda*. The Hague: IFLA.

IFLA (International Federation of Library Associations and Institutions). (2015b). *Libraries can Drive Progress across the Entire UN 2030 Agenda.* The Hague: IFLA.

IFLA (International Federation of Library Associations and Institutions). (2017). Toolkit: Libraries, Development and the United Nations 2030 Agenda. The Hague: IFLA.

IFLA (International Federation of Library Associations and Institutions). (2018). *The Santiago Declaration on Access to Information to Achieve Sustainable Development in Latin America and The Caribbean.* Retrieved from https://www.ifla.org/files/assets/hq/topics/libraries-development/documents/ifla-febab-lac-declaration-en.pdf (Accessed 12 January 2020).

IFLA (International Federation of Library Associations and Institutions). (2020). *Access to Information to Achieve Sustainable Development in Latin America and the Caribbean.* Retrieved from https://www.ifla.org/publications/node/81792. (Accessed 12 January 2020).

Igbinovia, M.O. (2016). Libraries as Vehicle to Sustainable Developmental Goals (SDGs): Nigerian's Current Status and Outlook. *Library 16 Hi Tech News*, 5, 16-17. doi:https://doi.org/10.1108/LHTN-03-2016-0010

Igbinovia, M.O. (2017). Librarians' Involvement in Cross-disciplinary Research and its Implication for Sustainable Development Goals (SDGs): The Nigerian Experience. *Library Review*, 66(4/5), 251-265. doi:https://doi.org/10.1108/LR-09-2016-0078

Igbinovia, M.O., & Osuchukwu, N.O. (2018). Predictors of Knowledge Sharing Behaviour on Sustainable Development Goals among Library Personnel in Nigeria. *International Federation of Library Associations and Institutions,* 44(2), 119–131. DOI:https://doi.org/10.1177/0340035218763445.

Jain, P., & Jibril, L. (2018). Achieving Sustainable Development Through Libraries: Some Preliminary Observations from Botswana Public Libraries. Satellite Meeting: Africa Section – Libraries as Centers of Community Engagements for Development, 22-23 August 2018, Kuala Lumpur, Malaysia.

Kear, R.L. (2018). Libraries, Development, and Implementation of the UN 2030 Agenda: A Regional Workshop held in Montego Bay, 16–18 February 2017. International Information & Library Review, 50(1), 60-62. DOI:https://doi.org/10.1080/10572317.2018.1422906.

Meschede, C., & Henkel, M. (2019). Library and Information Science and Sustainable Development: A Structured Literature Review. *Journal of Documentation*, 75(6), 1356-1369. doi:https://doi.org/10.1108/JD-02-2019-0021

Mulindwa, G.K. (2015). National Libraries in Africa: Refocusing Their Work to Ensure Delivery of Services. *Library Trends*, 64(1), 72-83. doi:https://doi.org/10.1353/lib.2015.0040

Nhamo, G. (2012). Participatory Action Research as Platform for Community Engagement in Higher Education. *Journal of Higher Education in Africa*, 10(1), 1-20.

Nhamo, G. (2019). Higher Education and the Energy Sustainable Development Goal: Policies and Projects from University of South Africa. In: G. Nhamo and V. Mjimba (eds.). *Sustainable Development Goals and Institutions of Higher Education*, pp. 31-48. https://doi.org/10.1007/978-3-030-26157-3_3.

Nhamo, G., Nhamo, S., Nhemachena, C. (2018). What Gets Measured Gets Done! Towards an Afro-barometer for Tracking Progress in Achieving Sustainable Development Goal 5. *Agenda*, 32(1), 60-75. https://doi.org/10.1080/10130-950.2018.1433365

Nhamo, G., Nhemachena, C., Nhamo, C. (2019a). Using ICT Indicators to Measure Readiness of Countries to Implement Industry 4. 0 and the SDGs.*Environmental Economics and Policy Studies,* 22, 315–337. https://doi.org/10.1007/s10018-019-00259-1.

Nhamo, G., Nhemachena, C., Nhamo, S. (2019b). Is 2030 Too Soon for Africa to Achieve the Water and Sanitation Sustainable Development Goal? *Science of the Total Environment*, 669, 129-139. https://doi.org/10.1016/j.scitotenv.2019.03.109

Njobvu, B. (2007). Zambian Libraries to Fight Poverty Through the Provision of Information to Help the Country Attain the Millennium Development Goals by 2015. *Zambia Library Association Journal*, 22(2), 56-65.

O'Keeffe, P. (2016). The Role of Ethiopia's Public Universities in Achieving the United Nations Sustainable Development Goals. *International Review of Education*, 62, 791–813. doi:https://doi.org/10.1007/s11159-016-9599-9

Okoye, I.B., Mbagwu, F.C., Eden, A., Umunnakwe, G.C. (2019). Sustainable Development Goals (SDGs) a Sine Qua Non for Improved Services in Public Library in Nigeria: Challenges and Prospects. *The Information Technologist: An International Journal of Information and Communication Technology*, 16(1), 11-25.

Okunlola, A.A., Oluwaniyi, J.T., Oyuedapo, R.O. (2018). A Survey of the Awareness of the 2030 Agenda: A Case of Libraries and Librarians in Osun State, Nigeria. *International Journal of Library and Information Science*, 10(2), 28-34. doi:https://doi.org/10.5897/IJLIS2017.0791

Peer, V., & Stoeglehner, G. (2013). Universities as Change Agents for Sustainability: Framing the Role of Knowledge Transfer and Generation in Regional Development Processes. *Journal of Cleaner Production,* 44, 85-95. https://doi.org/10.1016/j.jclepro.2012.12.003

Popoola, B.O. (2019). Involving Libraries in Improving Health Literacy to Achieve Sustainable Development Goal-3 in Developing Economies: A Literature Review. *Health Information and Libraries Journal*, 36, 111-120. doi:https://doi.org/10.1111/hir.12255

Samantaray, M. (2017). Role of Libraries in Quality Education for Achieving SDGs. *Journal of Advances in Library and Information Science*, 6(1), 31-35.

Sedlacek, S. (2013). The Role of Universities in Fostering Sustainable Development at the Regional Level. *Journal of Cleaner Production*, 48, 74-84. https://doi.org/10.1016/j.jclepro.2013.01.029

United Nations. (2015). *Transforming our World: The 2030 Agenda for Sustainable Development*. New York: United Nations Secretariat.

Vyas-Doorgapersad, S. (2011). ODL for Sustainable Development in India. *Progression*, 33(2), 51-71.

Wand, P. A. (2016). Information Centers and Socioeconomic Development in MENA: Finding a Quantitative Relationship. *International Federation of Library Associations and Institutions Journal,* 42(2), 109–117. doi:https://doi.org/10.1177/0340035216642457

11 Leadership Capacities Contributing to the Success of a Multi-Stakeholder Partnership in Eswatini

Wendy van Eyck and Yolande Steenkamp

Abstract

The United Nations (UN) recognises multi-stakeholder partnerships (MSPs) as necessary for implementing the Sustainable Development Goals (SDGs). Unfortunately, research into the leadership capacities that contribute to the successful realisation of an MSP project has been sparse, especially in an African context. This chapter uses a case study of a single phase of one MSP, Project Last Mile, in the Kingdom of Eswatini (Eswatini). Using an inductive approach, a content analysis of semi-structured interviews with four assigned leaders was undertaken. The interviews sought to understand what leadership capacities the leaders perceived contributed to the project realising targets of the SDGs. The themes of complementary collaboration, pragmatic optimism, unassuming influence, single-minded conviction, consistent trust-building and patient prioritisation emerged. While each of the leaders perceived these capacities to contribute to the successful delivery of the project, the data revealed a multidirectional phenomenon between individual capacities and team leadership capacities. The data also revealed an additional factor perceived to influence the successful delivery of the project in addressing SDG targets in an enabling environment. Further, the data suggests that the genders of the leaders may be a factor.

Keywords

SDGs · Leadership · Leadership capacities · Multi-stakeholder partnerships · Project Last Mile · Africa

1 Introduction and Background

Project Last Mile is an MSP pioneering a scalable model of capacity building through transferring the business and technical expertise of the Coca-Cola system to ministries of health in African countries. Despite billions invested to strengthen health systems in Africa, ministries of health have been unable to reach private sector benchmarks in terms of supply chain or strategic marketing (Yadav et al. 2013; Chaudhuri 2015). Consequently, in 2009, a delegation from the Global Fund to Fight AIDS, Tuberculosis and Malaria (the Global Fund) and the Bill & Melinda Gates Foundation approached The Coca-Cola Company to discover how its core competencies

W. van Eyck · Y. Steenkamp (✉)
Department of Business Management, University of Pretoria, Pretoria, South Africa
e-mail: wendy@believagency.com; yolande.steenkamp@up.ac.za

G. Nhamo et al. (eds.), *Sustainable Development Goals for Society Vol. 1*, Sustainable Development Goals Series, https://doi.org/10.1007/978-3-030-70948-8_11

might be used to support the distribution of life-saving medicines. The delegation had seen Coca-Cola products in remote villages in Africa, which had infrequent or no access to medicines and other health products and services. This prompted the delegation to discover why it was possible to find a product like Coca-Cola almost anywhere, but not life-saving medicines. In response, The Coca-Cola Company and its foundation partnered with the Global Fund and the Bill & Melinda Gates Foundation to form Project Last Mile to explore how private sector practices could be shared to strengthen public health systems. USAID entered the partnership in 2014. Following successful engagements in Tanzania, Ghana, Mozambique, Tanzania and South Africa, the Global Fund invited Project Last Mile to support the Ministry of Health and the National Emergency Response Council on HIV and AIDS in Eswatini in 2016 (Project Last Mile 2019a).

Project Last Mile's support in Eswatini involved building capacity through sharing the strategic marketing skills and network of the Coca-Cola system to generate demand for HIV prevention and health services among adolescent girls and young women (AGYW). In August 2017, Phase I (including *design and test* components) started with the appointment of a steering committee and working group. The design component aligned with the market research methodology approach The Coca-Cola Company would use. Based on the research findings, the MSP developed a marketing strategy called GirlChamp. This strategy involved creating safe spaces for AGYW to be informed and empowered about health issues which affect their health in a girls-only health club environment. It included healthcare workers by providing training to promote youth-friendly service to AGYW seeking healthcare services (Christie et al. 2018; Project Last Mile 2019b).

The *test* component ran GirlChamp Super Saturday events in three health facilities in the Manzini district during November–December 2018. Key to these pilots was community mobilisation through in-person dialogues with traditional leaders, clinic committees, parents and schools. AGYW were engaged through Project Last Mile-led consultations, radio advertising and social media campaigns. Project Last Mile expected 750 AGYW to participate in the three GirlChamp Super Saturday events. However, 1926 AGYW attended the event and registered with the Client Medical Information System for future health services. Additionally, 172 AGYW chose to participate in accessing same-day health services, including HIV testing. The Coca-Cola Foundation funded Phase II of the project to expand the campaign nationally because of the demonstrated impact and success of Phase I (Christie et al. 2018).

The work of Project Last Mile Eswatini supports the Ministry of Health towards realising SDG 3 (good health and well-being) through partnering to strengthen the means of implementation (SDG 17). SDGs are a set of goals announced in 2015 and developed by leaders from 193 countries to achieve sustainable development through economic, social and environmental dimensions (i.e. the "triple bottom line" focus on people, planet and profit). To realise one or more of the 169 targets of the SDGs requires individuals, organisations and governments to work together (United Nations 2015; Stibbe, Reid and Gilbert 2019). Following the Rio Summit of 2002, the UN recognised the necessity of collaboration by all sectors to achieve a global vision of sustainable development. This principle provided structure to the SDGs, with SDG 17 recognising partnerships as a core means of implementation for all the goals. The United Nations (Stibbe et al. 2019) has defined partnerships for the SDGs as follows:

> Partnerships for sustainable development are multi-stakeholder initiatives voluntarily undertaken by governments, intergovernmental organisations, major groups and other stakeholders, which efforts are contributing to the implementation of inter-governmentally, agreed development goals and commitments.

Each SDG is connected to targets to monitor progress. SDG 3 comprises 13 targets. The work of Project Last Mile Eswatini is contributing to targets 3.3 and 3.7 of the SDGs. The MSP is supporting a reduction in the number of new HIV infections (key indicator 3.3.1) through capability building in the Ministry of Health to create safe spaces for AGYW to access information and healthcare. The MSP contributed to target 3.7 by generating demand for sexual and reproductive healthcare services among women of reproductive age through the GirlChamp strategy. Project Last Mile thus successfully partnered with the Global Fund and Ministry of Health in Eswatini to facilitate an MSP which contributed to achieving targets of SDG 17.

SDG 17 focuses on partnerships for the goals through mobilising available resources in all sectors for financing, technology, capacity building, trade and policy, and institutional coherence. This SDG contains 19 targets, of which 17.9, 17.16 and 17.17 are relevant to this case study. Target 17.9 of the SDGs aims to "(e)nhance international support for implementing effective and targeted capacity-building in developing countries to support national plans to implement all the SDGs …" (United Nations General Assembly 2015). In addition, target 17.16 focuses on utilising MSPs "that mobilize and share knowledge, expertise, technology and financial resources, to support the achievement of the sustainable development goals in all countries, in particular developing countries" (United Nations General Assembly 2015). Finally, target 17.17 seeks to "(e)ncourage and promote effective public, public-private and civil society partnerships, building on the experience and resourcing strategies of partnerships" (United Nations General Assembly 2015).

If the drive to employ MSPs to realise the SDGs in sub-Saharan Africa is to be successful, understanding how leadership capacities contribute to the successful delivery of MSP projects is essential. Therefore, this study of Project Last Mile Eswatini seeks to explore the leadership capacities, which facilitate MSPs to successfully deliver projects that achieve SDG targets.

2 Literature Review

2.1 Overview of Leadership and Its Capacities

While there is little consensus on a single definition of leadership, many modern definitions recognise it as a process between leaders and followers who work together to achieve a shared goal, usually under the influence of the leader (Northouse 2016). The interrelationship between leaders and followers is receiving increasing focus, with relational leadership theory being at the forefront of emerging leadership theories. The relational approach departs from traditional theories by moving beyond the idea of leadership as a one- or two-way exchange between leaders and followers. It recognises that leadership occurs in many different everyday contexts and interactions, both formal and informal, and is consistently emerging and adapting to circumstances. It sees leaders as not removed from the context but positioned within social networks and practices. In short, relational leadership moves beyond understanding how leaders use relationships to align or cause a subordinate to do something to considering how leaders work with others within networks of relationships to achieve change (Uhl-Bien 2006; Cunliffe and Eriksen 2011; Carsten and Uhl-Bien 2015).

Uhl-Bien (2006) identifies two approaches to relational leadership. The *entity perspective* reflects the traditional concept of relationships as a connection between individuals. The *relational perspective* views leadership as an ongoing and shifting social construction process because the perceptions and thoughts of the person or organisation are continuously emerging. This is possible because power is shared throughout the organisation and not concentrated on a few individuals. Showing how leadership occurs in ordinary, everyday interactions, Cunliffe and Eriksen (2011) built on existing relational leadership theory and proposed it not as a theory or model but rather "a way of being-in-the-world that embraces an intersubjective and relationally-responsive way of thinking and acting". They found that relational leading is as much about the relationships leaders build between people as creating opportunities for living conversations.

Related to relational approaches is the emerging theory that shows how leadership processes occur in situations where multiple leaders arise based on the context in which leadership occurs (DeRue 2011; Dinh et al. 2014; D'Innocenzo et al. 2016). These theories approach leadership not as a top-down approach but as a complex, adaptive process which involves a series of leading and following interactions (DeRue 2011). Moreover, these theories depart from thinking about leadership capacities in terms of the individual and prefer to understand them in terms of a whole team (Day et al. 2004).

Another emerging approach viewing leadership as a process is responsible leadership. It emphasises the importance of forming and maintaining good relationships with all stakeholders within and outside an organisation (Maak and Pless 2006). Responsible leadership focuses on how the needs of a wide range of internal and external stakeholders can be considered and balanced, while creating resonance, trust and social capital, to realise the shared aim of responsible action (Maak 2007; Waldman 2011).

Women and leadership are also a growing field, broadly split into two opposing views. The first, more dominant view, claims that gender affects leadership styles. The second denies any great difference in leadership styles between genders, proposing that women leaders have rejected traditional feminine roles and characteristics (De La Rey 2005). Eagly and Johnson (1990) asserted that women adopt a democratic leadership style, while men often assume autocratic styles. Further, Eagly and Johannesen-Schmidt (2001) observed that women tend towards communal attributes like affection, empathy and nurturing, often displayed through supporting each other and contributing to solving relational problems. Men are likely to display dominance, risk-taking and competition (Eagly and Johannesen-Schmidt 2001). Stanford et al. (1995) found that women leaders are inclined to follow a highly participatory team-based management approach. Further, Appelbaum, Audet and Miller (2003) observed that the descriptors of consideration, transformational, participative, socio-expressive and people oriented were most often attributed to leadership approaches of women. Men's approaches to leadership were most often described using terms such as structure, transactional, autocratic, instruction giving and business oriented.

De La Rey (2005) proposes that the differences in leadership between genders are not based on biographical differences but on a different point of departure which makes women leaders inclined to display leadership consistent with responsible and relational leadership. She argues for a new way of leading that values human rights, diversity and sustainability, and holds that women might be well positioned for pioneering alternative leadership styles. Hare et al. (1997) and Rosener (2011) found that the characteristics of women leaders, such as sharing power, inclusion and participation, are suited to transformational leadership.

So far, this review has focused on leadership theories developed in the Global North. It now turns to leadership literature in sub-Saharan Africa. A 60-year review of African leadership literature found that, with very few exceptions, the emphasis until the 2000s was on leadership in the political sphere. From then onwards, the focus broadens as more African scholars and female researchers participate in research on leadership and gender, leadership styles and connection between leadership and management. Despite these shifts, the authors note the need for further research, specifically in the areas of African leadership theory and leadership in the private sector (Fourie et al. 2017). This literature review of leadership theories and capacities in Africa confirms this need.

A few researchers (Van der Colff 2003; Malunga 2006; Bolden and Kirk 2009) have discussed the importance of leadership development from within Africa that supports transformation at all levels through inclusive and participative processes in sub-Saharan Africa. Much of the research in the postcolonial era centres around the concept of *Ubuntu* leadership. Notably, Haruna (2009) says that the "great man" or centralised leadership model that many parts of Africa have adopted in the postcolonial era restricts community participation and disregards the contributions of families, communities and networks in achieving collective goals. Haruna argues that community, collaboration and cooperation are hallmarks of the lived reality in sub-Saharan Africa. Therefore, he observes that a community-based leadership approach would be consistent with cultural value systems and facilitate broader social change and transformation. Interestingly, this echoes deconstructions of the "great man" approach to leadership in precolonial Southern Africa by African historians. Weir (2008), for example, has offered more complex explanations to the leadership of Zulu chiefs that also takes account of the influence of chiefly females and communal dynamics.

Limited research exists on the capacities required by leaders in Africa to solve complex problems. However, in her survey of MBA students, Kurasha (2016) found that the respondents perceived the following characteristics of leaders to be critical for achieving sustainable development: (a) *the ability to change thinking and practice*; (b) *the ability to treat followers equally* and promote inclusiveness; (c) *the motivation to close the gender gap* by empowering women; (d) *a willingness to depict feminine attributes and traits* of cooperation and understanding like gentleness, empathy and sensitivity; (e) *the ability to embrace pluralistic knowing and seek union*; and (f) *the ability to create and sustain relationships*, which in an African context includes accommodating the heterogeneity and diversity of people in a region.

While the literature on leadership in Africa is sparse, studies into women leaders in Africa are even scarcer. The literature that does exist is focused on how women leaders are perceived in business (Ndinda and Okeke-Uzodike 2012), educational (Odhiambo 2011; Lumby and Azaola 2014), religious (Makoro 2007; Wakahiu and Salvaterra 2012) and political (Gouws and Kotzé 2007; Adams 2008; Gouws 2008) spheres. The review reflects a severe gap in research on how traditional or emerging African leadership approaches may contribute to realising the SDGs.

2.2 Linking Leadership Capacities to SDGs

Most observers of the drivers in realising the SDGs recognise the need for leadership. However, the role of leadership capacities in implementing complex development agendas is still unclear. Further, there is limited literature that describes the leadership capacities required to successfully implement an MSP to realise the SDGs. The following section will review existing research into leadership capacities and the SDGs.

Stibbe et al. (2019) propose that there are three basic types of partnerships for the SDGs and classify these on the Partnership Spectrum. The first category is partnerships which "leverage/exchange". These are partnerships that exist because both organisations recognise the complementarity of their collaboration and exchange knowledge and expertise to further their goals. The second category is partnerships that "combine/integrate", where organisations share a strategic goal and believe that they will achieve more working together than separately. The final category is partnerships which "transform", and here organisations collaborate to tackle complex challenges through system transformation.

Pattberg and Widerberg (2016) identify that different types of leadership by individuals and organisations are required at different stages over the course of an MSP. Gray (2007) identifies several tasks leaders need to undertake to develop successful partnerships for sustainable development (creating a vision; assembling and uniting partners; understanding the problem and designing a process; conflict management; brokering; and entrepreneurship) but does not relate this to the capacities leaders need to achieve these tasks.

A notable departure from the focus on the role or type of leadership required in MSPs to achieve the SDGs appears in the work of Fowler and Biekart (2017). They found that a special type of person—an interlocutor—is needed for SDGs to be implemented successfully. The authors pinpoint that the interlocutor must lead through "exerting influence without authority", which they place under the attribute of "leadership and conflict management". Additionally, Fowler and Biekart propose that an interlocutor has attributes of trustworthiness and trust-building; the ability to recognise and adapt to changes in the system; an awareness of how structures of power affect decision-making; commitment for the long haul; the ability to translate ideas across sectors; and the ability to think independently while remaining accountable.

James (2008) focuses on African non-governmental organisational leaders and the leadership capacities needed in their cultures and contexts. His research sought to understand what makes African civil society leaders change their behaviour. He recognised that when civil society leaders shift from a bureaucratic leadership style to a participative leadership style, they yielded the best results. James recommends that leadership development, and therefore leaders in Africa, needs to reflect the cultural, social and family context in which they occur. However, he observes that many of these recommendations are also popular in contemporary Western literature.

While literature on leadership in MSPs covers the types of leadership and tasks individuals and organisations require at various stages of MSPs, this is not linked to the leadership capacities. Further, we were unable to uncover any literature that looked at how leadership capacities might help attain the SDGs. We could also find no literature on the role of gender and leadership capacities. This review indicates the lack of research into the types of leadership capacities needed to realise the SDGs through MSPs.

3 Research Design

We chose an exploratory single case study design to explore the perceptions assigned leaders in the MSP had about the leadership capacities they believed contributed to successfully achieving targets of the SDGs. For the purposes of this study, capacities refer to the sum of embedded traits, values, characteristics, competencies, behaviours and practical knowledge an individual leader possesses.

A purposive sampling approach was used to select research participants who could give a suitable depth of information to answer the research question (Simons 2009; Zhang and Wildemuth 2009). Therefore, we purposefully selected the participants because of their assigned leadership roles within Project Last Mile Eswatini Phase I between August 2017 and December 2018. Consequently, the assigned leaders were not a random sample of leaders from Project Last Mile. However, as a purposive sample, the selected participants could provide a unique insight into which leadership capacities contributed to the success of the project. We recruited participants via email, requesting them to take part in an interview either face to face or telephonically.

The four leaders who participated represented three countries, were male and female, comprised different races and covered a 20-year age range. The assigned leaders hold different leadership roles and responsibilities in Project Last Mile Eswatini and manage stakeholders at various levels. While there is no assigned hierarchy, leaders report to one another, creating an assumed hierarchy. The leaders perform their tasks through in-person and virtual meetings. The leaders do not work in the same location but are in four different cities across three countries.

As a result of the geographically spread locations, three interviews took place over Zoom and one in person. A single round of interviews with the leaders conducted by one researcher was undertaken. The interviews with the leaders lasted approximately 60 min and were recorded on an audio recorder. We selected semi-structured interviews to increase the reliability and allow us to collect complete and consistent sets of information about perceptions and attitudes across interviews. The interviews followed a standard structure to explore which leadership capacities each leader identified in themselves and other assigned leaders. For example, the following questions were included in the interview guide: (a) As someone who followed other leaders before becoming a leader, what did you learn about leadership that you apply in your position as a leader in Project Last Mile? (b) Think about the other leaders in Project Last Mile Eswatini. What are the different traits and behaviours that you admire in the leadership styles of each of them?

The approach to data analysis was neither purely inductive nor deductive in nature, but a combination of both. This allowed for structure and flexibility to construct theoretical notions and frameworks in a deductive way while exploring data inductively with an open mind. Each interview was transcribed in its entirety. Next, we carried out preliminary coding in Microsoft Word by using the review function to find and label significant statements and phrases about leadership capacities in each transcript. Statements and phrases were labelled from the data, using either the exact words of the participants, a new phrase or established word to label the phenomena. These preliminary codes were then converted from Microsoft Word into a table in Microsoft Excel with a bespoke macro. The preliminary data was sorted and analysed—combining some of the codes—to decide on final codes. Meticulous records were kept as codes were reconsidered and the code sheet was modified. Once the full set of data had been coded using the final codes, the researcher analysed the data and divided the final codes into categories based on common properties. The researcher grouped the quotations from each of these categories together, searching for linkages, frequency and commonalities. Through multiple interrogations of the data categories, themes emerged.

4 Presentation and Discussion of Findings

The analysis revealed several leadership capacities that facilitated Project Last Mile Eswatini's successful realisation of one or more SDG targets. The first capacity (complementary collaboration) was a shared approach to leadership held by all the leaders. The remaining capacities (pragmatic optimism, openness to learning, single-minded conviction, consistent trust-building, patient prioritisation) represent common traits, values and characteristics that the leaders all perceived to have contributed to the success of the project. This distinction between team and individual capacities will be considered in the discussion.

4.1 Capacities

4.1.1 Complementary Collaboration

Significantly, all the leaders believed that Project Last Mile Eswatini was successful because of the sum of all the leadership capacities the leaders contributed. Not only did each leader acknowledge the contributions of the other leaders, but they also named their ability to share the leadership roles in different contexts as key to the success of the project. The Director of Delivery described this:

> to achieve this is about the collaboration. Not one leader has driven this whole thing from start to finish. So, emphasising that we all bring different leadership qualities that are very complementary to each other [is] what's enabled this to be successful.

The careful selection of the leaders of Project Last Mile Eswatini contributed to an enabling environment for the leaders to support one another. The Project Director acknowledged intentionally selecting individuals who would have "chemistry" with the other leaders in the project, and described a situation where he had to intervene and restructure the team due to a personality clash. He commented,

> my greatest success as a leader is picking the team … the fact that I haven't had to be as involved in Eswatini speaks for the process working, because it means the right people were put onto the team …

The women leaders felt that some of the success of the collaborative approach was due to traits usually assigned to women. The Subject Matter Expert said,

> I think women are potentially better in the space than men because women are by their nature more collaborative, and team players. So, I had no necessity to be the boss, much as I was the leader.

The only male leader made no observation about how gender might impact the team's collaborative approach to leadership.

4.1.2 Pragmatic Optimism

The leaders displayed the ability to identify and align with existing relationships, projects, human resources and infrastructure to produce pragmatic solutions which fitted the context of the problem. The Director of Delivery said:

> I am always trying to look at an opportunity and see the glass half full instead of glass half empty, but in doing that, with the lens of what's feasible, with the lens of, 'Yes, it's awesome. These are all the things we could achieve. But let's be realistic in our context, and create optimism, an opportunity around what works where we are'.

A further capacity that the leaders employed was an understanding of when to push and when to step back. For this, they drew on their lived experiences of implementing similar programmes. This maturity allowed them to understand that the goal of Project Last Mile Eswatini was to deliver a solution that improved the current situation, not to provide a perfect solution. The example of consultants that have extensive strategic know-how but little implementation experience was used. This example illustrated that without awareness as a leader that the solution must be practical and contextual, there will be limited alignment between the ideal scenario and what a programme is trying to achieve. This ability to apply pragmatic optimism as a leader also called for being comfortable with imperfection. The Subject Matter Expert explained:

> It's a question of how much can you improve the existing situation, it's not how you make the existing situation perfect, because it won't be, there's too many structural, economic, and social issues that will impact that. The way you get around that is certainly with experience. People that have worked in developing markets get that.

4.1.3 Openness to Learning

Each of the leaders approached the project open to learning and with the recognition that everyone came to the table with something to offer. This required them to exercise their capacities of humility, self-awareness and desire to evolve and grow as individuals. The Subject Matter Expert said, "This is not about proving how much you know, this is about using how much you know to be able to get people on the ground to understand how much they know".

There was a shared understanding that to avoid alienating other stakeholders on the project, the leaders needed to ensure that they did not debase existing knowledge or experts. This necessitated the leaders humbling themselves and acknowledging their lack of expertise in certain areas. Behaving in this way aided the leaders with developing relationships based on equality rather than superiority. This was a necessary capacity, as the nature of this MSP meant that the sole way to lead was through influence. The Project Last Mile leaders possessed no institutional authority over the various stakeholders. This behaviour demanded a high level of self-awareness regarding what they did not know, so they would understand when to "listen and learn" and when to share their expertise.

Finally, the leaders demonstrated the capacity of being open to learning through their willingness to evolve within their roles and their readiness to experiment to find solutions. This meant the leaders were willing to learn new things and accept new partners that could help them improve. This also extended to being willing to try out new approaches and solutions, even at the risk of failure. The Director of Delivery shared that "this marketing approach was totally new. We didn't know what we were doing, every day was a learning opportunity…we tested this every day together".

4.1.4 Single-Minded Conviction

Project Last Mile introduced an innovative approach to the strategic marketing landscape of public health in Eswatini. As a result, the leaders experienced various forms of resistance from different stakeholders in implementing the model. The leaders persevered because they believed in the power of the model they were pioneering. They described the capacity that enabled them to resist going with the status quo as conviction and linked this to the ability to establish and communicate firm boundaries about what was beneficial to achieving the goal and what was a distraction. An enabling environment within Project Last Mile allowed for single-mindedness in their convictions. The Director of Delivery explained:

> We're able to constantly push back because we believe so strongly in the power of this model. And, I've always felt in Project Last Mile empowered to say no. And that I think, is a very strong value that I appreciate. And that has helped us stay the course and not taking on things that just are not in our wheelhouse.

4.1.5 Consistent Trust-Building

While relationship building occurred in different ways between the leaders in this study, they each highlighted trust as capacity that built relationships. The Country Coordinator explained,

> I wouldn't be able to achieve [in this project] without being honest and being trusted and being trustworthy. ... The consistency in delivery, the consistency in ensuring that we are aligned, and the consistency in ensuring that the project grows really holistically. You know, it's some of the things that when you look at and you say, 'Ah, you know what, this is an honest person. I like their work ethic, and have demonstrated a strong work ethic.' That is all you can really work off of really when it comes to trust.

Trust emerged through observing other leaders consistently showing themselves to be trustworthy. The Project Director observed,

> The big word for me with trust is consistency. Consistency and trust go hand in glove, because they can be derailleurs of trust. And the thing is, you want to make sure that you build relationships that have enough goodwill in them that they can survive a couple of trust breakages.

One possible area of trust breakage is conflict, and here the leaders applied the capacities of honesty and respect to navigate conflict and further invest in their relationships. The Subject Matter Expert said, "I think being honest with people in a gentle way is important. So, if you do disagree, then you should disagree, and you should present the reason why that is". However, the leaders did not take on conflict for conflict's sake, but to get the job done. The Country Coordinator added, "If things are not going right, in our own special way, without offending the other, we just are able to voice it out, and make sure that we clear things off, and we move forward".

To demonstrate their trust in one another, the leaders found opportunities to lift one another up. The Director of Delivery described this as:

> When I'm in meetings [or] when I'm presenting the work we do, I specifically say [the Country Coordinator] did this, and [the Subject Matter Expert] does this. And they're here in this meeting for their perspectives, to share, to contribute. It's about including everybody's voice, because they've earned it, and they have a different, valuable perspective to bring to the table.

This approach established a reciprocal environment of trust where each leader believed that the other leaders would do everything possible to help them succeed in realising the project.

4.1.6 Patiently Prioritising

The leaders perceived their capacity for patience in and with the process to have contributed to the project's success. Interestingly, all the leaders commented that this was learnt from other leaders. The Subject Matter Expert described learning this from a former boss at a private sector company:

> I think that good African men leaders develop an amazing capacity for patience and wisdom. And when I was working for [my former boss], he never used to move fast enough for me because I'm also quite action orientated. And still, to this day, I would say that he could have moved faster. But I learned a lot around his approach.

In addition, the Director of Delivery felt that she had grown the capacity for patience while working at Project Last Mile:

> One of the leadership qualities that I have been able to kind of absorb is the ability, although it's still not my nature, but the awareness around patience. Patience is not a leadership virtue I had previously [laughs]. But there is a balance and a grace that I've learned from [the Project Director] to not seeing everything in extremes and also being comfortable with a little bit of uncertainty.

The leaders were careful to note that patience did not suggest a lack of progress and were careful not to confuse motion with progress. Patience was linked to the leaders' capacity to prioritise, as the Director of Delivery explained, "I think the ability to not see every problem as a massive crisis, but kind of prioritise what actually is important, what's urgent, and what is just life and will pass". The leaders believed that this ability to patiently prioritise led to greater impact. The Subject Matter Expert said, "You can't boil the ocean. You only have one set of resources. You have to focus on doing fewer, better things, right, with more impact".

The concept of patiently prioritising emerged as connected to the leaders' capacity for resilience. The Project Director explained, "an important part of resilience is like peace in the journey. It's not having knee-jerk reactions to situations like that, or giving up too early because you feel emotionally unstable about the situation". This ability to stay in the course and see the bigger picture amid uncertainty and pushback from stakeholders allowed the leaders to utilise patience to make progress towards realising the project goals.

4.2 Discussion

The analysis of the data revealed a multidirectional phenomenon that we did not anticipate. The leaders perceived that their ability to drive a successful project was a result of the sum of all the leadership capacities each leader contributed individually. In turn, team and group leadership dynamics lead to the emergence of individual capacities. This aligned with previous research by Uhl-Bien (2006) and Day et al. (2004) who conceptualised leadership in terms of leadership capacities that belong to the whole team rather than to individuals. As a result, while the themes reflect the capacities of individual leaders, they also need to be considered as the capacities that may drive the emergence, formation and implementation of team leadership capacities.

Thus, the data suggests that traditional theories of leadership, such as transformational leadership, which emphasise the individual leadership capacities that a leader brings to a team, are on its own insufficient for describing the full range of leadership capacities required to facilitate an MSP. While we observed patterns that correlated with transformational leadership theory—a clear, common vision and placing the greater good ahead of personal gain—the theory did not explain leadership capacities as a team phenomenon.

An interesting aspect of emerging literature in the field of relational leadership which relates to this data is the focus on how power is shared rather than concentrated on a few individuals (Uhl-Bien 2006). Within this study, the leaders demonstrated this through their emphasis on how the success of the project was due to the combination of their leadership capacities and teaming approach to leadership. Notably, the leaders felt uncomfortable claiming that it was one individual leader that brought about the success of the project. The data in this study provides an example of the capacities individual leaders need to achieve for leadership to emerge from a group rather than individuals.

The data shows that the enabling environment allowed the phenomenon of leadership capacities emerging from within the team of leaders to occur. This data suggests that the enabling environment was a result of deliberate choices about the composition of the leadership team, organisational culture and values, and individual leadership capacities the leaders brought to the project. The supportive, organisational culture empowered the leaders to exercise their individual leadership capacities, which in turn led to the emergence of shared leadership capacities. These concepts reflect studies on responsible leadership theory by recognising that leadership is a complex, adaptive process that occurs through forming and maintaining good relationships (Maak and Pless 2006; Maak 2007).

Another unexpected factor that emerged from the findings was how the leaders perceived that gender contributed to strengthening the ability of the leaders to combine their leadership capacities as a team. Three female leaders led this phase, with the single male leader acting as a thought partner when necessary and as a relationship manager of stakeholders at the highest level. This corresponds with the findings of Eagly and Johnson (1990) who found the strongest evidence of gender differences in leadership styles related to the tendency for women to adopt a democratic rather than an autocratic style of leadership. Further research must establish whether these characteristics are leadership capacities influenced by gender or the practice of emerging leadership styles as suggested by several authors (Hare et al. 1997; De La Rey 2005; Rosener 2011). This aspect of leadership capacities and gender is still underdeveloped in literature. However, the findings seem to confirm the dominant view that gender affects leadership styles and therefore capacities (De La Rey 2005). The female leaders displayed communal attributes identified by Eagly and Johannesen-Schmidt (2001), such as being supportive of one another and contributing to solving conflict. In addition, existing research into gender and leadership has found that women tend to follow a more team-based approach to leadership, including sharing power (Stanford et al. 1995; Hare et al. 1997; Appelbaum et al. 2003; Rosener 2011).

Any discussion of leadership in the context of an MSP in Africa must consider how it relates to literature from African scholars. Since most of the leaders were female, there may also be a correlation between gender and values of ubuntu as proposed by Ngunjiri (2016). Further, the findings from this study seem to back up existing research that leadership for development from within Africa needs to be an inclusive and participative process that reflects the idea of ubuntu (Van der Colff 2003; James 2008; Haruna 2009; Ncube 2010). What makes this finding interesting is that only three of the four leaders were born in Southern Africa and only one of the leaders grew up in a culture with a community-based leadership approach. As a result, it is unclear whether this leadership style is unique to an African context or individuals raised in traditional African culture. However, this implies that the traits of community, collaboration and cooperation that Haruna (2009) argues are hallmarks of leadership in Africa can be transferred to other cultures and may fulfil an important role in realising global goals. It also suggests that the community-based leadership approach proposed by Haruna (2009) which develops leadership as community, collaboration and cooperation contains capacities that can be learnt by leaders.

Another interesting element related to the literature on leadership in an African context was the focus on patience as an important team capacity. Not only was this a capacity which the leaders felt was transferable and learnt, but it was also a capacity the leaders described as learnt from African leaders. This is a value that has been linked to ubuntu values (Prinsloo 2000). This suggests that ubuntu values and value-based leadership contribute to realising goals in collaborative environments.

Finally, these findings seem to challenge the research of Fowler and Biekart (2017) that a special type of person is needed in MSPs to achieve SDGs. This study found that the multidirectional phenomenon of leadership capacities as an exchange between the group and individuals drove the success of the MSP to realise targets of the SDGs. This finding suggests that the type of person is less important than how individuals function together. While individual capacities increase the likelihood of team collaboration, the team dynamics also cause individual leadership capacities to emerge. This once again reflects some of the emerging literature on relational leadership that suggests that leadership emerges based on various human and non-human factors that make up the given context (DeRue 2011; Dinh et al. 2014; D'Innocenzo et al. 2016). This may be important in practice, as it speaks to how MSPs recruit leaders and resource teams to successfully deliver a project.

5 Conclusion

This study reported the findings of an exploratory case study that sought to answer the question: Which leadership capacities drive an MSP to successfully deliver a project that contributes towards achieving SDG targets? To explore this question, we interviewed four assigned leaders of an MSP successfully contributing to realising targets of SDG 3 in Eswatini. Through a content analysis of semi-structured interviews, six themes around the leadership capacities that appeared to drive the success of the Project Last Mile Eswatini emerged. These were complementary collaboration, pragmatic optimism, unassuming influence, single-minded conviction, consistent trust-building and patient prioritisation. The data revealed a multidirectional phenomenon whereby the individual capacities served to enable the emergence of team leadership capacities and vice versa.

These findings suggest the need for further research on how relational and collaborative leadership theories might contribute to the realisation of the SDGs, as well as to establish whether women leaders are practising new styles of participatory leadership and how this may contribute to MSPs. Another area to be considered for future research is whether the gender composition of leadership teams working towards realising MSPs affects the outcomes of such projects.

Recognising which leadership capacities contribute to the successful delivery of MSP projects is essential for realising SDGs in sub-Saharan Africa. In practice, a better understanding of the leadership capacities which facilitate MSPs to successfully deliver projects that realise one or more targets of the SDG can drive momentum towards achieving these global goals.FundingThis work was supported by the Baden-Württemberg-STIPENDIUM.

References

Adams, M. (2008). Liberia's election of Ellen Johnson-Sirleaf and women's executive leadership in Africa. *Politics & Gender* 4: 475-484.

Appelbaum, S.H., Audet L. and Miller J.C. (2003). Gender and leadership? Leadership and gender? A journey through the landscape of theories. *Leadership & Organization Development Journal* 24: 43-51.

Bolden, R. and Kirk P. (2009). African leadership: Surfacing new understandings through leadership development. *International Journal of Cross Cultural Management* 9: 69-86.

Carsten, M.K. and Uhl-Bien M. (2015). Follower beliefs in the co-production of leadership. *Zeitschrift für Psychologie*.

Chaudhuri, A. (2015). Supply chain innovations in healthcare. In: Singh V.K. and Lillrank P. (ed.), *Innovations in healthcare management: Cost-effective and sustainable solutions*: 261-272. Florida: CRC Press.

Christie, S., Linnander E., Brault M., Curry L. and Mertens K. (2018). *eSwatini Phase 1 Monitoring & Evaluation Report. Reporting Period: August 2017 – September 2018*. New Haven, CT, Yale Global Health Leadership Initiative, Yale University

Cunliffe, A.L. and Eriksen M. (2011). Relational leadership. *Human relations* 64: 1425-1449.

D'Innocenzo, L., Mathieu J.E. and Kukenberger M.R. (2016). A meta-analysis of different forms of shared leadership–team performance relations. *Journal of Management* 42: 1964-1991.

Day, D.V., Gronn P. and Salas E. (2004). Leadership capacity in teams. *The Leadership Quarterly* 15: 857-880.

De La Rey, C. (2005). Gender, women and leadership. *Agenda* 19: 4-11.

DeRue, D.S. (2011). Adaptive leadership theory: Leading and following as a complex adaptive process. *Research in Organizational Behavior* 31: 125-150.

Dinh, J.E., Lord R.G., Gardner W.L., Meuser J.D., Liden R.C. and Hu J. (2014). Leadership theory and research in the new millennium: Current theoretical trends and changing perspectives. *The Leadership Quarterly* 25: 36-62.

Eagly, A.H. and Johannesen-Schmidt M.C. (2001). The leadership styles of women and men. *Journal of Social Issues* 57: 781-797.

Eagly, A.H. and Johnson B.T. (1990). Gender and leadership style: A meta-analysis. *Psychological Bulletin* 108: 233.

Fourie, W., van der Merwe S.C. and van der Merwe B. (2017). Sixty years of research on leadership in Africa: A review of the literature. *Leadership* 13: 222-251.

Fowler, A. and Biekart K. (2017). Multi-stakeholder initiatives for sustainable development goals: The importance of interlocutors. *Public Administration and Development* 37: 81-93.

Gouws, A. (2008). Obstacles for women in leadership positions: The case of South Africa. *Signs: Journal of Women in Culture and Society* 34: 21-27.

Gouws, A. and Kotzé H. (2007). Women in leadership positions in South Africa: The role of values. *Politikon* 34: 165-185.

Gray, B. (2007). The process of partnership construction: Anticipating obstacles and enhancing the likelihood of successful partnerships for sustainable development. *Partnerships, governance and sustainable development. Reflections on Theory and Practice*: 27-41.

Hare, A.P., Koenigs R.J. and Hare S.E. (1997). Perceptions of observed and model values of male and female managers. *Journal of Organizational Behavior: The International Journal of Industrial, Occupational and Organizational Psychology and Behavior* 18: 437-447.

Haruna, P.F. (2009). Revising the Leadership Paradigm in Sub-Saharan Africa: A Study of Community-Based Leadership. *Public Administration Review* 69: 941-950.

James, R. (2008). Leadership development inside-out in Africa. *Nonprofit Management and Leadership* 18: 359-375.

Kurasha, P. (2016). Characterising Leadership for Sustainable Development in Africa: A Case Study.

Frontiers of Marketing Research (ISSN-2456-3919) 1: 2-6.

Lumby, J. and Azaola M.C. (2014). Women principals in South Africa: Gender, mothering and leadership. *British Educational Research Journal* 40: 30-44.

Maak, T. (2007). Responsible leadership, stakeholder engagement, and the emergence of social capital. *Journal of Business Ethics* 74: 329-343.

Maak, T. and Pless N.M. (2006). Responsible leadership in a stakeholder society–a relational perspective. *Journal of Business Ethics* 66: 99-115.

Makoro, T. (2007). The political emancipation of women in South Africa and the challenge to leadership in the churches. *Studies in World Christianity* 13: 53-66.

Malunga, C. (2006). Learning leadership development from African cultures: A personal perspective. *INTRAC PraxisNote* 25: 1-13.

Ncube, L.B. (2010). Ubuntu: A transformative leadership philosophy. *Journal of Leadership Studies* 4: 77-82.

Ndinda, C. and Okeke-Uzodike U. (2012). Present but absent: Women in business leadership in South Africa. *Journal of International Women's Studies* 13: 127-145.

Ngunjiri, F.W. (2016). "I Am Because We Are" Exploring Women's Leadership Under Ubuntu Worldview. *Advances in Developing Human Resources* 18: 223-242.

Northouse, P.G. (2016). *Leadership: Theory and practice.* Thousand Oaks: Sage.

Odhiambo, G. (2011). Women and higher education leadership in Kenya: A critical analysis. *Journal of Higher Education Policy and Management* 33: 667-678.

Pattberg, P. and Widerberg O. (2016). Transnational multistakeholder partnerships for sustainable development: Conditions for success. *Ambio* 45: 42-51.

Prinsloo, E.D. (2000). The African view of participatory business management. *Journal of Business Ethics* 25: 275-286.

Project Last Mile (2019a). *How we approach marketing health services in developing countries.* Port Elizabeth: Project Last Mile.

Project Last Mile (2019b). *Project Last Mile Impact Report.* Port Elizabeth: Project Last Mile.

Rosener, J.B. (2011). Ways women lead. In: Werhane P.H. and Painter-Morland M. (ed.), *Leadership, Gender, and Organization*: 19-29. New York: Springer.

Simons, H. (2009). *Case study research in practice.* Los Angeles: Sage Publications.

Stanford, J.H., Oates B.R. and Flores D. (1995). Women's leadership styles: a heuristic analysis. *Women in Management Review* 10: 9-16.

Stibbe, D., Reid S. and Gilbert J. (2019). *Maximising the Impact of Partnerships for the SDGs.* The Partnering Initiative and UN DESA.

Uhl-Bien, M. (2006). Relational Leadership Theory: Exploring the social processes of leadership and organizing. *The Leadership Quarterly* 17: 654-676.

United Nations (2015). *Partnerships for Sustainable Development Goals: A legacy review towards realizing the 2030 Agenda.* United Nations.

United Nations General Assembly (2015). *Transforming our world: the 2030 Agenda for Sustainable Development.* United Nations, UN General Assembly

Van der Colff, L. (2003). Leadership lessons from the African tree. *Management Decision* 41: 257-261.

Wakahiu, J. and Salvaterra M. (2012). Sustainable leadership: Lessons and implications of a leadership development program for women religious in Africa. *Journal of Pan African Studies* 5: 150-167.

Waldman, D.A. (2011). Moving Forward with the Concept of Responsible Leadership: Three Caveats to Guide Theory and Research. In: Pless N.M. and Maak T. (ed.), *Responsible Leadership*: 75-83. Dordrecht: Springer Netherlands.

Weir, J. (2008). Challenging heroic masculinity: leadership myths of nineteenth century King Shaka Zulu.

Yadav, P., Stapleton O. and Van Wassenhove L. (2013). Learning from Coca-Cola. *Stanf Soc Innov Rev* 11: 51-55.

Zhang, Y. and Wildemuth B.M. (2009). Qualitative analysis of content. *Applications of Social Research Methods to Questions in Information and Library Science* 308: 319.

Part IV

Service Delivery and the Attaining SDGs at the Local Government Level

The Global Water Partnership-South America and the Transboundary Implementation of Integrated Water Resources Management (SDG Target 6.5)

12

Alejandra Mujica, Corina Piaggio, Leandro Raúl Díaz, Juan Bussi, Aldo Palacios Courret, and Carlos Hiroo Saito

Abstract

The Global Water Partnership (GWP) has been working towards the attainment of the Sustainable Development Goals (SDGs), especially SDG 6 target 6.5. This was so given the interrelationship between the anthropogenic and the natural water cycles, supported by the water security concept. GWP-South America (GWP-SAM) regional activities and the results of its efforts supporting South American countries are presented. Data was collected from country reports and compared to the United Nations reports on SDG progress. Additional data was gathered during consultative workshops and other events that were held. The work also sought to determine different institutional arrangements, enabling environment and selected management instruments. Data tables and maps were produced comparing country performances relative to each of the indicators identified. It emerged that the GWP promoted supporting regional activities, based on participatory processes and the mainstreaming IWRM. This also involved matters of scale and the vision of national and provincial stakeholders involved in IWRM. A strategic plan was proposed to meet the necessity of knowledge development and capacity building aimed at combating social inequalities and value the cultural and social diversity of the societies involved.

A. Mujica · C. Piaggio
Global Water Partnership-South America, Montevideo, Uruguay
e-mail: alejandra.mujica@gwpsudamerica.org; corina.piaggio@gwpsudamerica.org

L. R. Díaz
Global Water Partnership-Argentina and Universidad Nacional de Tucumán, San Miguel de Tucumán, Argentina
e-mail: ldiaz@herrera.unt.edu.ar

J. Bussi
Global Water Partnership-Uruguay and Universidad de la República, Montevideo, Uruguay

A. P. Courret
Global Water Partnership-Chile and Sinergias, Región de Los Lagos, Chile
e-mail: ahpc@sinergias.cl

C. H. Saito (✉)
Global Water Partnership-Brazil and University of Brasilia, Institute of Biological Sciences and Centre for Sustainable Development, Asa Norte, Brasília-DF, Brazil
e-mail: saito@unb.br

G. Nhamo et al. (eds.), *Sustainable Development Goals for Society Vol. 1*, Sustainable Development Goals Series, https://doi.org/10.1007/978-3-030-70948-8_12

Keywords

IWRM · SDG 6 · South America · Water security · Sustainability · Management instruments

1 Introduction

One of the United Nations (UN) Sustainable Development Goals (SDGs) that mobilise the world community is SDG 6 that deals with ensuring the availability and sustainable management of water and sanitation for all (McCrackena and Meyer 2018). Saito (2017) concluded that water is the most highlighted issue during the 2005–2015 and 2018–2028 decades identified by the United Nations. Water is also highlighted as the greatest risk facing humanity by the World Economic Forum Report (WEF 2015), and it is embedded in almost all the SDGs as an inseparable part of the development of all sectoral agendas. Given the fear of getting lost across all the SDGs in terms of implementation gaps, Shah (2016) highlights that a dedicated 'Water Goal' (SDG 6) had to be included among the 17 SDGs.

Additionally, the strategic existence of SDG 6 represents an expansion of one of the Millennium Development Goals (MDGs), which was focused on drinking water and basic sanitation (MDG 7, Target 7.C). SDG target 6.5 has been found to be a powerful opportunity to rethink interrelationship between the anthropogenic and the natural water cycles that is supported by the water security concept (Lautze and Manthrithilake 2012).

Globally, the UN seeks to support countries in monitoring water issues within the framework of the 2030 Agenda for Sustainable Development (AfSD) that is supported by the UN-Water Integrated Monitoring Initiative for SDG 6 and the Global Water Partnership (GWP). The GWP is a global action network established by the Swedish International Development Agency (SIDA), the United Nations Development Programme (UNDP) and the World Bank in 1996. Since its inception, the GWP has grown to encompass over 3000 partner organizations in 179 countries, with 68 accredited Country Water Partnerships and 13 Regional Water Partnerships. In the South American region, GWP-South America (GWP-SAM) encompasses 7 Country Water Partnerships (CWP) involving the following countries: Argentina, Brazil, Chile, Colombia, Peru, Uruguay and Venezuela. From these countries, there are more than 350 members.

GWP coordinates the SDG 6 Integrated Water Resources Management (IWRM) Support Programme (SDG6-SP) that aims to support country and regional responses to SDG target 6.5. This is considered an entry point to accelerate the implementation of SDG 6 and other water-related goals. In South America, GWP-SAM has been working on this by supporting the definition of the baseline for SDG indicator 6.5.1 in different countries and also how to address SDG indicator 6.5.2. Addressing SDG indicator 6.5.2 is done in partnership with the United Nations Economic Commission for Europe (UNECE) and the United Nations Educational, Scientific and Cultural Organization's (UNESCO) International Hydrological Programme (IHP).

Gain et al. recognize that one of the major challenges for SDG 6 implementation would be the transboundary waters and challenges pertaining to water security. Water security is conceptualized as a function of 'availability', 'accessibility to services', 'safety and quality' and 'management'. The SDG indicator 6.5.2 is also strategic because it is expected that it can stimulate cooperation among countries on shared surface and groundwater resources, and this can extend these practices to other issues (McCrackena and Meyer 2018).

This chapter presents a diagnosis of SDG target 6.5 for GWP-SAM countries. A background section concerning SDG target 6.5 follows this introduction. In sequence, the research design is briefly described. The main findings and discussion section focus on the status of SDG indicators 6.5.1 and 6.5.2 for the seven countries belonging to GWP-SAM, its regional activities mainstreaming SDG target 6.5 and the main results of the GWP-SAM efforts in South American countries.

An overall regional conclusion about the present situation of both SDG indicators 6.5.1 and 6.5.2 and the main challenges for SDG target 6.5 is presented, with the necessary strategies to advance both SDG 5 target indicators under review.

2 The Literature Review

The origin of IWRM can be traced to the last century. The New Delhi Statement, Global Consultation on Safe Water and Sanitation (1990), addressed as the first guiding principle the protection of the environment and safeguarding of health through the integrated management of water and liquid and solid wastes (UN 1990). The Dublin Statement on Water and Sustainable Development (1992) established that governments should improve their capacity to implement the full range of activities for integrated water resources management, and these demanded a holistic approach that linked social and economic development, with protection of natural ecosystems (UN 1992). This was the precursor to incorporating IWRM Agenda 21 of the United Nations Conference on Environment and Development (UNCED). This was about improving water resources management by connecting the many different water services and providing good governance, appropriate infrastructure and sustainable financing. It was in this context that the GWP was founded to foster IWRM.

Snellen and Schrevel (2004) argue that IWRM should be viewed as an instrument for sustainable development. Koudstaal et al. (1992) argue that the word 'integrated' should go beyond traditional concepts and practices of coordination among water management agencies, the interaction between groundwater and surface water, or even a planning approach considering strategies and impacts. It needs to consider the carrying capacity rather than just planning, and it should encompass interactions instead of focusing on technicalities.

A great effort was put to monitor global progress on IWRM. The SDG indicator 6.5.1 is a composed index resulting from a combination of an assessment on four key dimensions of IWRM namely enabling environment, institutions and participation, management instruments and financing. These four components are measured on a scale of 0–100, using 33 questions in a self-assessed country questionnaire (Table 12.1). It

Table 12.1 Four key dimensions of the IWRM for SDG indicator 6.5.1

	1. Enabling environment	2. Institutions and participation	3. Management instruments	4. Financing
National level	• Policy • Law • Plans	• Authorities • Cross-sectoral coordination • Capacity • Public participation • Business participation • Gender objectives	• Availability monitoring • Water-use management • Pollution control • Ecosystem management • Disaster management	• Budget for investment • Budget for recurring costs
Subnational	Policy	Gender objectives	Data and information sharing	• Subnational or basin budget for investment • Revenues raised
Basin/aquifer/ local	Basin/aquifer management plans	• Basin/aquifer organizations • Local public participation	• Basin management Instruments • Aquifer management Instruments	
Federal countries only	Provincial water law	Provincial authorities		
Transboundary	Management arrangements	• Organizational arrangements • Gender objectives	Data and information sharing	Financing for cooperation

Source: Authors: Based on UN-Environment (2018, p. 10)

takes into account the diversity of users and uses of water, and data for the production of this indicator is collected through a questionnaire and responses are consolidated through consultations between relevant stakeholders.

For each question, the degree of IWRM implementation is ranked according to the following six thresholds: 0, 20, 40, 60, 80 and 100 (Table 12.2). Question scores in each dimension are averaged to give a dimension average for each of the four, rounded to the nearest whole number. The four key dimension averages are then averaged to calculate the final indicator 6.5.1 score for each country, on a scale of 0–100. The lines represent the different scale levels of water resources management, from national authorities to local ones, with the addition of a transboundary cross-national level (the last line).

Table 12.2 Thresholds and their meaning for the degree of IWRM implementation for SDG indicator 6.5.1

Degree of implementation (0–100)	General profile based on overall IWRM implementation
Very high (91–100)	The vast majority of IWRM elements are fully implemented; objectives consistently achieved; plans and programmes periodically reviewed and revised
High (71–90)	Policy objectives are consistently achieved. IWRM objectives of plans and programmes are generally met, and geographic coverage and stakeholder engagement are generally good
Medium-high (51–70)	Capacity to implement elements of IWRM is generally adequate. IWRM is being used by the majority of relevant authorities to guide their work, and elements are generally being implemented under long-term programmes
Medium-low (31–50)	Elements of IWRM are generally institutionalized, and implementation is underway. Plans are based on IWRM, approved by government and starting to be used by authorities to guide work
Low (11–30)	Implementation of elements of IWRM has generally begun, but with limited uptake across the country, and potentially low engagement of stakeholder groups. Plans may exist, but are not based on IWRM
Very low (0–10)	Development of elements of IWRM has generally not begun, or has stalled and is not progressing

Source: Based on UN-Environment (2018, pp. 12 and 16)

The SDG indicator 6.5.1 has highlighted the importance of financing to IWRM, and this differs from the traditional approach for IWRM, which used to consider the enabling environment; institutions and participation, which means institutional arrangements; and management instruments as the key process, and financing (the fourth column) was part of the enabling environment, as it could be seen at GWP IWRM Toolbox (Thalmeinerova et al. 2017).

Although SDG indicator 6.5.1 has been receiving much more attention, the issue related to SDG indicator 6.5.2 has been managed much earlier in world history. This started from local water basin simple agreements to an upper level and more sophisticated instruments. The Atlas of International Freshwater Agreements by United Nations Environment Programme and Oregon State University identifies 400 water agreements adopted since 1820 (UNEP/OSU 2002). The 1966 Helsinki Rules on the Uses of the Waters of International Rivers is considered a pioneering effort on rules governing international rivers. It was adopted by the International Law Association (ILA) but although it had no formal status, it inspired the creation of the Convention on the Law of Non-Navigational Uses of International Watercourses, adopted by the United Nations on 21 May 1997. In Europe, the UNECE Convention on the Protection and Use of Transboundary Watercourses and International Lakes (Water Convention) was adopted on 17 March 1992 and entered into force on the 6 October 1996. It has inspired many agreements, most notably the 1994 Convention on the Cooperation for the Protection and Sustainable Use of the Danube River. The regional success of the UNECE Water Convention led its opening up for accession by all United Nations member states.

It is also important to note that transboundary waters were characterized by conflicts. The drainage structure would lead to upstream countries affecting downstream territories belonging to other countries (Gottgens et al. 1998; Valbo-Jørgensen et al. 2008). Dams could promote changes in flow patterns of a river, affecting the sediment transportation and flooding pulses. Waterways downstream could affect upstream morphology. Environmental impacts could encompass decrease in land-use productivity due to loss of natural sediment depositions or lack of water for irrigation. Biodiversity can be affected leading to loss of fish productivity (Bakker 2009; Rivera 2015; Villar 2016; Petersen-Perlman et al. 2017).

Achieving transboundary agreement around principles[1] of equitable and reasonable utilization and participation, obligation not to cause significant harm, general obligation to cooperate, regular exchange of data and information, and notification concerning planned measures with possible adverse effects is a long and hard journey: each situation is unique, which means there is no single path. It demands commitment from all governments and political will, at all levels (UN-Water 2008). Cooperation may be reached by local practices apart from the national level state agreements, and geopolitics and regional history can influence these processes.

Differences on state economic levels, scientific and technological infrastructure, threshold values for water quality and identification of contamination can influence the advancement of agreements (Brels et al. 2008; Petersen-Perlman and Fischhendler 2018). The observed experiences and case studies on the governance of transboundary waters inform the general preferences for models based on the simple coordination of actions, guided by the domestic regulatory framework of involved countries. According to Souza et al. (2014), shared management is not adopted and countries seek to establish institutional formats such as mixed commissions that do not harm their sovereignty over water resources. To measure the progress of cross-border cooperation in accordance with goal 6.5, indicator 6.5.2 was adopted. The indicator is defined as the 'percentage of the area of the transboundary basin with an operational agreement for cross-border cooperation' as follows:

$$\text{SDG indicator } 6.5.2(\%) = \left[(\text{A}+\text{C})/(\text{B}+\text{D})\right]*100$$

where [A] is the total surface area of transboundary basins/sub-basins of rivers and lakes covered by operational arrangements within the territory of the country in km^2; [B] is the total surface area of transboundary basins of rivers and lakes within the territory of the country in km^2; [C] is the total surface area of transboundary aquifers covered by operational arrangements within the territory of a country in km^2; and [D] is the total surface area of transboundary aquifers within the territory of a country in km^2 (UNECE and UNESCO 2018, p. 19).

SDG indicator 6.5.2 is a starting point for the diagnosis that includes mapping the number of transboundary waters (superficial water basins and groundwater) to consider the existence of shared waters and the lack of cooperative processes. This can seem easy but much of the time, policymakers and diplomats do not know about it. The next section presents the research design.

3 Research Design

This work is based on data gathered from official documents. The documents include the country reports to the United Nations regarding SDG progresses, and the United Nations reports on SDG target 6.5 implementation. A report published by the United Nations Environment Programme presented the global baseline for SDG 6 indicator 6.5.1 (UN-Environment 2018) with the overall degree of IWRM implementation. This report analyses how 172 countries put in place IWRM initiatives based on data supplied by countries. This global report did not contain data for all the countries, and missing data was searched in country reports or other internal documents. For example, Uruguay produced its report much later. Additionally, some countries presented divergent data.

[1]Convention on the Law of the Non-Navigational Uses of International Watercourses, UN General Assembly Resolution A/RES/51/229, 8 July 1997, https://www.un.org/ga/search/view_doc.asp?symbol=A/res/51/229

Aiming to answer questions seeking to identify the region's weaknesses, disaggregated data for the SDG indicator 6.5.1 from each country were sought for the decomposed indicator addressing the following key pillars: enabling environment, institutional arrangements, management instruments and financing. The same was done for SDG indicator 6.5.2. In this case, the UN Progress on Transboundary Water Cooperation: Global baseline for SDG indicator 6.5.2 (UNECE and UNESCO 2018) was the first step of the data research. Collected data were organized and presented in tables and maps to help in quick visualization and countries' comparison. The nature of data and the quality of report concerning scales and hierarchy of water basins were analysed to complement diagnosis and produce recommendations.

4 Main Findings and Discussions

4.1 Addressing SDG 6.5.1 in South America

The UN-Environment (2018) identifies a delicate situation which reinforces the decision of GWP-SAM to dedicate more efforts on the attainment of SDG 6.5.1. Furthermore, Venezuela and Uruguay did not present data for the 2018 global report. Drawing from the countries which reported, the overall situation in South America is as follows (Table 12.3): medium level of IWRM implementation SDG indicator 6.5.1 score (Brazil), medium-low level of IWRM implementation SDG indicator 6.5.1 score (Argentina, Colombia) and low level of IWRM implementation SDG indicator 6.5.1 score (Chile, Peru).

The countries also presented different data sets. For example, Brazil had certain disaggregated data (dimension 1-60, dimension 2-64, dimension 3-44, dimension 4-34, final score 51) in the annex of the UN-Environment Programme Report about the global baseline for SDG 6 indicator 6.5.1 (UN-Environment 2018) and in another document produced by the National Agency of Water (ANA 2019) they present a different disaggregated data (dimension 1-80, dimension 2-55, dimension 3-40, dimension 4-40, final score 53.8).

These results were anticipated by GWP-SAM as an initial evaluation of the regional weakness 1 year before. However, where are the region's weaknesses? What are the disaggregated data for the SDG indicator 6.5.1?

Countries did not present the same level of information in their National Voluntary Reviews to the United Nations. Even this categorization by the United Nations should be taken with caution because the difference between Brazil and Colombia is rather insignificant, yet they were put in different profiles of IWRM implementation. If each component is analysed separately, it is possible to see that Colombia reached medium-high degree of implementation in three components and Brazil only got the same degree of implementation in two components, but with higher scores. The IWRM implementation level is presented for each of its components separately in Fig. 12.1.

Table 12.3 Decomposition of SDG indicator 6.5.1 for each Country Water Partnership of GWP

	1: Enabling environment	2: Institutional arrangements	3: Management instruments	4: Financing	Final score	IWRM implementation SDG 6.5.1 level
Argentina	39	48	34	32	38	Medium-low
Brazil	60	64	44	34	51	Medium-high
Chile	18	26	19	28	23	Low
Colombia	55	55	53	38	50	Medium-low
Peru	34	26	34	24	30	Low
Uruguay	40	42	36	19	34	Medium-low
Venezuela	No data	No data	No data	No data	No data	No evaluation

Source: Authors, based on UN-Environment (2018) and Uruguay (2018)

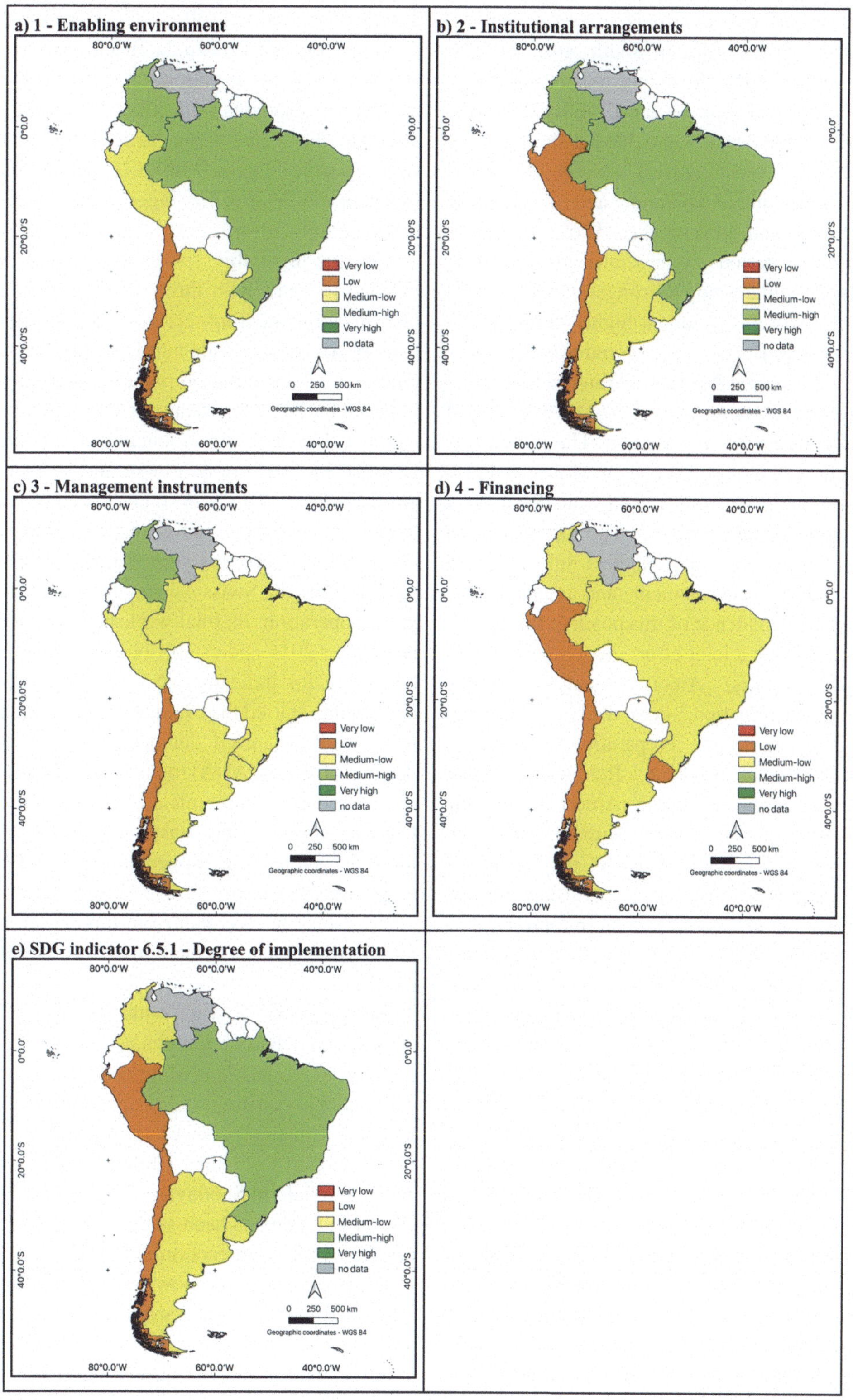

Fig. 12.1 Map of GWP-South America countries with their degree of implementation for each component of SDG indicator 6.5.1. (**a**) Enabling environment. (**b**) Institutional arrangements. (**c**) Management instruments. (**d**) Financing. (**e**) SDG indicator 6.5.1: Degree of implementation

The results show that financing scores are the lowest for all countries in South America. The low level of IWRM financing means that these countries, which were so-called underdeveloped countries some decades ago, have a short history of economic growth to invest in water and sanitation infrastructure and planning and management instruments and governance. Those countries with a longer period of democratic governance in sequence could develop a stronger enabling environment, with a consequent higher sub-score in this issue (see Brazil's case scored 80). The existence of a National Agency of Water in this country reinforces its commitment to SDG 6 in general and SDG target 6.5 in particular. The joint effort to create a National Agency of Water in Peru, from Brazilian experience, may put Peru on a good path to attain SDG 6.5. The fact that these countries presented their highest sub-scores in both enabling environment and management instrument is evidence of this positive future.

The GWP, in a joint effort involving its South America, Central America and Caribbean regional coordination, organized a Technical Workshop on Project Preparation focused on Transformational Climate Resilience Water Project Concepts in Latin America and the Caribbean for the Green Climate Fund, on September 3–5, 2019, at Panama City. The aim of the workshop was to prepare governmental and non-governmental organizations in these regions to increase fundraising and strengthen the financing dimension of IWRM, which was the weakest dimension of all South American countries in their SDG indicator 6.5.1. Nine countries from South America were present at the workshop, namely Colombia, Ecuador, Peru, Brazil, Bolivia, Paraguay, Uruguay, Chile and Argentina.

What then can be done in the region to mainstream SDG 6 in general and SDG target 6.5 in particular? GWP-SAM had focused on (1) countries, which had difficulties to send data for global report because it is important to engage them in a global path, and (2) helping countries to improve their IWRM performance; that is, countries with lower scores of IWRM should move to the next level of performance. A series of workshops on IWRM supported by GWP IWRM Toolbox were held as follows: 2015 in Brazil and 2016 in Chile and Venezuela. In addition, a workshop to design local policies on integrated management of droughts took place in Argentina in 2016. In this context, the next step demanded some specific targets, with at least one country focusing on a specific lagging situation that deserved more efforts.

Considering this scenario, GWP-SAM decided to work with three countries that were late in implementing certain IWRM aspects, namely Argentina, Chile and Uruguay. The three countries received the preparatory workshop for baseline for SDG 6 indicator 6.5.1: Argentina on October 19, 2017, in Buenos Aires; Chile on October 30, 2017, in Santiago; and Uruguay on April 18, 2018. This was so because other countries were already establishing their baselines. For example, Peru had started in October 2016 supported by the Swiss Agency for Development and Cooperation. Its final workshop was done in November 2016 and established an overall baseline score for indicator 6.5.1 of 30/100. Brazil had already started the evaluation and the country has a well-organized database and a National Agency of Water (ANA) to take care of this issue. In a recent communication, the Brazilian Government developed baseline data for indicator 6.5.1 presenting movement in the IWRM score since 2010 from 44.2 to 53.8 in 2016 (ANA 2019). Nevertheless, data from this report is different from that presented in UN-Environment (2018).

Recent social conflict outbreaks in Chile in 2019 raised debates about the necessity to review their social model, their water code and also their own national constitution law in a more inclusive and social justice perspective. According to the Final Report of the Chilean Stakeholders Workshop held in Santiago on October 30, 2017, 'Water in Chile has been seen and treated by the private sector as an economic resource and most of the water rights are now private, making any integrated management hard to implement from the public sector'.

This same Chilean stakeholders workshop proposed interesting issues for global debate about SDG indicator 6.5.1. They argued in favour of the necessity to 'change from IWRM to Integrated Water Management (IWM). This meant considering the ecosystemic approach as a good and service and not as a resource, which implies an economic perception and implicates markets' (Chile 2017, p. 3).

4.2 Addressing SDG 6.5.2 in South America

The UN Progress on Transboundary Water Cooperation (UNECE and UNESCO 2018) shows a bad situation regarding progress towards SDG indicator 6.5.2 in South America. There are many countries that did not make their presentations and were excluded from the global report. There are also countries with reports and data that need further clarification. The overall performance in transboundary cooperation is low, and the situation for international groundwater cooperation is worse compared to artificial transboundary water basins. A set of regional data for SDG indicator 6.5.2 is presented in Table 12.4 and Fig. 12.2. The degree of implementation for each component is based on the range of scores presented earlier.

Similar to the work on the progress towards SDG indicator 6.5.1, the first GWP-SAM objective was to know where exactly are the region weaknesses related to transboundary cooperation and how cooperation is spatially distributed. What are the disaggregated data for the SDG indicator 6.5.2 (artificial transboundary waters or international groundwaters)? The first step was to obtain the data from each country for the decomposed indicator. It emerged that there is weak performance regarding the attainment of SDG indicator 6.5.2 for all South America compared to SDG indicator 6.5.1. There is a lack of data from country reports in general, and the aquifers component is the weakest part compared to surface water bodies component.

This SDG indicator 6.5.2 has an additional difficulty compared to SDG indicator 6.5.1. Country reports could present conflicts or differences in data interpretation. For example, Brazil reported that the Guarani Aquifer (shared by Brazil, Argentina, Paraguay and Uruguay) was an agreement and this raised the country score. However, Uruguay indicated that there was no agreement to manage it as a shared aquifer, and this lowered its national score. For surface water bodies, Maroni is presented as a shared water basin among Brazil, Suriname and French Guiana by UNEP/OSU (2002), but not by Brazil (ANA 2019). In this regard, it can be concluded that Paraguay has no Country Water Partnership connected to GWP, although there are some isolated institutional members. This therefore presents an opportunity to reinforce positions by organizing the countries to get the status of a Country Water Partnership. Both Paraguay and Bolivia are strategic in this sense, because they are geographically located in a more central position in the continent and they have a very important role to play in transboundary cooperation.

The work also reveals that the main reference point adopted is the UNEP/OSU (2002), which has a list of transboundary water basins as shown in Table 12.5. However, there are a list of other transboundary water basins in national reports,

Table 12.4 The set of South American SDG indicator 6.5.2

	Rivers and lakes component (%)	Aquifers component (%)	SDG indicator 6.5.2 final score (%)	Degree of implementation
Argentina	No data	No data	No data	No data
Brazil	98.2	0	62.4	Medium-high
Chile	0	0	0	Very low
Colombia	1.1	No data	No data	No data
Peru	14.1	No data	No data	No data
Uruguay	100.0	0	54.7	Medium-high
Venezuela	7.0	0	3.5	Very low

Source: Authors, based on UNECE and UNESCO (2018) and Uruguay (2018).

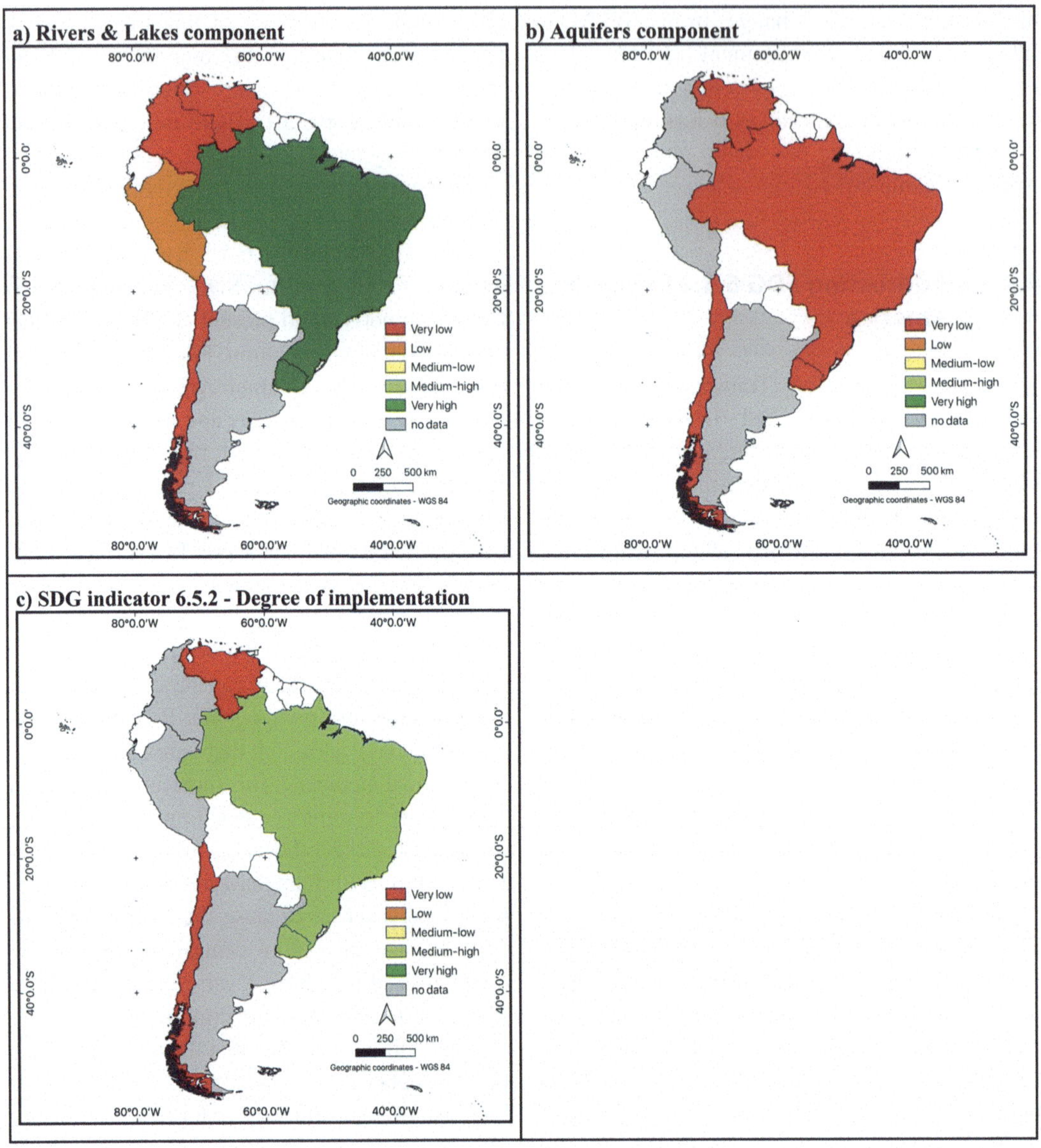

Fig. 12.2 Latin America and the Caribbean cooperation on transboundary water, river and lake basins, and aquifers, and countries where further clarification is still needed. (**a**) Rivers and lakes component. (**b**) Aquifers component. (**c**) SDG indicator 6.5.2: Degree of implementation. Source: Authors, based on UNECE and UNESCO (2018, p. 41)

which demands careful attention. This is so because these basins cannot simply be part of larger water basins. Thus, the main difficulty is related to the scale of analysis and reporting. National reports consider sub-basins as transboundary water basins due to a local/national view. For example, River Plate basin can be divided into Paraguay, Parana and Uruguay basin. Yet in some reports (Wolf et al. 1999; UNEP/OSU 2002), River Plate basin is the transboundary water basin to be considered, while in others like Brazil's report (ANA 2019), what appears is each of the sub-basin. Another example concerns the Morona and Pastaza water basins. They belong to the Amazon water basin, although they are reported as transboundary water basin with

Table 12.5 Transboundary water basins according to UNEP/OSU (2002)

Water basin name	Countries involved
Amacuro	Venezuela, Guyana
Amazon	Brazil, Bolivia, Colombia, Guyana, Peru, Venezuela, Ecuador
Aviles	Argentina, Chile
Aysén	Argentina, Chile
Baker	Argentina, Chile
Barima	Venezuela, Guyana
Cancoso/Lauca	Bolivia, Chile
Carmen Silva/ Chico	Argentina, Chile
Catamayo-Chira	Peru, Ecuador
Catatumbo	Colombia, Venezuela
Chuy	Brazil, Uruguay
Comau	Argentina, Chile
Corantijn/ Courantyne	Guyana, Suriname
Cullen	Argentina, Chile
Essequibo	Venezuela, Guyana, Suriname
Gallegos-Chico	Argentina, Chile
Jurado	Colombia, Panama
La Plata	Argentina, Bolivia, Brazil, Paraguay, Uruguay
Lagoon Mirim-Lagoon Patos	Brazil, Uruguay
Laguna Filaret	Argentina, Chile
Lake Fagnano	Argentina, Chile
Lake Titicaca-Poopo System	Chile, Peru, Bolivia
Maroni	Brazil, Suriname, French Guiana
Mataje	Colombia, Ecuador
Mira	Colombia, Ecuador
Oiapoque/ Oyupock	Brazil, French Guiana
Orinoco	Brazil, Colombia, Venezuela
Palena	Argentina, Chile
Pascua	Argentina, Chile
Patia	Colombia, Ecuador
Puelo	Argentina, Chile
Rio Grande (South America)	Argentina, Chile
San Martin	Argentina, Chile
Seno Union/ Serrano	Argentina, Chile
Tumbes-Poyango	Peru, Ecuador
Valdivia	Argentina, Chile
Yelcho	Argentina, Chile
Zapaleri	Argentina, Chile, Bolivia
Zarumilla	Peru, Ecuador

their own agreements.[2] All these problems emerge due to the procedure of accounting to the SDG indicator 6.5.2 in the country reports and how the countries perceive the existence of transboundary water basin. The local perspective guiding national reports is responsible for some ruptures of the hierarchy of water basin structures. For example, the Quaraí basin is part of the Uruguay basin, but sometimes it is mentioned as an additional transboundary water basin. Hence, there is a mix of different hierarchical watercourses (ANA 2019). Agreements can exist in different sub-basins involving different countries, and an absence of agreement of the entire large basin can lead to different regional accountability of transboundary water agreements. Villar et al. (2018) argue in favour of the necessity of multi-scale and multi-actor perspective as transboundary water conflicts are not limited exclusively to interstate conflicts. Hence there should be better understanding of the development process of state decision-making or its influences on intranational power dynamics.

Another problem relates to contradictory information. For example, the Essequibo water basin (ANA 2019) is not mentioned as part of Brazilian territory, yet as per UNEP (2007), this water basin is shared by Brazil, Guyana, Suriname and Venezuela. This situation can cause confusion to readers because UNEP (2007) did not consider that the frontier between Brazil and Guyana is based on the drainage divide of the Amazon basin (for the Brazilian side) and the Essequibo and Corentyne basins (for the Guyana side).

There are also problems related to water transfer canals. For example, Canal Mauri is a canal shared by Peru and Chile, which transfers water from the Mauri River watershed (which is a tributary of Desaguadero) into the Caplina River for irrigation purposes. This canal replaced the for-

[2] https://aplicaciones.senagua.gob.ec/reslotaip2017/rendcuent/PLANTA%20CENTRAL/FASE%201/RESPALDOS/GESTIÓN/Respaldos%20Gestión%20Social/Respaldo%20gestión%20social/Acuerdo_CB_GIRH_Cuencas_Transfronterizas.pdf

mer Canal Uchusuma built in 1820. Canal Mauri is also considered a transbasin diversion, which transports water from the Lake Titicaca watershed to the Pacific slope.

Transboundary agreements also depend on the absence of territorial dispute between countries. In South America, there is a case of absence of transboundary water basin agreement, which depends on a previous definition of territory limits such as in the case of the Essequibo water basin. The Guayana Essequibo case is a territorial dispute dating back 200 years between Venezuela on one side and the British Empire and Guyana on the other. Venezuela claims territory up to the western bank of the Essequibo River. According to Nikolić (2018), this is an example of flagrant breaking of *uti possidetis* principle of international law and the author believes that the British acknowledgment of the borderlines in 1825 will probably reinforce legal arguments on the side of Caracas in this dispute. This type of historical background context cannot be captured in numbers and indexes of SDG 6.5.2 implementation.

The continuous effort to organize data and mainstream transboundary cooperation among South American countries has gained momentum after the initiatives of the Peru Government (the Ministry of Foreign Affairs, in coordination with the Ministry of Agriculture and Irrigation, through the National Water Authority of Peru) and joint efforts of different international organizations including GWP-South America, to lead a regional process on this issue. There was an International Symposium on Transboundary Waters in Latin America, held in Lima, Peru, from November 8th to 9th, 2018. The aim was to promote effective water governance and implementation of actions to guide and facilitate cooperation through the strengthening of regional capacities. Several questions were posed for debate such as the following: (1) How to define if the agreements are operational or not? (2) What should be the role of the foreign ministries in the negotiation/promotion of the agreements, the necessity to leverage countries' knowledge and expertise, and the resources and networks to identify key issues related to SDG 6.5? (3) What is the region's take on the necessity to widen transboundary water debates to other knowledge domains such as ichthyology, climatology and anthropology? (4) What is the importance of considering management plans outside the water basins? The external influences may threaten water security in the Pantanal region due to atmospheric flows known as low-level jets (Vernekar et al. 2003) associated with deforestation and land-use change processes in the Amazon biome (which is outside Paraguay water basin where the Pantanal is situated) (Bergier et al. 2018; Schulz et al. 2019).

These issues were proposed to develop a systemic integrated view of transboundary water cooperation to go beyond the past activities more focused on capacity building in International Water Law, with a total of ten courses or workshops from 2014 to 2018 in the continent promoted on behalf of GWP-SAM. This initiative is remarkable considering Hussein et al. (2018)'s proposal to include informal, formal and technical talks as a preoperational arrangement phase of the progress in SDG indicator 6.5.2. The authors argue that reports demonstrated how nonformal activities such as technical meetings have contributed to develop water cooperation and effectively resulted in the birth of water agreements. To this end, there were meetings that took place in 2019 including the XVII Brazilian Congress of Limnology and the Second Ibero-American Congress of Limnology that took place in Florianopolis, Brazil, from August 4 to 9, 2019, organized by Brazilian Association of Limnology (ABLimno). The XXIII Brazilian Congress of Water Resources took place in Foz do Iguaçu, Brazil, from November 24 to 28, 2019, and was organized by Brazilian Association of Water Resources (ABRH). Both meetings were supported by the Global Water Partnership-Brazil and ANA. The first event focused on an ecological approach with scientific basis for transboundary water in South America and the second event focused on a broad management approach. Furthermore, GWP-SAM was present at the Convention of the Protection and Use of Transboundary Watercourses and International Lakes Task Force on Water and Climate—tenth meeting (Geneva, Switzerland, May 2019).

All these initiatives represented an increase on the regional capacity building and they reinforced the importance to generate intersectoral dialogues within national governments and with other governments, mainly in the case of transboundary waters. This is completely aligned with the International Decade for Action 'Water for Sustainable Development' (2018–2028) framework, which demands support on knowledge generation and dissemination to reach the decade objectives.

One of the main findings from the mentioned events relates to less attention given to aquifers. McCrackena and Meyer (2018) inform of a gap between transboundary surface water and groundwater when they compared operational cooperation of waters worldwide. The authors identify the lack of formal and operational cooperation over shared groundwater as resulting from the limited scientific and technical knowledge regarding their location, extent and other physical characteristics. Thus, the organization of formal database is essential to improve regional performances on transboundary groundwater agreements and thus on the SDG indicator 6.5.2. There is also a necessity to build the infrastructure to collect such data in many low- and middle-income countries as one of the greatest challenges to walk on the 2030 AfSD path (Nature 2020). Due to this weakness, the second workshop of the Intergovernmental Council of the International Hydrological Programme (IHP) of UNESCO to follow up indicator 6.5.2 of the SDGs was taken in Montevideo, Uruguay (June 2018). The focus was on the 'aquifers' component. A list of aquifers in South America is presented in Table 12.6.

The absence of well-organized databases became a central point to a better coordination of IWRM, including transboundary water basin management. Thus, the ultimate regional goal related to this issue is to organize a multiscale GIS database platform containing transboundary rivers and lakes as much as transboundary aquifers. This means the region will encompass both SDG indicators from target 6.5. Such a database should include watershed shapefiles in different hierarchical scales, territory extension, existence

Table 12.6 List of aquifers in South America

Aquifer code (UNESCO 2009)	Aquifer name	Countries involved
1S	Choco-Darién	Colombia-Panama
2S	Táchira-Pamplonita	Colombia-Venezuela
3S	La Guajira	
4S	Grupo Roraima	Brazil-Guyana-Venezuela
5S	Boa Vista-Serra do Tucano-North Savanna	Brazil-Guyana
6S	Zanderij	Guyana-Suriname
7S	Coesewijne	
8S	A-Sand/B-Sand	
9S	Costeiro	Brazil-Guayana Francesa
10S	Tulcán-Ipiales	Colombia-Ecuador
11S	Zarumilla	Ecuador-Peru
12S	Puyango-Tumbes-Catamayo-Chira	
13S	Amazonas	Bolivia-Brazil-Colombia-Ecuador-Peru--Venezuela
14S	Titicaca	Bolivia-Peru
15S	Pantanal	Bolivia-Brazil-Paraguay
16S	Agua Dulce	Bolivia-Paraguay
17S	Ollagüe-Pastos Grandes	Bolivia-Chile
18S	Concordia/Escritos-Caplina	Chile-Peru
19S	Aquidauana-Aquidabán	Brazil-Paraguay
20S	Caiuá/Bauru-Acaray	
21S	Guaraní	Argentina-Brazil-Paraguay-Uruguay
22S	Serra Geral	
23S	Litoráneo-Chuy	Brazil-Uruguay
24S	Permo-Carbonífero	
25S	Litoral Cretácico	Argentina-Uruguay
26S	Salto-Salto Chico	
27S	Puneños	Argentina-Bolivia
28S	Yrendá-Toba-Tarijeño	Argentina-Bolivia-Paraguay
29S	El Cóndor-Cañadón del Cóndor	Argentina-Chile

Source: Authors, based on UNESCO (2009, p. 95)

of agreements and water management plans, and additional layers such as country limits. The multiscale data storage should start from local/national point of view to a widened perspective to management instruments. This is an independent effort to coordinate regional actions because we cannot count on the country's voluntary report only and information is the basic source of rational mobilization.

This is part of the priority issues established by GWP-SAM Strategic Plan. The following aspects are proposed to be covered: water solutions for the Sustainable Development Goals, climate resilience through water, transboundary water cooperation and necessity to develop knowledge and capacity building (mainly supported on GWP IWRM Toolbox), all of them guided by the commitment to combat social inequalities and value the cultural and social diversity (Saito 2019). It is expected that through this, GWP-SAM can increasingly help and support regional and national processes, based on participatory approach. This approach leads to an empowered vision of national and provincial stakeholders involved in IWRM, through the registration of the challenges and potential solutions to be implemented in the field of water governance.

5 Conclusions

The work shows that countries in South America have different backgrounds and data management systems regarding SDG target 6.5 requirements. Transboundary water presents several challenges in the region, including data gathering and organization. Financing is also a key issue to implement SDG target 6.5. Institutional arrangements, enabling environment and management instruments are also different. The strength of GWP-SAM in the region is based on a set of efforts to collect and organize data, and develop independent initiatives to improve capabilities by developing knowledge and scientific literacy mainly supported on GWP IWRM Toolbox. These initiatives were performed by a coordinated work involving GWP-SAM Regional Coordination, GWP Country Water Partnerships and local GWP partners and members to advance SDG 6.5 in their own countries. GWP-SAM has also been increasingly supporting regional and national processes based on participatory approaches. These approaches lead to an empowered vision of national and provincial stakeholders involved in IWRM. The challenges and potential solutions are identified and these assist in improving implementation of strategies aimed at water governance and attainment of SDG target 6.5 by 2030. GWP-SAM intends to serve as a guiding instrument for water policy in general, allowing the SDG target 6.5 to be reached correctly and progressively in South American region.

References

ANA (2019). ODS 6 no Brasil: visão da ANA sobre os indicadores. Brasília: Agência Nacional de Águas.

Bakker, M.H.N. (2009). Transboundary river floods: examining countries, international river basins and continents. *Water Policy, 11*(3), 269-288.

Bergier, I., Assine, M.L., McGlue, M.M., Alho, C.J.R., Silva, A., Guerreiro, R.L., Carvalho, J.C. (2018). Amazon rainforest modulation of water security in the Pantanal wetland. *Science of the Total Environment, 619-620*, 1116-1125. https://doi.org/10.1016/j.scitotenv.2017.11.163

Brels, S., Coates, D., Loures, F. (2008). *Transboundary water resources management: the role of international watercourse agreements in implementation of the CBD.* CBD Technical Series n. 40. Montreal, Canada: Secretariat of the Convention on Biological Diversity.

Gottgens, J.F., Fortney, R.H., Meyer, J., Perry, J.E., Rood, B.E. (1998). The case of the Paraguay-Paraná waterway ("Hidrovia") and its impact on the Pantanal of Brazil: a summary report to the society of wetlands scientists. *Wetlands Bulletin (The Society of Wetlands Scientists), September*,12-18.

Hussein, H., Menga, F., Greco, F. (2018). Monitoring Transboundary Water Cooperation in SDG 6.5.2: How a Critical Hydropolitics Approach Can Spot Inequitable Outcomes. *Sustainability, 10*(10), 3640. https://doi.org/10.3390/su10103640

Koudstaal, R., Rijsberman, F.R., Savenije, H. (1992). Water and sustainable development. *Natural Resources Forum, 16*(4), 277-290. https://doi.org/10.1111/j.1477-8947.1992.tb00859.x

Lautze, J., & Manthrithilake, H. (2012). Water security: old concepts, new package, what value?

Natural Resources Forum, 36(2), 76-87. https://doi.org/10.1111/j.1477-8947.2012.01448.x

McCrackena M., & Meyer, C. (2018). Monitoring of transboundary water cooperation: Review of Sustainable Development Goal Indicator 6.5.2 methodology. *Journal of Hydrology, 563*, 1-12 https://doi.org/10.1016/j.jhydrol.2018.05.013

Nature (2020). Get the Sustainable Development Goals back on track (Editorial). *Nature, 577*(7788), 7-8. https://doi.org/10.1038/d41586-019-03907-4.

Nikolić, A. (2018). The Guayana Esequiba Conflict: Key historical frameworks and legal issues. *The Review of International Affairs (Belgrade, Serbia), LXIX*(1169), 70-82. https://www.diplomacy.bg.ac.rs/pdf/ria/2018/RIA-1169_2018.pdf. Accessed 24 Apr 2020.

Petersen-Perlman, J.D., Veilleux, J.C., Wolf, A.T. (2017). International water conflict and cooperation: challenges and opportunities. *Water International, 42*(2), 105-120. https://doi.org/10.1080/02508060.2017.1276041

Petersen-Perlman, J.D., & Fischhendler, I. (2018). The weakness of the strong: re-examining power in transboundary water dynamics. *International Environmental Agreements: Politics, Law and Economics, 18*(2), 275-294. https://doi.org/10.1007/s10784-018-9387-z

Rivera, A. (2015). Transboundary aquifers along the Canada-USA border: Science, policy and social issues. *Journal of Hydrology: Regional Studies, 4* (Part B), 623-643.

Saito, C.H. (2017). Quais seriam as Questões Globais que desafiam a Educação Ambiental? Para além do modismo, uma análise sistemática e uma visão sistêmica. *Revista Eletrônica do Mestrado em Educação Ambiental, v. Especial*, 4-24. https://doi.org/10.14295/remea.v0i0.7138

Saito, C.H. (2019). Global Water Partnership e as ideias-chave em sua nova Estratégia Global 2020-2025. Rega - *Revista de Gestão de Água da América Latina, 16*(e13), 1-10. https://doi.org/10.21168/rega.v16e13.

Schulz, C., Whitney, B.S., Rossetto, O.C., Neves, D.M., Crabb, L, Oliveira, E.C., et al. (2019). Physical, ecological and human dimensions of environmental change in Brazil's Pantanal wetland: Synthesis and research agenda. *Science of the Total Environment, 687*, 1011-1027. https://doi.org/10.1016/j.scitotenv.2019.06.023

Shah, T. (2016). *Increasing water security: the key to implementing the Sustainable Development Goals*. TEC Background Papers n.22. Stockholm: Global Water Partnership.

Snellen, W.B., & Schrevel, A. (2004). *IWRM, for sustainable use of water; 50 years of international experience with the concept of integrated water resources management*. Background document to the FAO/Netherlands Conference on Water for Food and Ecosystems. Alterra-report 1143. Wageningen: Alterra,. https://edepot.wur.nl/30428. Accessed 24 Apr 2020

Souza, M., Veloso, F.T., Santos, L.B., Caeiro, R.B.S. (2014). Governança de recursos comuns: bacias hidrográficas transfronteiriças. *Revista Brasileira de Política Internacional, 57*(2), 152-175. https://doi.org/10.1590/0034-7329201400309

Thalmeinerova, D., Fábrega, J. R., Guan, Y., Janusz-Pawletta, B., Kileshye-Onema, J. M., & Saito, C. H. (Eds.), (2017) IWRM Toolbox Teaching Manual. Stockholm: Global Water Partnership. http://www.gwp.org/globalassets/global/toolbox/references/iwrm_teaching_manual.pdf. Acessed 18 Nov 2019.

UN (1990). *New Delhi Statement, Global Consultation on Safe Water and Sanitation*. UN-GA, A/C.2/45/3, 11 October 1990. http://www.ielrc.org/content/e9005.pdf. Accessed 06 Aug 2020.

UN (1992). *The Dublin Statement on Water and Sustainable Development. International Conference on Water and the Environment*, January 31, 1992 in Dublin, Ireland. http://www.un-documents.net/h2o-dub.htm. Accessed 24 Apr 2020

UN-Environment (2018). *Progress on integrated water resources management - Global baseline for SDG 6 Indicator 6.5.1: degree of IWRM implementation*. Geneva, Switzerland: UNEP.

UN-Water (2008). *Transboundary Waters: Sharing Benefits, Sharing Responsibilities*. Thematic Paper. Zaragoza: United Nations Office to Support the International Decade for Action 'Water for Life' 2005-2015 (UN-IDfA).

UNECE & UNESCO (2018). *Progress on Transboundary Water Cooperation - Global baseline for SDG indicator 6.5.2*. Paris, France: United Nations and UNESCO.

UNEP (2007). *Vulnerabilidad y Resistencia Hidropolíticas en Aguas Internacionales: América Latina y el Caribe*. Nairobi, Kenya: UNEP Press.

UNEP/OSU. (2002). *Atlas of International Freshwater Agreements*. Nairobi, Kenya: UNEP Press.

UNESCO (2009). *Atlas of Transboundary Aquifers: Global Maps, Regional Cooperation, and Local Inventories*. Paris, France: UNESCO-International Hydrological Programme.

Uruguay. (2018). *Informe Voluntario Nacional - Uruguay 2018*. Montevideo, Uruguay: Presidencia de la República Oriental del Uruguay.

Valbo-Jørgensen, J., Marmulla, G., Welcomme, R.L. (2008). Migratory Fish Stocks in Transboundary Basins - Implications for Governance, Management and Research. In: V. Lagutov (Ed.) *Rescue of Sturgeon Species in the Ural River Basin. NATO Science for Peace and Security Series C: Environmental Security* (pp. 61-86). Basel: Springer Nature.

Vernekar, A.D., Kirtman, B.P., Fennessy, M.J. (2003). Low-Level Jets and Their Effects on the South American Summer Climate as Simulated by the NCEP Eta Model. *Journal of Climate, 16*(2), 297-311. https://doi.org/10.1175/1520-0442(2003)016%3C0297:LLJATE%3E2.0.CO;2

Villar, P.C. (2016). International cooperation on transboundary aquifers in South America and the Guarani Aquifer case. *Revista Brasileira de Política Internacional, 59*(1), e007. https://doi.org/10.1590/0034-7329201600107

Villar, P.C., Ribeiro, W.C., Sant'Anna, F.M. (2018). Transboundary governance in the La Plata River basin: status and prospects. *Water International, 43*(7): 978-995. https://doi.org/10.1080/02508060.2018.1490879

WEF. (2015). *Global Risks 2015, 10th edition*. http://www3.weforum.org/docs/WEF_Global_Risks_2015_Report15.pdf. Accessed 24 Apr 2020

Wolf, A.T., Natharius, J.A., Danielson, J.J., Ward, B.S., Pender, J.K. (1999). International River Basins of the World. *International Journal of Water Resources Development, 15*(4), 387-427. https://doi.org/10.1080/07900629948682

Sustainability Reporting Through UNGC at Unisa: Opportunities and Challenges in Mainstreaming GRI Standards and the SDGs

13

Godwell Nhamo

Abstract

As late as 2018, academic publications have still been recommending more research on sustainability reporting in higher education. This chapter focuses on how the University of South Africa (Unisa), an open-distance e-Learning institution, reports sustainability matters under the United Nations Global Compact (UNGC) and examines the opportunities and challenges in mainstreaming the Global Reporting Initiative (GRI) Standards and Sustainable Development Goals (SDGs). The study is premised on lived experiences and participatory action research (PAR). Findings reveal that Unisa has made progress in sustainability reporting. It further emerged that there is evidence of reporting on sustainability matters as reflected in both Unisa's annual and UNGC reports. However, with the advent of twin globally recognised sustainability mainstreaming and reporting frameworks such as the Sustainable Development Goals (SDGs), especially through the higher education and colleges SDG Accord and Global Reporting Initiative's (GRI) Sustainability Standards, Unisa is awakening to the need to mainstream these to enable both implementation and reporting. Hence, the latest UNGC report embeds the Unisa SDGs Localisation Declaration of November 2019. The chapter recommends a hybrid sustainability reporting model. The work further recommends that both the GRI Practical Guide and the Global Alliance's SDG Accord be used in localising the SDGs across Unisa to enhance implementation and reporting.

Keywords

Unisa · UN Global Compact · GRI Sustainability Standards · SDG Accord · SDGs

1 Introduction

The University of South Africa's (Unisa) understanding of sustainability implies "the long-term maintenance of responsibility, which has environmental, economic, and social dimensions, and encompasses the concept of stewardship and the responsible management of resource use" (Unisa 2012, p. 23). Unisa further emphasises that the spaces of sustainability and sustainable development should not be about environmental issues exclusively. It should also embrace the core institutional functions of teaching and learning as well as research and community engagement.

G. Nhamo (✉)
Exxaro Chair in Business and Climate Change, Institute for Corporate Citizenship, University of South Africa, Pretoria, South Africa
e-mail: nhamog@unisa.ac.za

G. Nhamo et al. (eds.), *Sustainable Development Goals for Society Vol. 1*, Sustainable Development Goals Series, https://doi.org/10.1007/978-3-030-70948-8_13

Given that environmental reporting has been neglected in many organisations as part of sustainability reporting for a long time, this chapter is biased towards tracing activities in this regard.

To report effectively in this generation, Unisa should be aware of other reporting platforms that have a bearing on its current framework based on the United Nations Global Compact (UNGC) to which it is a signatory. Unisa ratified the UNGC on 17 January 2007 and it is possibly the only institution that has done so in Africa. However, the emergence of the 2030 Agenda for Sustainable Development (AfSD) and its aligned 17 Sustainable Development Goals (SDGs) (United Nations 2015) has forced the entire world, including Unisa, to rethink sustainability reporting. The SDGs remain an expanded agenda from the Millennium Development Goals (MDGs) whose sole rallying point is not to leave anyone behind (Nhamo et al. 2018, 2019), with the aim to transform the world (United Nations 2015). Coming closer to home, the Global Alliance's 2017 SDG Accord presents another relevant platform for sustainability reporting by Unisa (Global Alliance 2017). The other set of reporting frameworks challenging Unisa's framework are the GRI's Sustainability Standards complemented by both the GRI's SDG Campus and the Practical Guide. The SDG Campus is a tool developed by the GRI in collaboration with the UNGC and the World Business Council for Sustainable Development (WBCSD) soon after the ratification of the SDGs (GRI 2016). This was followed by a further refinement by the GRI and UNGC that resulted in the Practical Guide (GRI 2018; UNGC 2018).

From the aforementioned, it becomes clearer that Unisa's sustainability reporting framework as informed by the UNGC is now inadequate as the UNGC template focuses on the ten principles (UNGC 2017). Zemanová and Druláková (2020a) and Mattera et al. (2020) highlight that since 2016 the UNGC has been aligning its work to the SDGs. However, Zemanová and Druláková (2020b, p. 1) note that "despite the enormous scholarly interest in the UNGC since the very beginning, its impact on the integration of the SDGs into the business activities, risk management and reporting of its participants remains understudied".

Given the foregoing, this chapter raises an objective to assess and recommend how Unisa, as a higher education business entity, can best mainstream other sustainability reporting frameworks, namely the SDG Accord, GRI Sustainability Standards and SDG Campus, into the UNGC. Alternatively, one may ask the question: Is the current UNGC reporting framework used by Unisa in its sustainability reporting adequate in the face of changing sustainability reporting in universities and colleges in the context of the 2030 AfSD and its 17 SDGs? The response will be clearly a big NO!

The rest of the chapter is organised as follows: the next section looks at the literature. This is followed by the methodology and the presentation and discussion of findings, with the last section being the conclusion.

2 A Literature Survey and Background

This literature review is based on the global sustainability reporting frameworks highlighted in the introduction. An effort will be made to discuss these frameworks with the view of identifying opportunities and challenges for integrating these frameworks into an already existing Unisa sustainability reporting framework based on the UNGC.

2.1 The UNGC and GRI Sustainability Reporting Frameworks

The UNGC was launched in 2000 with the objective of addressing sustainability and other social issues in company reporting. It advocates for the alignment of company operations and strategies with ten universally accepted principles in the areas of human rights, labour, environment and anti-corruption (UNGC 2017). The UNGC is now being viewed as another platform to act in delivering the SDGs. It is also seen as a leadership platform seeking the "development, implementation and disclosure of responsible corporate practices" (UNGC 2017, p. 36). It remains the biggest corporate sustainability initiative globally. It has been ratified by over 9500 companies, 3000 non-business signatories and 70 local networks in more than 160 countries (UNGC 2017). The finer details concerning the UNGC principles are reflected in Table 13.1.

The ten principles in Table 1 are informed by a range of global declarations that were in place then which include the Universal Declaration of Human Rights, the International Labour Organization's Declaration on Fundamental Principles and Rights at Work, the Rio Declaration on Environment and Development and the United Nations Convention Against Corruption (UNGC 2017). What is of interest is the fact that the UNGC principles are complemented by the SDGs. There is an alignment with a cluster of environmental SDGs (SDGs 12–15) for principles 7–9; matters of equity, peace and security (SDGs 10 and 16) for principles 1–2 and 10; and sustainable jobs (SDG 8) for principles 3–6. This provides an opportunity for institutions such as UNISA to mainstream SDGs (UNDG 2016) in UNGC-based reporting frameworks.

Table 13.1 The ten principles of the UNGC

Dimensions	Principles
Human rights	1. Businesses should support and respect the protection of internationally proclaimed human rights, and 2. make sure that they are not complicit in human rights abuses
Labour	3. Businesses should uphold the freedom of association and the effective recognition of the right to collective bargaining. 4. The elimination of all forms of forced and compulsory labour. 5. The effective abolition of child labour. 6. The elimination of discrimination in respect of employment and occupation
Environment	7. Businesses should support a precautionary approach to environmental challenges; 8. undertake initiatives to promote greater environmental responsibility; and 9. encourage the development and diffusion of environmentally friendly technologies
Anti-corruption	10. Businesses should work against corruption in all its forms, including extortion and bribery

Source: UNGC (2017, p. 36)

To bring the UNGC principles into alignment with the SDGs, the UNGC (2017) introduced a campaign called "Making Global Goals Local Business". This drive is informed by the fact that the SDGs are implemented at national and local levels, more than on a global scale. Since the launch of the SDGs in September 2015, local networks of the UNGC have been hosting activities on the ground. In South Africa, for example, there were a series of events to raise awareness of SDGs and to show local businesses how to engage their stakeholders to scale up implementation. Such engagements were also targeted at supporting nation building, transformation and social cohesion leading to enhanced movements towards the eradication of poverty (SDG 1), inequality (SDG 10) and unemployment (SDG 10). To facilitate further engagements with the SDGs from companies, the UNGC in partnership with the GRI, the WBCSD and other key stakeholders developed the SDG Campus, the Women's Empowerment Principles (WEP) Gender Gap Analysis Tool and the SDG Industry Matrix. In order to attain the ambitious 2030 AfSD, the UNGC put in place a portfolio of Action Platforms. The Action Platforms are embedded in the ten principles of the UNGC and each Action Platform convenes key stakeholders that include business, global compact local networks, civil society, governments, etc. The structure of the Action Platforms is presented in Fig. 13.1.

Under the Reporting SDGs Action Platform, there are twin dimensions, namely pathways to low-carbon and resilient development that link up with SDGs 7, 9 and 13 and decent work in global supply chains responding to SDGs 1, 3 and 5 (UNGC 2017). As for the Breakthrough Action Platform, there are also twin dimensions, namely health as everybody's business addressed

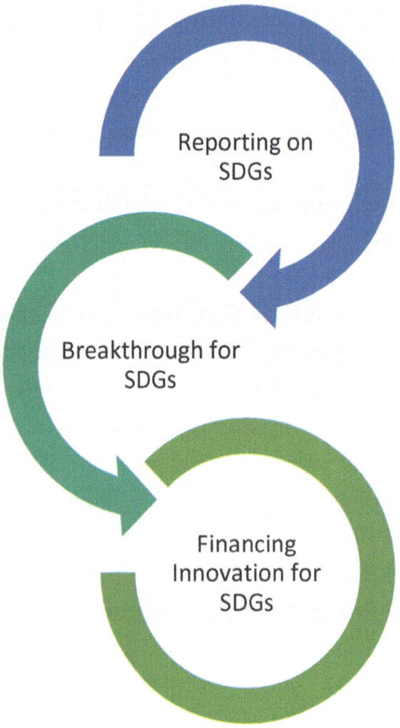

Fig. 13.1 UNGC action platforms. Source: Author, based on UNGC (2017, p. 21)

by SDGs 2–6, 8, 10 and 11 and water security through stewardship responding to SDGs 2, 3, 6 and 13. The last Action Platform dealing with financing innovation sits on other twin dimensions involving business for humanitarian action and peace linking up to SDGs 1, 3–6, 8, 10 and 16 and the justice and strong institutional aspects that speak to SDGs 1, 8–10, 16 and 17. What emerges from the explanations and parcelling out of the SDGs herein are overlaps and complements reflecting that the SDGs are intertwined and that they are best viewed from the one-in-all and all-in-one approach (Nhamo et al. 2018).

Another initiative from the UNGC is the Blueprint for Business Leadership on the SDGs (UNGC 2017). This initiative is designed to inspire every business regardless of its size to embark on leading action to support the attainment of the 2030 AfSD and its Global Goals. The common leadership qualities that include ambition, collaboration, accountability and consistency are applied in terms of business strategy, supply chains, partnerships and related matters.

In its 2018 progress report, the UNGC assessed how businesses were adopting the ten principles and evaluated actions in delivering the SDGs. It emerged that 80% of the companies surveyed were taking actions on the SDGs (UNGC 2018). The UNGC further reported that they had partnered with the GRI to encourage the business community to communicate their sustainability actions more effectively. To this end, there was a joint publication entitled "Integrating the Sustainable Development Goals into Corporate Reporting: A Practical Guide" (GRI 2018; UNGC 2018). This publication followed another earlier proclamation that was released in a document entitled "SDG Campus" carried out by the two organisations in conjunction with the WBCSD (GRI 2016). From the survey, the UNGC asked businesses to indicate SDG-related activities they were targeting. The findings are presented in Fig. 13.2.

From the 17 SDGs, the top 3 that businesses indicated to be involved in included SDG 3 (health), SDG 5 (gender) and SDG 8 (decent work). Those SDGs in which there were limited activities from businesses (bottom three) included SDG 14 (life in oceans), SDG 15 (life on land) and SDG 2 (poverty eradication).

As for the Practical Guide highlighted earlier, this has been aligned with the GRI Sustainability Standards and the United Nations Guiding Principles for Business and Human Rights (UNGC 2018). From the Practical Guide, there are three major steps indicated required by companies to act on SDGs that include defining priority SDG targets (Step 1), measuring and analysing (Step 2) and reporting, integrating and implementing (Step 3). The three steps are further divided into three sub-steps as shown in Fig. 13.3.

Although all steps in Fig. 13.3 are crucial, it is necessary to zero in briefly to check what the requirements for principled prioritisation are. The GRI (2018) and UNGC (2018) consider principled prioritisation as a process that includes the consideration of risks to people and the environment and beneficial SDG-related products, services and investments. These give us the first and second entry points in this compartment. The first entry point focuses on the contribution of a

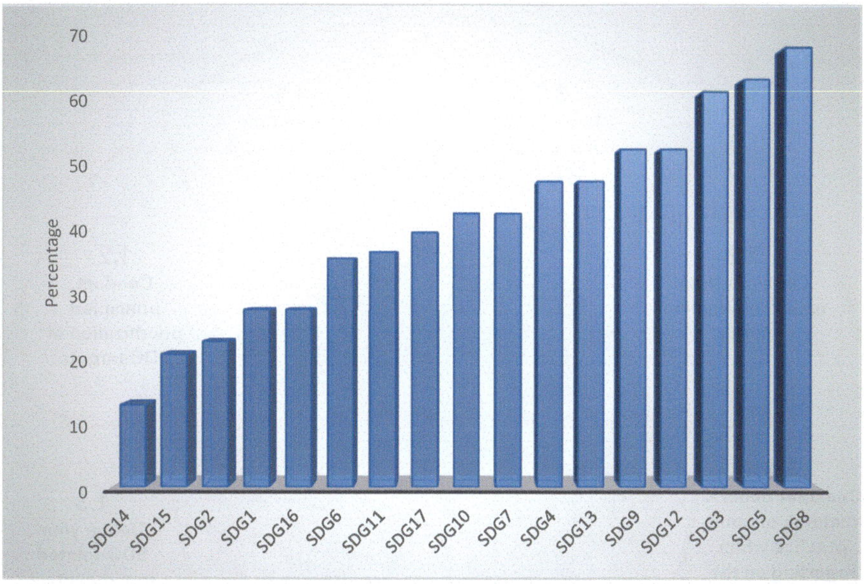

Fig. 13.2 Ranking in terms of targeted SDGs from corporate activities. Source: Author, based on UNGC (2018, p. 7)

business in achieving the 2030 AfSD through fulfilling its responsibility to address potential and actual negative impacts on people and the environment linked to business operations and value chains. The second entry point looks at additional contributions that can be made to achieve the SDGs through the application of their knowledge, skills and other capabilities that benefit people and the environment.

The GRI Sustainability Standards come in twin major thematic areas: (1) universal standards and (2) topic-specific standards (GRI 2016). There are three universal standards, namely the Foundation (GRI 101), which is the starting point for using the GRI Standards; the General Disclosure (GRI 102) that focuses on reporting the contextual information about an organisation; and the Management Approach (GRI 103) that looks at reporting the management approach for each material topic. The topic-specific standards are aligned and founded on the three pillars of sustainable development, namely social, economic and environmental pillars. The GRI 200 series addresses economic matters, the GRI 300 series focuses on the environment and the GRI 400 addresses the social dimensions of the sustainable development.

The GRI Standards further presents reporting principles in defining report content that include stakeholder inclusiveness, sustainability context, materiality and completeness (GRI 2016). In addition, there are also reporting principles for defining report quality borrowed from general accounting fundamentals. These report quality principles include accuracy, balance, clarity, comparability, reliability and timeliness.

2.2 The SDG Accord for Universities and Colleges

In 2017, the Global Alliance drew up the SDG Accord (SDGA) as a university and college sector response to the SDGs (Global Alliance 2017). In coming up with the SDGA, the stakeholders realised the unique role the university and college sector plays in capacitating billions of youths as agents of change. Hence, harnessing the energies brought about by the SDGs in IHE mandates of teaching and learning, research and development as well as community engagement was fundamental. In fact, among the 17 SDGs there is SDG 4, which is dedicated to "Ensure inclusive and equitable quality education and promote lifelong

Fig. 13.3 Steps to integrate SDGs into corporate reporting. Sources: GRI (2018) and UNGC (2018, p. 5)

learning opportunities for all" (United Nations 2015, p. 14). Given that all the SDGs are intertwined, investments in inclusive and equitable quality education will have spin-offs to improve health (SDG 3) and gender equality (SDG 5), which is also one of the targets under SDG 4.3: "By 2030, ensure equal access for all women and men to affordable and quality technical, vocational and tertiary education, including university" (United Nations 2015, p. 17); raise awareness on water, sanitation and hygiene (WASH) as enshrined in SDG 6; bring capacity to understand and address challenges of climate change (SDG 13); conserve life in the oceans (SDG 14); and ensure biodiversity (SDG 15). In fact, universities and colleges also train agriculture graduates that directly address issues pertaining to hunger and food security (SDG 2), develop entrepreneurs and engineers involved in the energy sector (SDG 7), design future sustainable cities (SDG 11) and work towards factories that result in sustainable consumption and production (SDG 12). Universities and colleges further prepare adaptable graduates for industry and society (SDG 9), thereby facilitating engagement as employees and/or entrepreneurs, both aspects that directly address matters pertaining to job creations (SDG 8). Lastly, lawyers and defence forces deal with matters of peace and security (SDG 16). As is emerging from the aforementioned, the education SDG may as well qualify as the greatest of all the SDGs.

Coming back to the SDGA, the Global Alliance (2017, p. 3) presents twin statements resembling the purpose of the Accord:

- First, the Accord is an opportunity to inspire, celebrate and advance the critical role that education has in delivering the SDGs and the value it brings to governments, businesses and wider society.
- Second, the Accord is a commitment learning institutions are making to one another: to do more to deliver the goals, to report on each signatory's progress annually and to do so in a way which shares the learning with others both nationally and internationally.

In terms of reporting, the education and college sector SDG reporting metrics are due for presentation annually during the United Nations High Level Political Forum (HLPF) in New York. The SDGA is open for signature from institutions, organisations, individuals and student organisations (Global Alliance 2017) and the details are presented in Box 13.1.

The fact that universities and colleges that are signatories to the SDGA must report annually remains a challenge—not only in terms of paperwork but also in terms of measuring significance of movement towards SDGs' implementation. For institutions such as Unisa that already report to the UNGC annually, this can bring about fatigue. However, if handled in a mainstreamed manner that integrates, localises and lands the SDGA into the UNGC framework, this can be a great opportunity to address multiple reporting platform through a single report—the expanded UNGC report.

At the time of writing this chapter, there had been two reporting cycles for the SDGA (Global Alliance 2018, 2019). It is clear from the SDGA 2019 Report that there has been significant progress in terms of drawing reporting baselines. The 2019 SDGA Report presents findings from a survey of 110 institutions regarding their readiness to engage the SDGA. From these, 51 institutions responded from 19 countries. There was action regarding climate change and education as these were considered priorities. However, positive developments and actions were also taking place across all 17 SDGs. The global commitment to

Box 13.1: The SDG Accord

Recognising that we are at different stages in our journey towards a sustainable future, with differing strategic priorities reflecting our strategic ambitions, areas of expertise and organisational cultures

This Accord calls upon us, the world's universities and colleges:

To embed the Sustainable Development Goals into our education, research, leadership, operations, administration and engagement activities.

We, the Accord Signatories, recognise:

- The indivisible and interconnected nature of the universal set of Goals—people, prosperity, planet, partnership and peace.
- That, as educators, we have a responsibility to play a central and transformational role in attaining the Sustainable Development Goals by 2030.

And we thereby assert:

That as leaders or individual practitioners, academics, students or researchers, we will:

- Align all major efforts with the Sustainable Development Goals, targets and indicators, including through our education, research, leadership, operational and engagement activities.
- Aim to involve members from all key stakeholder groups in this endeavour, including students, academics, professional staff, local communities and other external stakeholders.
- Collaborate across cities, regions, countries and continents with other signatory institutions as part of a collective international response.
- Using our own unique ways give an account to both local and global communities on our progress towards the Sustainable Development Goals, informing others and sharing our learning.
- Report on how our institution contributes to the Goals annually and what more can we do.

Source: Global Alliance (2017, p. 4)

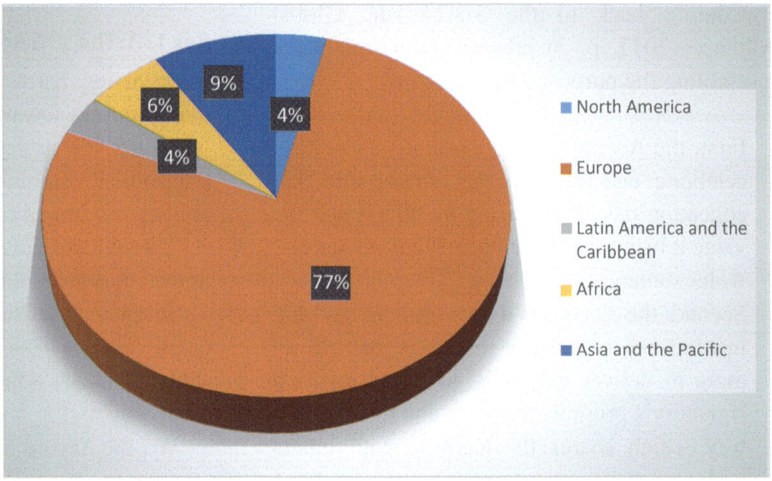

Fig. 13.4 Global distribution of SDGA signatories, 2019. Source: Author, data from the Global Alliance (2019, p. 5)

the SDGA has grown, with 1045 signatories in 85 countries (Global Alliance 2019), and the distribution is shown in Fig. 13.4. The lion's share of signatories (that includes endorsing partners, institutions and individuals) was from Europe with a total count of 808 signatories.

Among the key findings were that over 70% of SDGA members had mapped activities to the SDGs either entirely or partially (Global Alliance 2019). In addition, an estimated 62%, a 50% increase from 2018 reporting (Global Alliance 2018), had proclaimed their work on SDGs publicly. The major focus and areas of strength in SDG mainstreaming were in policy and strategy commitments from top management and/or leadership, staff and governance frameworks. Furthermore, 71% of the surveyed institutions were of the view that the SDGs remained a strategic framework, enabling dialogue with top management. The reporting period showed that "Over the last 12 months, the SDGs that institutions outlined they have the biggest impact on was SDG 4 Quality Education, SDG 3 Good Health and Wellbeing, and SDG 5 Gender Equality. When asked what they would be prioritising over the next 12 months, 72% said SDG 13 Climate Action, 70% said SDG 4 Quality Education and 58% said SDG 3 Good Health and Wellbeing" (Global Alliance 2019, p. 7). A question regarding the priority SDGs in the 2020 reporting period was also raised, and the findings are reflected in Fig. 13.5.

What is of interest are parallels between the SDGA and the UNGC activities regarding the Global Goals. SDGs 2, 14 and 15 are found at the bottom of the ladders of both initiatives. This is worrying because SDGs 14 and 15 are in the traditionally neglected environmental sustainability space (Nhamo 2014). Furthermore, SDG 3 finds favour in the top three of both lists. However, whereas the SDGA has SDG 13 as its top priority followed by SDG 4, the UNGC has SDG 8 followed by SDG 5.

The 2019 Report further considered challenges that universities and colleges were facing in dealing with SDGs through the SDGA and possible next moves for the next reporting period. Among the challenges as ranked from least to highest are the lack of guidance from the United Nations, no adequate practice to learn from, a lack of guidance from own governments, a lack of engagement from students, a lack of engagement from departments, limited understanding of the SDGs within the institution, a lack of funding and a lack of staff capacity (Global Alliance 2019). As for the next steps in the 2020 reporting period, the following matters have been ranked from least to highest priority: investing funds into SDG activities, sharing learning on SDGs, increasing research on SDGs, planning to increase partnerships working on SDGs, integrating SDGs into the campuses and operations, mapping activities currently taking place on SDGs (baselines), creating activities to raising awareness of SDGs across the entire institution, incorporating SDGs at a strategic level and

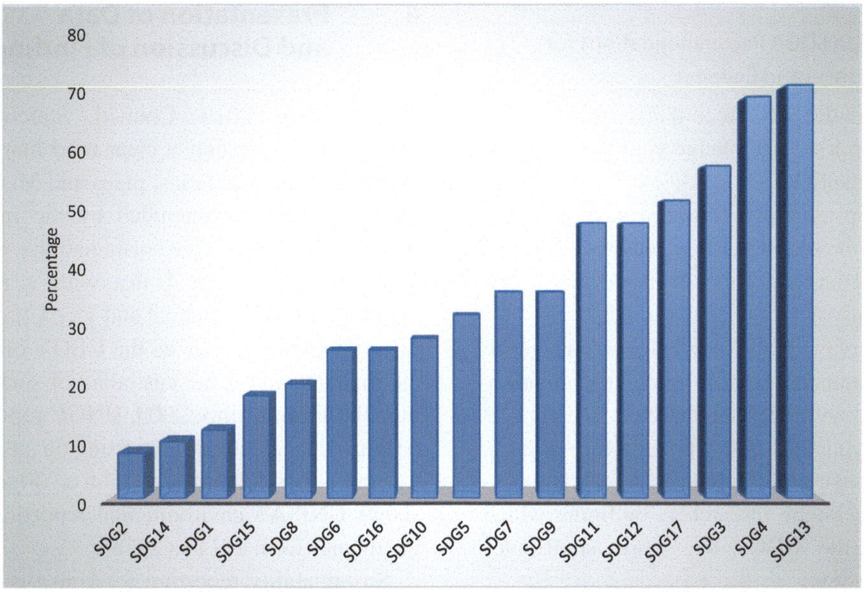

Fig. 13.5 SDGs to be prioritised in the 2020 reporting period. Source: Author, based on Global Alliance (2019, p. 7)

engaging students on SDGs through teaching and learning (Ibid.). A number of recommendations were presented for various levels and institutions starting with the United Nations, through governments, students and ultimately universities and colleges. Details of eight recommendations for the institutions are shown in Box 13.2.

In line with the SDGA, during the 2019 HLPF the Global Alliance, representing a partnership of 7000 universities and colleges from six continents, pronounced a Climate Emergency (UN Environment 2019, p. 1). From the declaration, the stakeholders agreed to work on a three-point plan focusing on the following key deliverables to scale up SDG 13 implementation:

i. Committing to going carbon neutral by 2030 or 2050 at the very latest
ii. Mobilising more resources for action-oriented climate change research and skill creation
iii. Increasing the delivery of environmental and sustainability education across curricula, campus and community outreach programmes

The next section focuses on methodological dilemma deliberations.

3 Methodology

This chapter focuses on how Unisa, an open-distance e-Learning institution, reports sustainability matters under the UNGC framework. The objectives set were (1) to determine the materiality of sustainability reporting by Unisa; (2) to establish the extent of sustainability reporting by Unisa under the UNGC; and (3) to establish the extent to which Unisa's sustainability reporting is localising the GRI Sustainability Reporting Standards and the SDG Accord into its reporting framework. The methodology draws from three main methods, namely lived experiences and application of the participatory action research (PAR), critical document and discourse analysis (CDDA) and key informant interviews. The CDDA is applied to a set of key documents that includes Unisa UNGC Reports (2012–2017) and the Unisa Annual Reports (2012–2017). Other documents of importance include a range of policies in the sustainability space such as the Unisa Environmental Sustainability Policy (Unisa 2016), Unisa Energy and Carbon Policy (2016) and related plans and strategies, as well as the Unisa SDGs Declaration (2019). The lived

Box 13.2: SDGA Recommendations for Universities and Colleges

1. Sign the SDG Accord at a senior leadership level and pledge your commitment to the SDGs.
2. Map and report formally on your contribution to the SDGs. Create benchmarks and targets. This ensures that you are transparent and aspirational.
3. Increase your knowledge exchange on sustainability, both interdepartmentally and with other institutions.
4. If your institution undertakes research, change the research application to necessitate researchers outlining which of the SDGs the work contributes towards.
5. If your institution looks to update or begin new processes, incorporate the SDGs, particularly in strategies, policies and updates to the curriculum or course review process.
6. Look for innovative ways to increase staff and student capacity to address the SDGs.
7. Incorporate the SDGs explicitly across the whole curriculum. It has relevance to every single course and module offered, and knowledge of sustainable development and its applicability to everyday scenarios is a much-coveted graduate attribute for employers.
8. Offer more funded extracurricular opportunities for students to get involved with sustainability initiatives.

Source: Global Alliance (2019, p. 17)

experience dates as far back as 2009 when the author joined Unisa as a staff member and initiated several projects until the time of completing this chapter. In addition, since the study was for Unisa, the necessary institutional permission and also ethics clearance were obtained. These paved the way to engage with both the management and other staff.

4 Presentation of Data and Discussion of Findings

The UNISA 2016 Council Statement on Sustainability presents a clear road map regarding the landmark policies, plans and Memoranda of Understanding concluded by the institution (Unisa 2016). Therefore, sustainability reporting is institutionalised at Unisa with a reporting organogram. The Principal and Vice Chancellor's office (which also acts as the CEO's office) has the mandate and is the custodian of such reporting. To this end, since 2009, UNGC reports have been produced annually reporting on sustainability and other initiatives at Unisa. Figure 13.6 shows UNISA's environmental reporting strategic themes from 2011 to 2018.

Sustainability reporting for Unisa is elevated to two major platforms, namely the annual reports in the form of a Council Statement on Sustainability and the UNGC's ten principles. The 2012 Council Statement of Sustainability starts by referring to the year 2007, the year in which Unisa became a signatory to the UNGC (Unisa 2012). The UNGC framework requires Unisa to report its activities based on ten UNGC principles that are parcelled out into three main thematic focus areas, namely finance people and the planet. In terms of governance, sustainability matters are directed from the Principal and Vice Chancellor's office in the portfolio of the Vice Principal: Advisory and Assurance Services.

Unisa has an opportunity to select a number of material GRI Sustainability Standards that are also linked to the SDGs and the ten principles of the UNGC. From this orientation, a picture such as one presented in Table 13.2 could emerge. In the table, the selected GRI Sustainability Standards have been aligned to respective SDGs and the UNGC as appropriate for ease of reference. It is important to note that the proposed alignment is not from Unisa-wide stakeholder engagement, but a lived experience from the author at Unisa, and could be of value as the stakeholders continue to be engaged.

From the GRI Environmental Standards identified, Unisa has done a significant work as such.

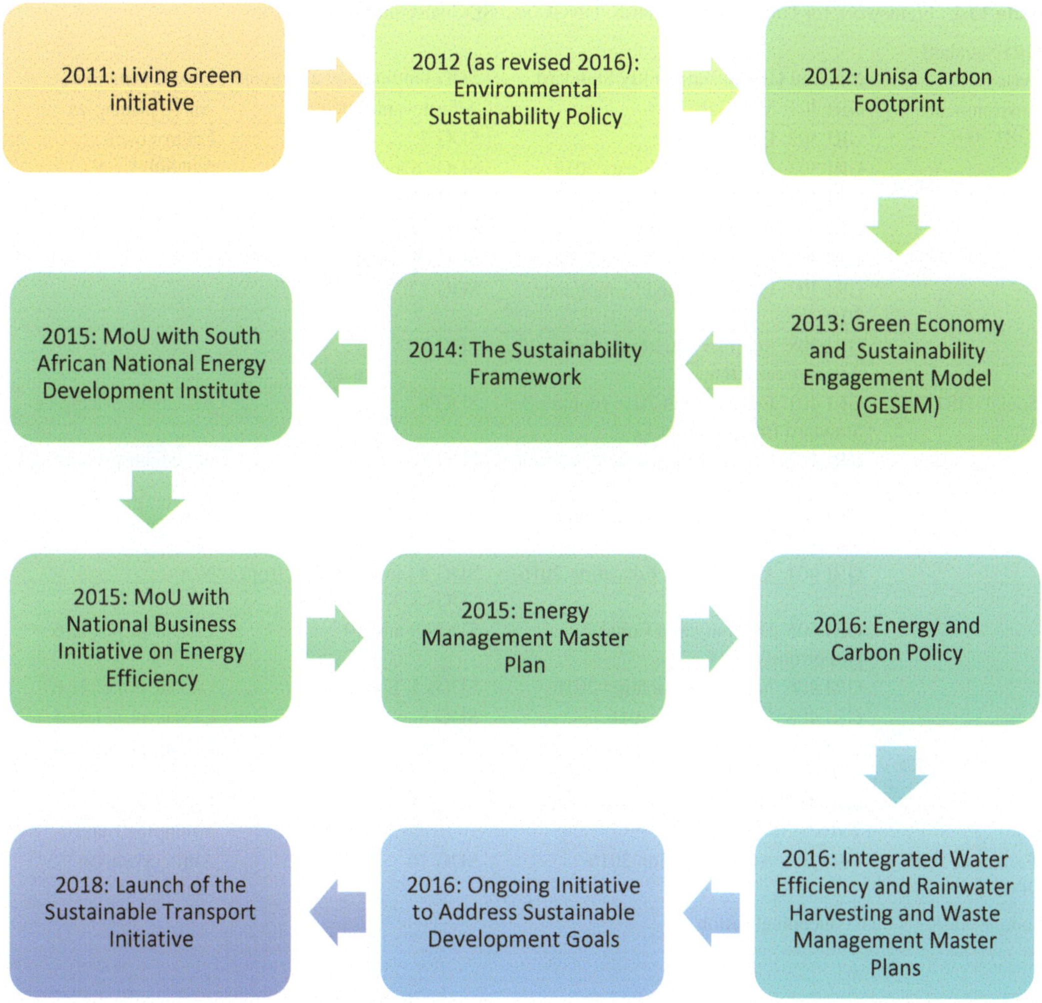

Fig. 13.6 Unisa's environmental reporting strategic themes between 2011 and 2018. Source: Author

The starting point was that Unisa ratified the Unisa Living Green campaign in 2011. From this campaign, a more comprehensive long-term road map code-named the Green Economy and Sustainability Engagement Model (GESEM) was developed. Approved on 22 January 2012, the GESEM is the brainchild of the author, the reason why part of the methodology is PAR. In 2016, the Unisa Council also approved the Energy and Carbon Policy (Unisa 2016). This Policy is also supported by the Energy Master Plan finalised a year earlier (Unisa 2015). Both the Policy and Master Plan rest on three main pillars of action, namely carbon management, sustainable energy transition and energy efficiency. Regarding carbon emissions, Unisa produced its carbon footprint benchmark in 2012 (Unisa 2012). Concerning environmental compliance, Unisa annually works on its environmental risk register driven by the Risk and Compliance Directorate. Unisa's sustainability reporting journey in the context of UNGC, GRI and SDGs will remain incomplete without sharing perspectives on supplier environmental assessments. In 2019, a one-stop SDG for society library online resource platform was created. For Unisa, the Environmental Sustainability Policy of 2012 (as revised in 2016) embeds the need for Unisa to move towards sustainable and green procurement. In fact, some of the points raised in the

Table 13.2 Framework for Unisa GRI Standards, UNGC and SDG sustainability reporting

GRI Standard Series	Material GRI Sustainability Standard	SDG landing and alignment	UNGC principles
Environment (GRI 300)	GRI 301: Materials 2016	All relevant SDGs	All 10 principles
	GRI 302: Energy 2016	SDG 7	Environment principles 7–9
	GRI 303: Water and Effluents 2018	SDGs 6 and 14	
	GRI 304: Biodiversity 2016	SDG 15	
	GRI 305: Emissions 2016	SDG 13	
	GRI 306: Effluents and Waste 2016	SDGs 6, 14 and 15	
	GRI 307: Environmental Compliance 2016	SDG 12	
	GRI 308: Supplier Environmental Assessment 2016	SDG 12	
Social (GRI 200)	GRI 401: Employment 2016 (containing Standard Interpretation 1)	SDG 8	All labour principles
	GRI 402: Labour/Management Relations 2016	SDG 8	All labour principles
	GRI 403: Occupational Health and Safety 2018	SDG 3	N/A
	GRI 404: Training and Education 2016	SDG 4 (and elements from SDGs 3, 8 and 13)	N/A
	GRI 405: Diversity and Equal Opportunity 2016	SDGs 8 and 10	Labour principle 6
	GRI 406: Non-discrimination 2016	SDGs 5, 8 and 10	Labour principle 6
	GRI 408: Child Labour 2016	SDG 8	Labour principle 5
	GRI 409: Forced or Compulsory Labour 2016	SDG 8	Labour principle 4
	GRI 412: Human Rights Assessment 2016	SDGs 4, 8	Human rights principles 1 and 2
Economic (GRI 200)	GRI 205: Anti-corruption 2016	SDG 16	Anti-corruption principle 10

Source: Author's own conceptualisation

three-point plan from the climate emergency declaration resonate well with Unisa. As already highlighted, Unisa has already agreed to a pathway to reduce its carbon footprint and work on renewable energy, especially solar energy (Nhamo et al. 2019). The university has also been on a drive to use both financial and human resources, including research and skill development towards climate action (Nhamo et al. 2019).

From the aforementioned discussion, it is clear that both opportunities and challenges exist for Unisa in mainstreaming and landing the SDGA and GRI Sustainability Standards into its current reporting framework and systems based on the UNGC. The key opportunity is that both the GRI and UNGC have been working together since the ratification of the SDGs to harmonise the ten principles of the UNGC with the GRI Sustainability Standards. Furthermore, the two organisations worked with the WBCSD to come up with the SDG Campus, a document showing how companies and organisations could land SDGs in their strategies and operation systems. But the key challenge comes in that the SDGA is being promoted globally as the main framework for landing and reporting on SDGs in universities and colleges. The SDGA is further linked directly to the HLPF of the SDGs, an aspect that brings pride to those fulfilling this requirement.

As in many instances, there is the general resentment from staff regarding anything viewed as adding to their workload. In this case, landing the SDGs into the UNGC to come up with what could be termed the UNGC+ annual report harmonising the UNGC, GRI Sustainability Standards and the SDGA could be viewed as adding to the existing workload. With such perceptions, current Unisa staff may request to be incentivised through additional rewards or capacity, an aspect that universities may not have. Furthermore, should Unisa get involved with the SDGA, it implies another separate annual report for the HLPF. This is an element covered by proposing the UNGC+ report, a comprehensive and expanded report that will embed the requirements of the SDGA inherently as well. Technically, Unisa will then prepare the UNGC+ report that will be dismantled into the UNGC, and the SDGA HLPF reports every year. The GRI Sustainability Standards will also be cross cutting into these twin reports as they will be used to determine materiality in reporting. The UNGC+ report will also provide information for reporting sustainability matters to the Unisa Council and for the preparation of the Council Sustainability Statement.

There has been movement towards the UNGC+ model as reflected in the Unisa UNGC report of 2019. The report embraced the SDGs Localisation Declaration, which commits Unisa to localise the SDGs across all its operations and core mandates including teaching and learning, research and innovation, and community engagement (Unisa 2019). A whole section (Section 5) of the 2018/2019 Unisa UNGC report is dedicated to the Unisa SDGs Localisation Declaration. Looking ahead, more work needs to be done to realise the implementation of the SDGs Localisation Declaration ratified on 29 November 2019. The following paragraphs discuss the emerging findings in line with the work that has been taking place elsewhere concerning the SDGs, sustainability reporting, the UNGC and other initiatives.

The progress made by Unisa in implementing the SDGs across its core mandates was observed earlier by Mawonde and Togo (2019). However, the authors noticed that several challenges existed, including limited funding. In addition, being an ODeL institution, the university battles to involve students in the projects. The authors also agreed to the direction that Unisa took through the SDGs Localisation Declaration to embrace most, if not all, of the SDGs. Drawing from the Swedish case on integrating sustainability into higher education systems, Argento et al. (2020) are of the view that different disciplines should be acknowledged and used as an asset to understand how the 17 SDGs are interconnected. However, integrating the SDGs and the UNGC is also likely to face the challenge of other initiatives, especially those from business schools aligned to the United Nations Responsible Management Education (UN-PRME) (Haertle et al. 2017). The UN-PRME has been reoriented to speak to the 2030 Agenda for Sustainable Development and the SDGs since 2016 (Storey et al. 2017).

Fleacă et al. (2018) consider aligning SDGs to strategy in higher education, a perspective the authors concede is difficult. In the authors' view, there is:

> Lack of capacity of higher education institutions to integrate the principles and practices of sustainable development into all aspects of education and learning, which hampers the capability to act as an entrepreneurial university. Embarking on the path of sustainable development goals (SDGs) requires HEI to design, launch, implement, and customise specific processes architecture to govern the advance of the sustainability approach (Fleacă et al. 2018, p. 2).

Drawing from Australian universities, Melles (2020) observes that sustainability reporting mainly aligns to environmental management, with the desire to reduce harmful greenhouse gases associated with SDG 7, SDG 12 and SDG 13 at the centre. However, there was another challenge that Unisa should be on the lookout for, that is, the lack of transparency and modest accomplishments in campus sustainability if compared to the corporate sector.

5 Conclusion

What emerges from this chapter is that there are a number of contesting sustainability reporting frameworks already in use that are applicable to institutions of higher education, such as the University of South Africa. Those covered by this work are the ten principles of the UNGC, the GRI Sustainability Standards and the SDGA. Of late, the SDGA is being promoted as the ultimate sustainability reporting framework for universities and colleges through the Global Alliance based in London, the United Kingdom. However, given that Unisa has been reporting sustainability matters under the UNGC for a long time, in fact since 2009, the facts on the ground are that one of the most feasible ways to move forward is to mainstream the GRI Sustainability Standards and the SDGA into the UNGC. Fortunately, this endeavour has been simplified by the fact that the UNGC and the GRI are compatible through harmonisation of the SDGs and the ten principles. At the time of writing, there was evidence that Unisa had gathered the critical mass towards the mainstreaming of the SDGs and the GRI Sustainability Standards onto its UNGC reporting. Nevertheless, Unisa is faced with the conventional challenges of a lack of capacity with few staff working on such reporting on a full-time basis. Furthermore, low levels of awareness on the GRI, SDGs and SDGA provisions remain. It was only in 2018 that the scaling up in terms of awareness raising, especially on SDGs, kicked off with colleges (equivalent to faculties), SDG road maps were instituted and the SDGs for Society library online resource platform was created. The outcomes of such initiatives will be evaluated in the future.

References

Argento, D., Einarson, D., Mårtensson, L., Persson, C., Wendin, K., Westergren, A. (2020). Integrating sustainability in higher education: a Swedish case. International Journal of Sustainability in Higher Education, 21 (6), 1131-1150. https://doi.org/10.1108/IJSHE-10-2019-0292.

Fleacă, E., Fleacă, B., Maiduc, S. (2018). Aligning Strategy with Sustainable Development Goals (SDGs): Process Scoping Diagram for Entrepreneurial Higher Education Institutions (HEIs). Sustainability 2018, 10, 1032; doi:https://doi.org/10.3390/su10041032.

Global Alliance. (2017). SDG Accord: The University and College Sector's Collective Response to the Sustainable Development Goals. London: Global alliance.

Global Alliance. (2018). Annual report to the United Nations High-level Political Forum on Sustainable Development (HLPF) as part of the SDG Accord mandatory institutional reporting, New York, 11th July 2018. London: Global alliance.

Global Alliance. (2019). Annual SDG Accord Report 2019 Progress towards the Global Goals in the University and College sector, New York, July 2019. London: Global alliance.

GRI (Global Reporting Initiative), UNGC (United Nations Global Contact) and WBCSD (World Business Council for Sustainable Development). (2016).

GRI (Global Reporting Initiative). (2016). Consolidated set of GRI sustainability reporting standards 2016. Amsterdam: GRI.

GRI (Global Reporting System) and UNGC (United Nation Global Compact). (2018). Business Reporting on the SDGs: Integrating the SDGs into Corporate Reporting – A Practical Guide. New York: UNGC.

Haertle, J., Parkes, C., Murray, A., Hayes, R. (2017). PRME: Building a global movement on responsible management education. The International Journal of Management Education, 15, 66e72, https://doi.org/10.1016/j.ijme.2017.05.002.

Mattera, M. & Alba Ruiz-Morales, C. (2020). UNGC principles and SDGs: perception and business implementation. Marketing Intelligence & Planning, https://doi.org/10.1108/MIP-08-2018-0319

Melles, G. (2020). Sustainability Reporting in Australian Universities: Case Study of Campus Sustainability Employing Institutional Analysis. In: Leal Filho W., Borges de Brito P., Frankenberger F. (eds) International Business, Trade and Institutional Sustainability. World Sustainability Series. Springer, Cham. https://doi.org/10.1007/978-3-030-26759-9_56

Nhamo, G. (2014). From sustainable development through green growth to sustainable development plus, *International Journal of African Renaissance Studies,* 9(2), 20–38. https://doi.org/10.1080/18186874.2014.987953.

Nhamo, G., Nhamo, S. and Nhemachena, C. (2018). What gets measured gets done! Towards an Afro-barometer for tracking progress in achieving Sustainable Development Goal 5. Agenda, 32(1), 60-75. https://doi.org/10.1080/10130950.2018.1433365.

Nhamo, G., Nhemachena, C. and Nhamo, S. (2019). Is 2030 too soon for Africa to achieve the water and sanitation sustainable development goal? *Science of the Total Environment*, 669, 129-139. https://doi.org/10.1016/j.scitotenv.2019.03.109.

Mawonde, A. & Togo, M. (2019). Implementation of SDGs at the University of South Africa. International Journal of Sustainability in Higher Education, 20(5), 932-950. https://doi.org/10.1108/IJSHE-04-2019-0156.

Storey, M., Killian, S., O'Regan, P. (2017). Responsible management education: Mapping the field in the context of the SDGs. The International Journal of Management Education, 15(2), 93-103. https://doi.org/10.1016/j.ijme.2017.02.009.

UN Environment. (2019). Higher and Further Education Institutions across the globe declare Climate Emergency. Nairobi: UN Environment.

UNDG (United Nations Development Group). (2016). MAPS – A Common Approach to the UNDG's policy support to the SDGs – An update on implementation March 2016. Available at: https://www.un.org/ecosoc/sites/www.un.org.ecosoc/files/files/en/qcpr/doco-summary-brief-on-maps-march2016.pdf (Accessed 3 July 2019).

UNGC (United Nation Global Compact). (2017). Making Global Goals Local Business: A New Era for Responsible Business. New York: UNGC.

UNGC (United Nation Global Compact). (2018). United Nations Global Compact Progress Report 2018. New York: UNGC.

Unisa (University of South Africa). (2012). University of South Africa Annual Report 2012. Pretoria: Unisa Press.

Unisa (University of South Africa). (2015). Energy Master Plan. Pretoria: University of South Africa.

Unisa (University of South Africa). (2016). University of South Africa Annual Report 2016. Pretoria: Unisa Press.

Unisa (University of South Africa). (2019). University of South Africa United Nations Global Compact Communication on Engagement 2018/19. Unisa Press.

United Nations. (2015). Transforming our World: The 2030 Agenda for Sustainable Development. New York: United Nations Secretariat.

Zemanová, Š. & Druláková, R. (2020a). Mainstreaming Global Sustainable Development

Štěpánka Zemanová and Radka Druláková (2020b) Goals through the UN Global Compact: The Case of Visegrad Countries. Journal of Risk and Financial Management, 13, 41; doi:https://doi.org/10.3390/jrfm13030041.

The Leadership and Implementation of Sustainable Development Goals in Finnish Municipalities

14

Satu Lähteenoja, Kaisa Schmidt-Thomé, Jani Päivänen, and Emma Terämä

Abstract

Cities and municipalities play a vital role in accelerating the transition towards reaching the Sustainable Development Goals (SDGs). In an urbanised world, many of the most significant sustainability initiatives are carried out locally. This study discusses the results of action research run with 12 forerunning municipalities in Finland. The research identified the current baseline in the municipalities' cross-sectoral strategic leadership work related to sustainability and co-designed solutions for more long-term and coherent SDG leadership at the local level. The conceptual framework links the literature on sustainability transitions seen as systemic socio-technical change with the public administration scholarship. The study showed that embedding sustainability in the practice of local level leadership is a matter of both procedures and substance. The local level administration needs to know how to orchestrate its sustainable development efforts but also know which themes and actions to concentrate on in order to utilise the full local potential. As a result, this chapter presents three models of SDG leadership at the local level. A participatory process, where the SDGs are localised in a cross-sectoral manner, gives the opportunity to bring sustainable development to the very centre of strategy work at the local level and can be recommended.

Keywords

Agenda 2030 · SDGs · Municipalities · Strategic leadership · Local level policy · Cross-sectoral governance

1 Introduction and Background

Right after the Sustainable Development Goal (SDG) report by the UN Secretary-General was published in December 2014, a series of scholars came out with a 'communication' (Hajer

S. Lähteenoja (✉)
Aalto University School of Arts, Design and Architecture, Espoo, Finland

Demos Helsinki, Helsinki, Finland
e-mail: satu.lahteenoja@demoshelsinki.fi

K. Schmidt-Thomé
Demos Helsinki, Helsinki, Finland
e-mail: kaisa.schmidt-thome@demoshelsinki.fi

J. Päivänen
Finnish Consulting Group, Helsinki, Finland
e-mail: jani.paivanen@fcg.fi

E. Terämä
The Finnish Environment Institute, Helsinki, Finland

Ministry of Finance, Helsinki, Finland
e-mail: emma.terama@syke.fi

G. Nhamo et al. (eds.), *Sustainable Development Goals for Society Vol. 1*, Sustainable Development Goals Series, https://doi.org/10.1007/978-3-030-70948-8_14

et al. 2015), raising their concern about the 'cockpit-ism' of the SDGs. The SDGs were to be accepted in September 2015, and the scholars wanted to put an end to the illusion that solving global problems would be a matter of top-down leadership, belonging alone to governments and intergovernmental organisations (IGOs). What the scholars were calling for was the mobilisation of a new set of change agents, such as businesses, cities and civil society (Hajer et al. 2015, p. 1652). They saw that cities/municipalities, on which we will focus in this chapter, should be encouraged to push for transformation in resource efficiency and environmental quality to start with.

What has happened since the communication? Is there evidence of a new sense of agency, or even an emerging legacy in sustainability transitions, among the non-state, non-IGO actors? The work that we report on is about the organisational arrangements and the key content matters through which the local level administration in Finland may (or may not) be taking leadership in sustainability transitions. We have been searching for signs of far-sighted leadership and coherent policymaking. Such deeds could even come coupled with a certain local level cockpit-ism, considering that a strong self-governance of the municipalities is a central characteristic of the Finnish multilevel governance in international comparisons (e.g. OECD 2016).

We were interested in the different ways that municipalities can steer and accelerate the local sustainability transition. Our conceptual framework links the literature on sustainability transitions seen as systemic sociotechnical change (e.g. Grin et al. 2010) with the public administration scholarship. In particular, we are addressing the possibilities to accelerate transitions by embedding sustainability in the practice of local level leadership. This is a matter of both procedures and substance. The local level administration needs to know how to organise and orchestrate its sustainable development efforts but also know which themes and actions to concentrate on in order to utilise the full local potential. We examine whether Agenda 2030 can support making these crucial choices in the context of cross-sectoral, strategic leadership and whether the SDGs can be localised in a manner that accelerates systemic transition.

Our aim has been to identify, categorise and describe how a set of Finnish municipalities are (or are not) operationalising the challenge of becoming the local level leaders and implementation agents of a global agenda. In this chapter, we first present a literature review and the theoretical background, followed by presenting the research methodology, including research questions and selected action research approach. Then we move to the key results about the current situation in Finnish municipalities and present the leadership models that were developed during the joint journey with the partaking set of Finnish municipalities. Before the conclusions, we discuss the role that the national level government can have in supporting the local level work in reaching the global goals.

Agenda 2030, including the broader programme framework and the set of tools developed to achieve the SDGs, has received much attention both among policymakers and scholars (UN 2019; Gusmão Caiado et al. 2018). This also holds increasingly true at the level of local authorities (Krellenberg et al. 2019; Zinkernagel et al. 2018; Valencia et al. 2019; von Wirth et al. 2019). Bowen et al. (2017) argued that, globally, there are three governance challenges to implementing the SDGs: (i) cultivating collective action, (ii) making difficult trade-offs and (iii) ensuring that societal actors are held accountable. Whether these challenges also hold true at a local level governance is an interesting question. We expect that focusing on the municipalities of a Nordic country with a tradition of strong self-governance and comprehensive responsibilities in service provision can provide interesting insights from a context where contributing to sustainability transition can be seen as a local level duty (the Local Government Act 2015).

Within the recent literature on sustainability transitions, understood as a radical transformation of the sociotechnical systems (Grin et al. 2010;

Köhler et al. 2019), there is a growing body of work on accelerating the required transitions. In these studies, acceleration is used both in broader terms, when underlining the urgency to bring about systemic change by challenging the dominant configurations (Roberts et al. 2018; Loorbach 2010), and when breaking free from the systemic gridlocks (Klitkou et al. 2015). In a somewhat narrower sense, the term *acceleration* may be used when referring to a phase within a transition process (Rotmans et al. 2001).[1] Both framings of acceleration have also been present in local level studies. However, these studies have often focused on the niche level—for example, on 'transition initiatives' (Gorissen et al. 2018) or civil society groups (Hargreaves et al. 2011)—and have bypassed the role of local level administration in the acceleration process (Quitzau et al. 2013). While many have high hopes for cities and other local level authorities as enablers and collaborators (Mukhtar-Landgren et al. 2019; Frantzeskaki et al. 2017) or as agents of coalition building and as participants in innovative experimenting (Swilling and Hajer 2017), we argue that there is a need to look at the work *within* those administrations themselves. If it holds true that considerations on 'how agency can contribute to large scale systemic changes' still represent a 'largely uncharted territory' (Gorissen et al. 2018, p. 172), it certainly makes sense to focus on the agency that lies with the authorities before rushing into the myriad compelling initiatives led by many other actors (put on your own mask first and only then help others).

Within the public administration scholarship, Zeemering (2017, p. 136) argued that the emphasis has been on '*what* managers should do to pursue sustainability, but not *how* to manage in pursuit of sustainability'. Building on the observations of Laszlo and Zhexembayeva (2011) on corporations, Zeemering was concerned about the public-sector organisations failing to embed sustainability into their strategy and operations. Realising that surveys of local government sustainability initiatives reveal the uneven adoption of policies, he joined the scholars that 'question the extent to which sustainability is being embraced as a comprehensive framework for planning or local government management' (Laszlo and Zhexembayeva 2011, p. 137). We share this concern and want to address the challenge of finding a practical framework for thinking about the deployment of necessary policy tools within these organisations. As Zeemering put it, managers in local government are becoming aware of the policy tools associated with sustainability, but many still need to learn how to lead their organisation in order to put them in place. Organising the administrative procedure to back the leadership is not doomed to be strategically paralysing (Quitzau et al. 2013) or locked into a system innovation capture (Pel 2016). Whereas the initiatives of the niche-level actors are often expected to be the most creative accelerators of sustainability, in our view, the 'endogenous renewal' of the local regime, in the name of sustainability, is also very much a matter of creativity (Hoffman and Loeber 2016).

Whether we are expecting too much from the local level apparatus must be put into the perspective of the strong tradition in local level self-government that prevails in Finland, like it does in most Nordic countries (Sallinen et al. 2017). Both the formal power relations and the considerable resourcing through the tax income lead us to expect that the municipalities are well positioned to face the sustainability challenge. In leadership literature, we have found two interesting avenues that have informed our research process. Both of them emphasise integration—one, integration between the different administrative sectors (Morse 2010; Leponiemi 2019), and the other, integration between the openly political decision-making and the expertise-based preparatory work of the local officials (Helgeson 2018; Zeemering 2017; Hölscher 2019). In Sect. 3.2 we will discuss how these integration challenges play out across the municipal spectrum in Finland.

[1]The four different transition phases conceptualised by Rotmans et al. (2001) are predevelopment, take-off, acceleration (or breakthrough) and stabilisation.

2 Materials and Methods

We report on findings from a project carried out within the research activities that the Finnish Government coordinates in order to support national level decision-making.[2] The main objective for the government was to increase the understanding of the potential of local level administration to contribute to reaching the Agenda 2030 targets set for nation states. The project consortium proposed relying on an approach based on action research and directly involving a set of municipalities in the study. In the tradition of participatory action research (McIntyre 2007), the consortium also carried out a process of localising the global level ambitions—made explicit via the SDGs—with the local level actors.

Using the SDG framework at a local level requires the localisation of Agenda 2030. *Localisation* means that global goals are interpreted from the perspective of local circumstances and relevance.[3,4] We studied how the localisation could be organised in a way that supports the acceleration of sustainability transformations across the Finnish municipal self-government. The broader research problem was the following: How do the municipalities embed sustainability into their operations? Which areas of sustainability receive the most attention and why? Which leadership models can accelerate the local level sustainability transition? In more detail, we asked how the leadership of sustainable development is currently organised in Finnish municipalities and which models can be abstracted based on the action research done. We also studied the preconditions for successful acceleration in each model. In terms of key substance areas, we mapped the potential (by SDG) to leap forward in sustainability. And, when reporting our study to the Prime Minister's Office, we proposed how the Finnish Government can best support municipalities in strategic SDG implementation.

The action research consisted of two parts. In the first part, cross-sectoral teams from 12 participating municipalities[5] started to work with an analysis of the current situation of sustainable development in their local municipality. The analysis was conducted together with the researchers. It included a text analysis of existing local administration strategy documents and the SDGs, interviews with public administration officials in municipalities and a co-creation workshop where potential substance-specific priority areas were identified and compared with those of other municipalities. In the second part, five municipalities continued the work by analysing their current leadership structures with respect to sustainable development. Based on the analysis, each municipality hosted a workshop together with the researchers in order to discuss the strengths and weaknesses of their existing models and the potential for improvement. In addition, we organised two common events for all original 12 municipalities[5] where the observations were discussed, and much of the research material was produced. In a joint seminar at the semi-final stage, we used a tailor-made card game, through which the leadership models best suited for different municipality types were construed. Each municipality selected a set of cards presenting those elements that best illustrated their model and situation. Afterward, the models were finalised by the research group and assessed by a steering group with representation across several government sectors.

We employed action research in the sense that our municipal field study subjects were selected from a small sample that was known to us ('small-*n* statistics'). In this sense, we could characterise ourselves as undertaking 'qualitative research by insiders' (Zeni 1998). We found that typical ethical safeguards designed for an outsider

[2]The Finnish Government's analysis, assessment and research activities. https://tietokayttoon.fi/en/frontpage.

[3]Sustainable Development Solutions Network, 2016. https://sdgcities.guide/.

[4]Road map for localizing the SDGs: implementation and monitoring at subnational level. The Global Taskforce of Local and Regional Governments, 2016. UN-Habitat, UNDP.

[5]The municipalities were chosen based on applications. They differ in size, in the socio-demographic characteristics of the citizens and in geographical location.

doing a field experiment (random selection, control groups, removing the personal influence of the researcher) were not applicable. Similarly, the typical qualitative research design of anonymous responses would have defeated the purpose of the open communication and dialogue of our municipal field study. The ensuing ethical guidelines that were followed were that participants were considered free from risk as (a) they were first informed and given the general information about the study and what was expected of them at different stages; (b) by choosing to participate, they gave informed consent; (c) they could refuse to participate, and they could withdraw at any time, even after the research had begun; and (d) respondents' identities were not disclosed in any outputs, and response data was aggregated to above-municipal level whenever possible. Project output was subject to a review of the participants prior to public disclosure. In accordance with open data guidelines and funder's requirements, the results were made public as a technical report and a policy brief.

3 Results

3.1 The Current Situation at the Local Level

Even though Finnish municipalities have long traditions in sustainable development, they are still facing several challenges in integrating sustainability into the core of both their leadership and operations. Many Finnish municipalities were already active in the Local Agenda 21 work in the early 1990s and have continued to work towards the sustainability objectives since. Today, many municipalities consider sustainability or responsibility as one of their core values and show commitment to it in their strategies. Thus, sustainability seems to be widely accepted as a broad societal aim, notably in Finland in the recent Government Programme (Finnish Government 2019; Berg et al. 2019). Furthermore, municipalities are relatively aware of the key challenges they still face around sustainability. These include, for example, climate change mitigation and reduction of social disparities (Fig. 14.1). Sympathising with Le Blanc (2015), who has shown how some themes targeted by Agenda 2030 are not well connected with each other, we have chosen to present the local level targets in thematic groups that are nested in each other.

The main challenges related to implementing sustainable development at the local level in Finland are linked to coherence in strategy and leadership, in the interlinkages of different dimensions of sustainability (e.g. social, environmental and economic dimensions), in the different time frames of implementation and in the challenges of cross-governmental leadership and management (see also Berg et al. 2019; Lyytimäki et al. 2020). In all the municipalities studied, the short-term economic goals were prioritised over the longer term environmental goals in local decision-making. This is understandable as Finnish politics operates in 4-year electoral terms and policymakers tend to focus on issues where they hope to see results in a few years' time, preferably still within the ongoing term.[6] The policy coherence challenges (as well as the trade-off challenges between different goals) create further obstacles for strategic decision-making (Lyytimäki et al. 2020). A recent example of global debates related to interlinkages is the question of climate change mitigation and employment in the locations where coal and carbon-intensive industry, such as peat production or the steel industry, plays an important role in the local economy.

Summa summarum, 5 years after the ratification of Agenda 2030, most Finnish municipalities are still in the early stages of localising and implementing the SDGs and in adopting them into the core of their strategic leadership and management structures. This is true despite the early onset of the sustainability

[6] Kettunen et al. (2019) have also brought up the time issue from another perspective: the 'right time' to do long-term reforms often has a short time window within a 4-year term as the time both before and after the elections is often less consensual.

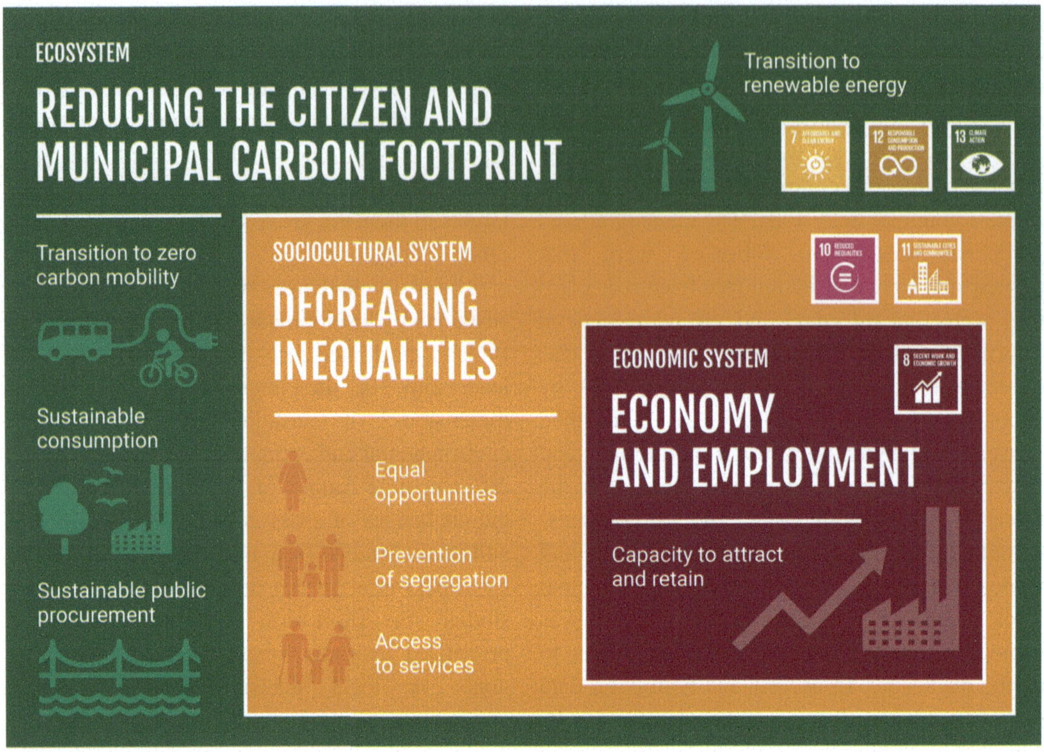

Fig. 14.1 Key sustainability topics at the local level in Finland

discourse in Finnish municipalities and society at large. Traditionally, sustainability has been considered to mostly be an environmental topic, and that is why any given leadership has also been restricted to the municipalities' environmental or technical services. As the SDGs cover all dimensions of sustainability, the topic becomes highly cross-sectoral, and the strategic leadership models of sustainable development need to be reconsidered.

3.2 Three Models for SDG Leadership at the Local Level

The partaking municipalities were at different stages, both in their sustainability work in general and in their local adaptation/localisation of SDGs. Some of the more well-resourced municipalities had already used SDGs in systematic Voluntary Local Reviews (VLRs), others had perhaps done one-time monitoring in a looser manner, yet others had only started discussions on how to use them. We wanted to produce an overview of the underlying features and explanations of their respective situations and on that basis form coarse models of 'SDG leadership'. Three different models for SDG leadership in municipalities were formed: *Guiding Stars*, the *Power of Networks* and *Active Individuals*. Next, we shall take a look at these models as structures and agencies, and at their respective strengths and weaknesses in terms of acceleration. Our task, then, was to (1) form a (tentative) model for leading and mainstreaming sustainability issues in municipal government and (2) propose ways in which municipalities could or should use SDGs to back up this work. First, we conducted a brief study of some of the models of municipal leadership proposed in the literature. The ideas of integrative leadership focusing on cross-sectoral work and partnership building (Morse 2010), leading decision-making in complex environments by structuring decisions (Helgeson 2018) and the model of transformative capacities, focusing on orchestrating these diverse local capacities (Hölscher 2019), all seemed relevant for leading sustainability transformations in complex organisations, but none covered the diversity of municipal constellations we were facing. It seemed clear that the diversity of Finnish municipalities in terms of population, economic base and municipal governmental resources required more than one leadership model. Based on these, we started to construct a couple of models that would somewhat cover the diverse field that we had identified by looking at the case study municipalities.

The first model, Guiding Stars, is a model of a municipality implementing sustainability goals in a very straightforward manner. Sustainability leadership is well resourced, operated by a dedicated sustainability team and aligned rather vertically top-down (Fig. 14.2). Its progress is (annually) monitored using a clear set of indicators/metrics. The work can be supported by cross-administrative programmes, impressive project portfolios and external funding. The SDGs can be linked directly to the municipal strategy, or they can be applied in a toolbox to guide the work in management teams, local council groups, etc.

The Guiding Stars model leans on strong internal expertise in the issues of sustainable development, expertise which is also utilised up to strategic management. The clear political mandate gives sustainability work a solid ground. Sustainability work is organised among a team, monitored regularly, and energised by direct feedback from management. The specific challenges and risks of the model relate to its top-down management: if messages from the management do not match experts' perceptions of the most important issues for the municipality, both performance and atmosphere may suffer. Therefore, the leaders must learn and nurture a culture of listening and letting their team members shine.

One of our case cities, City E, which we recognised as a clear-cut Guiding Star, was planning

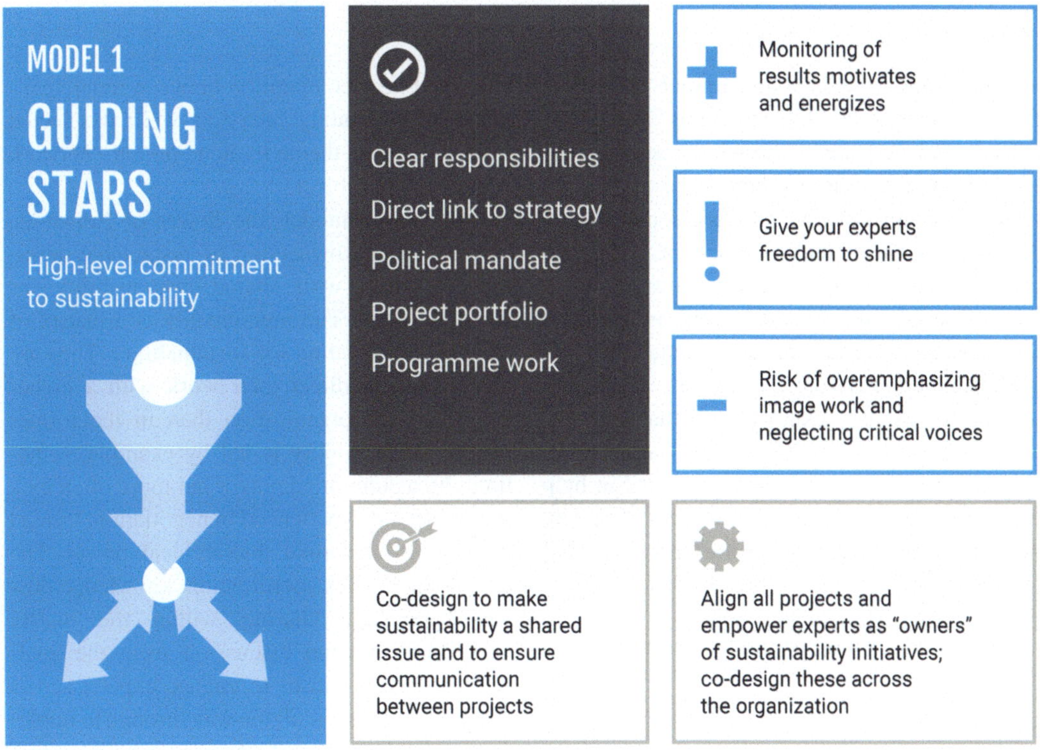

Fig. 14.2 The Guiding Stars model of sustainability leadership

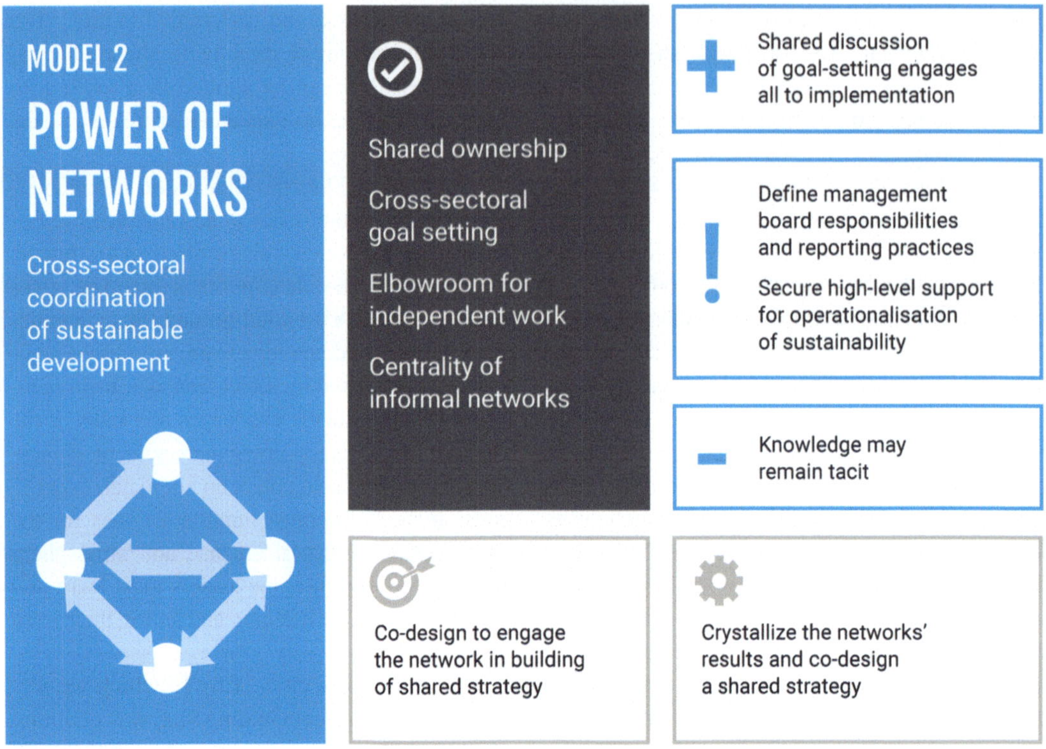

Fig. 14.3 The Power of Networks model of sustainability leadership

their first VLR, scheduled for summer–fall 2020.[7] They considered the SDG apparatus[8] as a 'magnificent and perpetually growing resource' for VLR. They saw VLR as a both horizontally and vertically communicative process, dependent on the commitment of top management. They laid great emphasis on high-level mandates, monitoring and resourcing. Adequate time is a crucial resource for face-to-face communication and workshopping horizontally among municipal sectors. It was also acknowledged that a review using and localising SDGs requires a considerable amount of time, but then again, it can help construct a common learning process. The city showcased an impressive project portfolio and perceived peer competition between municipalities as a motivating factor as well; it selected leadership as its theme in the temporary network formed in the project.

The second model, the Power of Networks, portrays shared ownership of sustainable development work, with shared responsibility for attaining goals. The municipality is a matrix of experts collaborating for sustainability. They are empowered to direct their work with a higher degree of self-organisation than in the former model. Sustainability is led by example, sometimes by a committed mayor but more often by a senior official experienced in negotiation, instead of leaning on formal, hierarchical power. The model emphasises coordination, knowledge flow and knowledge sharing within the matrix (Fig. 14.3). Shared discussion from the goal-setting stage promises to engage experts across the administration—all the way through to implementation. However, if processes and reporting responsibilities do not really support cross-

[7]A VLR declaration was launched by New York City in 2019 to encourage local and regional governments worldwide to formally report on their local progress towards SDGs.

[8]City E used the MayorsIndicators toolbox: https://mayorsindicators.com/

administrative work, it will be difficult to allocate the time, catch up with important content issues and so forth. Another challenge for the model is whether the actors are confident that they really are sharing the ownership. The difficulty of orchestrating the shared responsibility lies in its subtlety: Who ultimately has the responsibility to take things forward? Municipalities using this model must strive to develop effective management team practices and internal communication to embed a truly shared operating culture.

In one of our case cities matching the Power of Networks model, City T, a multidisciplinary cross-administrative group gathered to take part in the project workshops. Each member was tackling a range of social and environmental issues in their work and trying to find common ground in the SDGs. The members of the unofficial network were committed to sustainability goals, each from their own angle. While enjoying the freedom to work on these goals, they emphasised the need for management support and clarity in reporting. In SDGs, they saw an opportunity to identify key indicators, point out central goals and guarantee this support. In another city matching the Power of Networks model, a senior official used the term 'seduction' to point out the importance of interpersonal relationships when persuading people to engage in sustainable policy actions. In their discussions, the systematic use of SDGs was invoked, and a consensus was found that the SDGs should be used in managing the whole span of leadership, from sustainability goals to implementation and monitoring.

In the third model—Active Individuals—sustainable development work may not be formally organised at all. Still, the municipality's work and its stakeholder cooperation can include very-goal-oriented sustainability activities. These can be individual initiatives, for example, through the council, or activities outsourced to businesses or local/regional development companies. Entrepreneurs and civil society may also have certain informal responsibilities. The municipality may strategically decide to apply this model in order to allow sustainable development solutions to emerge in the local innovation ecosystem. However, bottom-up initiatives can be challenging to upscale to the system level.

This model, of course, depends on the commitment of active individual stakeholders specifically pursuing sustainable development (Fig. 14.4). However, grassroots initiatives and engagement are often targeted at some concrete improvement rather than being linked to more overreaching sustainability goals. The development of inclusion and local democracy does not necessarily guarantee a robust thrust towards sustainability transitions. The second challenge is to achieve a sufficiently cross-administrative status for sustainable development within the municipal organisation without the theme dissolving into an overarching but vague principle to which no content is attached. On the other hand, when working freely at the grassroots level, any individual—municipal officials, delegates and the 'fiery souls'[9] of civil society or business—can do a lot. This requires, however, an explicit commitment to deliberative democracy on the one hand, and an understanding of the municipality's position in these advances.

Representatives of our Case Municipality U, which had prominent Active Individuals features, acknowledged that sustainability was not adequately managed or 'owned'. Despite several cross-administrative management boards, the location and responsibilities concerning sustainability goals remained unclear. The municipality's representatives in our project, however, focused on the opportunities at hand. They sought ways to utilise SDGs to empower strong villages and active residential and other associations, and solidify a bottom-up approach to local environmental change. The municipality has launched an annual participative budgeting 'contest' where local initiatives are found and processed. SDGs could be used to assess and improve these initiatives. In the longer run, the municipality needs to solve the question of ownership and the coordination of resources for larger, longer term sustainability goals.

[9] 'Fiery souls' is a Scandinavian term for idealistic persons who often bear the bulk of the responsibilities (e.g. in non-governmental organisations).

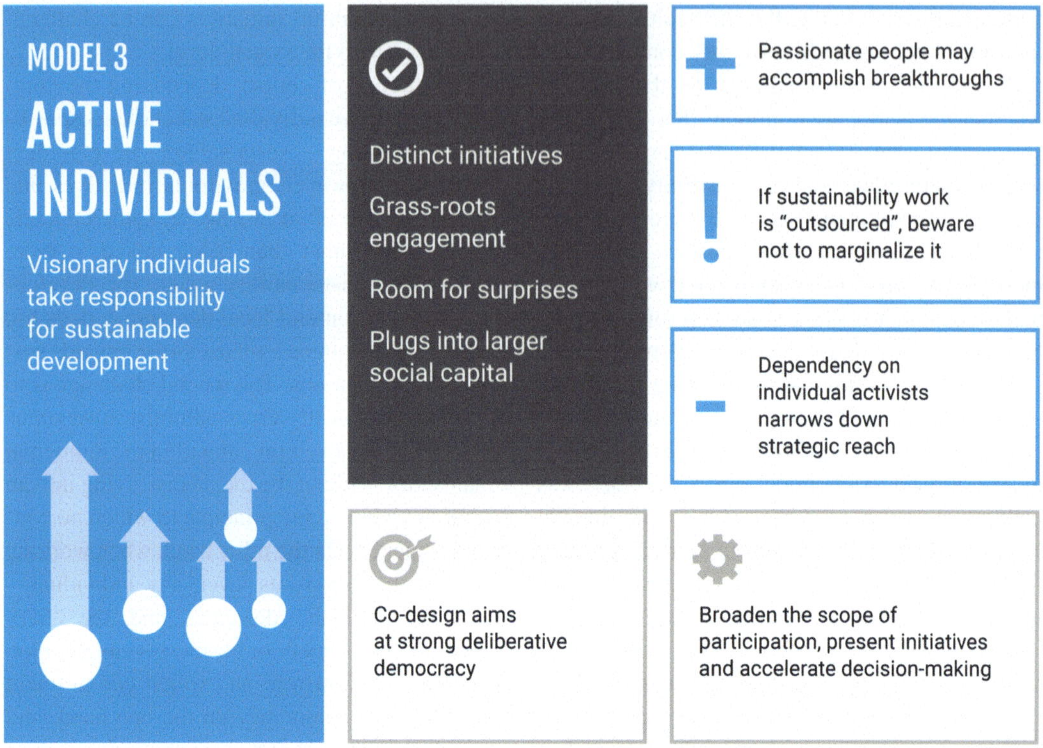

Fig. 14.4 The Active Individuals model of sustainability leadership

4 Discussion

4.1 Using SDGs for Strategic Municipal Leadership

Although the approaches to adapting and utilising SDGs varied a lot among our case municipalities, there were several common features. One is using long-term strategic goals—as opposed to single, 4-year time span political targets—for formulating more robust and cohesive goals, identifying key actions and monitoring success in approaching sustainability. In this vein, several Finnish municipalities/cities have already started to conduct VLRs of their progress towards reaching the SDGs. The city of Helsinki is one of these: while sustainable development is a long-standing priority for the city, preparing a VLR allowed it to connect that commitment to the SDGs and generate more knowledge about sustainability (City of Helsinki 2019).

Secondly, sustainability work means dealing with a broader time horizon, advancing the required long-term processes with perseverance, even stubbornness. This would be self-evident if it were not that the sustainability experts are dependent on external funding for projects and often for their own salaries. This 'project addiction' or treadmill chops long-term work into small, less strategically led pieces, particularly in smaller municipalities with few resources. When linked to strategic thinking, well-localised sustainability indicators (while always limited) provide a view over short-term projects. When understood properly, monitoring progress can function as an antidote to the project treadmill phenomenon.

Thirdly, SDGs provide the municipal sector with a shared language that can glue together experts working on diverse and seemingly incompatible challenges, ranging from saving the Baltic Sea to decreasing inequalities between demographic groups. While not solving any ethical questions per se, SDGs can be used to facilitate these often-challenging multidisciplinary discussions and analyse interlinkages—both synergies and trade-offs (Bowen et al. 2017). Thereby the framework should be helpful in prioritising resource use as well.

Another strength of the SDG framework was identified in its relative independence from partisan politics. Such politicisation was remarkably absent from the discussions and interviews, or at any rate, the dividing lines were not drawn on party political grounds. As Chigudu (2020) argued, progress on SDG 16 is critical to progress on the other SDGs and, therefore, it is imperative that countries vigorously attempt to overcome those challenges to meet the targets indicated for each goal. Some populist parties notwithstanding, the sustainability discourse has gradually, in the Finnish context at least, been mainstreamed into political discussions, nearly without a party colouring. The 4-year electoral term of municipal councils and chairpersons is a factor. Even so, by combining effective leadership in the big picture and co-designing the more detailed goals, leaders can establish a long- or medium-term programme for sustainability.

4.2 The Role of the Central Government in Supporting Municipalities

The relationship between the national government and local governments depends greatly on the national context. This ultimately pertains to national legislation on the matter but, in Finland, as in many other countries, increasingly to other policy instruments as well. *Collaboration, support, cooperation* and *guidance* are some of the terms that can be assigned to the nation state-local government interaction. Furthermore, increasingly important is also the willingness and ability to collaborate across local governments, a symbol of which is the growing number of national and international city networks.[10] Therefore, while it may be thought that a national government and local government relationship would follow a top-down approach, this is an increasingly dated view.

In Finland, municipal administration in local governments is based on the self-government of the municipality's residents. A municipal council, elected every 4 years, decides on the long-term goals of the municipality's activities and finances in its municipal strategy. In addition to the council, a municipality has to have a municipal board and a board of auditors. The council may also appoint other institutions, such as boards and chambers. The council elects the municipal director or mayor.

Finnish municipalities have the right to levy taxes, and the residents and users of services have the right to participate and influence the work of the municipality. The council must enable diverse and effective opportunities and means for participation. Residents (with no age limit) and the corporate entities and foundations operating in the municipality have the right to submit initiatives on matters concerning the municipality's activities.

Finnish municipal self-governance is anchored in the Constitution of Finland, and decision-making is based on the Local Government Act. The goals of the Local Government Act (see Fig. 14.5) can be viewed with an eye for the SDG framework, with, for example, 'safeguarding the financing of municipal tasks' having a direct link to the attainment of SDGs related to food (SDG 2), education (SDG 4), water and sanitation (SDG 6), work and growth (SDG 8) and so on. Therefore, it is in fact mandated for the local council to limit the risks and mitigate possible negative impacts related (almost directly) to a wide range of SDGs, and they should voluntarily do so for the rest.

The risk of not attaining SDGs in the local public domain is eventually felt by the municipality's inhabitants. It is the right of the individuals (by law) to take part in their local governance, and it is the council's obligation to provide the means to do so. It is clearly the obligation of the council and the mayor/municipal director to safeguard against later regret (i.e. to

[10] See, e.g., ICLEI: Local Governments for Sustainability and the Covenant of Mayors as examples of city networks.

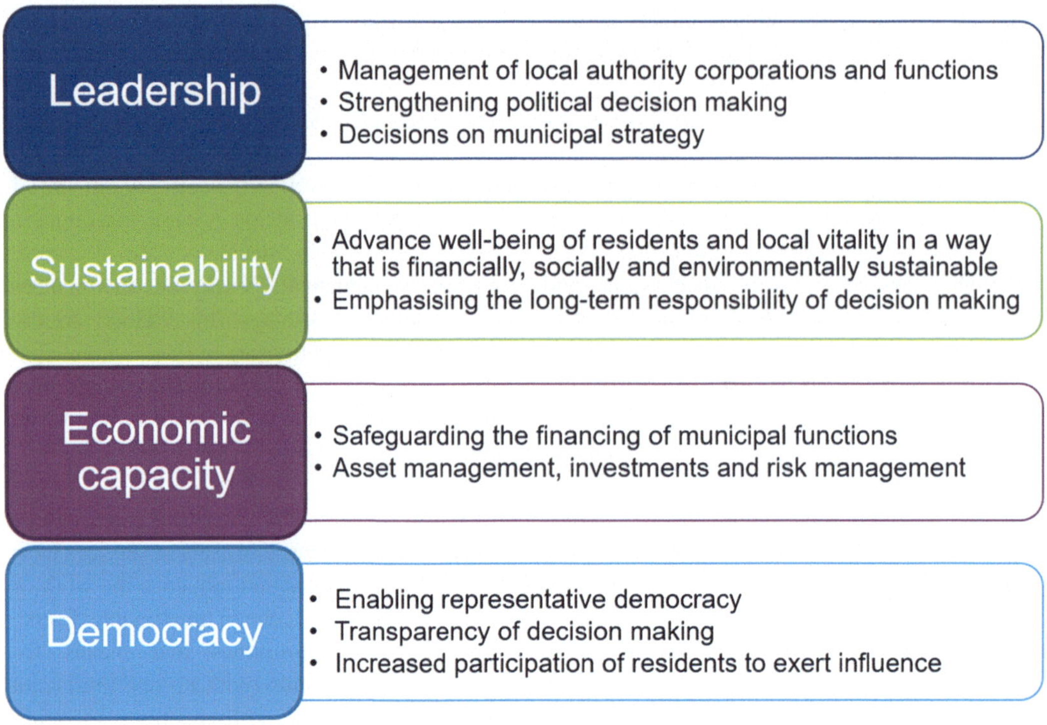

Fig. 14.5 The purpose of municipalities, according to the Local Government Act 410/2015

manage the risk on behalf of the residents and other local actors). At the same time, it is of great importance for the public sector to strive towards common goals. The SDG framework provides a good basis for cross-level governance objectives; however, it is important to specify the particulars that high-level target setting vs. localised target setting entails. This will also support the identification of responsibilities, timelines and, not least, opportunities for collaboration.

5 Conclusions

This study corroborated earlier findings on the crucial role of municipalities in the implementation of the SDGs (see Sect. 2). The mobilisation of change agents (cf. Hajer et al. 2015) committed to Agenda 2030 is very much ongoing at the local level, but differently so in the diverse municipalities. Sustainable development—as an overarching global goal, obliging nations—can be attributed to being a lawful right of citizens, governed by the local elected officials. Depending on the level of awareness and ambition, local level actors can have a strong positive influence on the sustainability of their region. They can significantly accelerate the transition towards achieving Agenda 2030 at the local level, which can then also empower policymakers at the national level to make decisions that support the development towards the common good.

The added value—as well as the challenge—in SDG implementation is its extremely cross-sectoral nature. The 17 goals almost entirely cover the responsibilities of the Finnish municipalities as described in the Local Government Act (410/2015). Therefore, if there is no or little high-level commitment to sustainable development, it is impossible to promote the topic in a strategically meaningful way. Similarly, it is impossible to push all the 17 goals forward at the same time, and in any case, the localisation of the goals is needed. Only when the localisation and local priority setting have been conducted and integrated into the strategic goals of the city can sustainable development become a guiding principle at the level of local decision-making. Lacking this, strategic leadership becomes difficult: How can networks of actors be coordinated if one's organisational goals are unclear? How can long-term planning be enacted if sustainable development is resourced through short-term projects, or in fact outsourced to other entities? As Huxley et al. (2019) have phrased it, 'progress will be limited if the institutional processes that drive action are not re-configured in line with sustainable local visions'.

Cross-sectoral leadership is a prerequisite to leading through and with the SDGs. The cross-sectoral leadership traditions and capacities vary between different municipalities, and there are no 'one-size-fits-all' solutions. In this study, we presented three models for SDG leadership at the local level. The Guiding Stars model represents a model of strong, high-level commitment and often ambitious goal setting. The Power of Networks model highlights the importance of coordination between different sectors and networks. The Active Individuals model shows how individuals and groups can start to act independently, even without high-level commitment. Sometimes good results at the grassroots level, however small, also support a gradual change in values and mindsets at the high level.

All of these models are abstractions; in reality, most municipalities have the features of different models in their governing and management structures. In the depicted study of 12 Finnish municipalities, it can be concluded that the biggest municipalities correspond most often to the first model, the middle-sized municipalities with the second model and the small municipalities with the third model. This, however, does not mean that generalisations should be made based on the size of the municipality. How such models of SDG leadership interact with a municipality's general management models requires more research and testing. It can, however, be expected that, just as different-sized organisations have different organisation structures, their embedding of SDG leadership into their practices should also be different. Joining Zeemering (2017), we see sustainability as having a strong government reform value. Local level sustainability work can best be accelerated if municipalities identify and adopt a suitable and empowering leadership model. Based on such sustainability leadership, local leaders can (and should) consciously use SDGs for shared strategy-making, for experts' motivation and for organisational learning.

References

Berg, A., Lähteenoja, S., Ylönen, M., Korhonen-Kurki, K., Linko, T., Lonkila, K-M., Lyytimäki, J., Salmivaara, A., Salo, H., Schönach, P., & Suutarinen, I. (2019). *PATH2030 – Evaluation of Finland's Sustainable Development Policy*. Publications of the Government's analysis, assessment and research activities 23/2019. Prime Minister's Office, Helsinki. http://urn.fi/URN:ISBN:978-952-287-655-3. Accessed 3 May 2020.

Bowen, K. J., Cradock-Henry, N. A., Koch, F., Patterson, J., Häyhä, T., Vogt, J., & Barbi, F. (2017). *Implementing the "Sustainable Development Goals": towards addressing three key governance challenges—collective action, trade-offs, and accountability. Current opinion in environmental sustainability.* Elsevier B. V. https://doi.org/10.1016/j.cosust.2017.05.002.

City of Helsinki (2019). From agenda to action: The implementation of the UN Sustainable Development Goals in Helsinki 2019. City of Helsinki. https://www.hel.fi/static/helsinki/julkaisut/SDG-VLR-Helsinki--2019-en.pdf. Accessed 15 May 2020.

Chigudu, D. (2020). Public sector corporate governance: Zimbabwe's challenges of strategic management in the wake of sustainable development. *Academy of Strategic Management Journal, 19*(1), 1–13. https://www.tandfonline.com/doi/abs/10.1080/14781158.2019.1667320.

Finnish Government (2019). *Programme of Prime Minister Sanna Marin's Government.* https://valtioneuvosto.fi/en/marin/government-programme. Finnish Government. Accessed 15 May 2020.

Frantzeskaki, N., Broto, V. C., Coenen, L., & Loorbach, D. (2017). *Urban Sustainability Transitions*. New York: Routledge. https://doi.org/10.4324/9781315228389.

Gorissen, L., Spira, F., Meynaerts, E., Valkering, P., & Frantzeskaki, N. (2018). Moving towards systemic change? Investigating acceleration dynamics of urban sustainability transitions in the Belgian city of Genk. *Journal of Cleaner Production, 173,* 171–185. https://doi.org/10.1016/j.jclepro.2016.12.052.

Grin, J., Rotmans, J., & Schot, J. (2010). *Transitions to Sustainable Development: New Directions in the Study of Long Term Transformative Change.* New York/London: Routledge. https://doi.org/10.4324/9780203856598.

Gusmão Caiado, R. G., Leal Filho, W., Quelhas, O. L. G., Luiz de Mattos Nascimento, D., & Ávila, L. V. (2018). A literature-based review on potentials and constraints in the implementation of the Sustainable Development Goals. *Journal of Cleaner Production 198*, 1276-1288. https://doi.org/10.1016/j.jclepro.2018.07.102.

Hajer, M., Nilsson, M., Raworth, K., Bakker, P., Berkhout, F., de Boer, Y., … Kok, M. (2015). Beyond cockpit-ism: Four insights to enhance the transformative potential of the Sustainable Development Goals. *Sustainability (Switzerland), 7*(2), 1651–1660. https://doi.org/10.3390/su7021651.

Hargreaves, T., Haxeltine, A., Longhurst, N., & Seyfang, G. (2011). *Sustainability transitions from the bottom-up: Civil society, the multi-level perspective and practice theory.* Working Paper – Centre for Social and Economic Research on the Global Environment. University of East Anglia. https://www.econstor.eu/obitstream/10419/48796/1/662352246.pdf. Accessed 17 August 2020.

Helgeson, C. (2018). Structuring decisions under deep uncertainty. *Topoi, 39*(2), 257–269. https://doi.org/10.1007/s11245-018-9584-y.

Hoffman, J., & Loeber, A. (2016). Exploring the micro-politics in transitions from a practice perspective: The case of greenhouse innovation in the Netherlands. *Journal of Environmental Policy and Planning, 18*(5), 692–711. https://doi.org/10.1080/1523908X.2015.1113514.

Huxley, R., Owen, A., & Chatterton, P. (2019). The role of regime-level processes in closing the gap between sustainable city visions and action. *Environmental Innovation and Societal Transitions, 33,* 115–126. https://doi.org/10.1016/j.eist.2019.04.001.

Hölscher, K. (2019). *Transforming urban climate governance: Capacities for transformative climate governance.* Erasmus University Rotterdam. http://hdl.handle.net/1765/118721.

Kettunen, P., Heino, H., & Sankala, I. (2019). Strategioista toiminnaksi? Kestävän kehityksen edistäminen paikallisella tasolla. *Focus localis, 47,* 1. In Finnish. https://research.utu.fi/converis/portal/Publication/40284020?lang=fi_FI.

Klitkou, A., Bolwig, S., Hansen, T., & Wessberg, N. (2015). The role of lock-in mechanisms in transition processes: The case of energy for road transport. *Environmental Innovation and Societal Transitions, 16,* 22–37. https://doi.org/10.1016/j.eist.2015.07.005.

Köhler, J., Geels, F. W., Kern, F., Markard, J., Onsongo, E., Wieczorek, A., … Wells, P. (2019). An agenda for sustainability transitions research: State of the art and future directions. *Environmental Innovation and Societal Transitions, 31,* 1–32. https://doi.org/10.1016/j.eist.2019.01.004.

Krellenberg, K., Bergsträßer, H., Bykova, D., Kress, N., & Tyndall, K. (2019). Urban sustainability strategies guided by the SDGs – A tale of four cities. *Sustainability (Switzerland), 11*(4), 1116. https://doi.org/10.3390/su11041116.

Laszlo, C., & Zhexembayeva, N. (2011). *Embedded sustainability: The next big competitive advantage.* Stanford, CA: Stanford Business Books.

Le Blanc, D. (2015). Towards integration at last? The Sustainable Development Goals as a network of targets. *Sustainable Development, 23*(3), 176–187. https://doi.org/10.1002/sd.1582.

Leponiemi, U. (2019). *Collective capacity – communality in small municipalities.* Doctoral dissertation, University of Tampere. In Finnish.

Loorbach, D. (2010). Transition management for sustainable development: A prescriptive, complexity-based governance framework. *Governance, 23*(1), 161–183. https://doi.org/10.1111/j.1468-0491.2009.01471.x.

Local Government Act (2015). 410/2015. Ministry of Finance, Finland. https://www.finlex.fi/en/laki/kaannokset/2015/en20150410.pdf. Accessed 2 May, 2020.

Lyytimäki, J., Lonkila, K.-M., Furman, E., Korhonen-Kurki, K., & Lähteenoja, S. (2020). Untangling the interactions of sustainability targets: synergies and trade-offs in the Northern European context. *Environment, Development and Sustainability, April 2020.* https://doi.org/10.1007/s10668-020-00726-w

McIntyre, A. (2007). *Participatory action research.* Los Angeles: Sage Publications.

Morse, R. S. (2010). Integrative public leadership: Catalyzing collaboration to create public value. *The Leadership Quarterly, 21*(2), 231–245. https://doi.org/10.1016/j.leaqua.2010.01.004.

Mukhtar-Landgren, D., Kronsell, A., Voytenko Palgan, Y., & von Wirth, T. (2019). Municipalities as enablers in urban experimentation. *Journal of Environmental Policy and Planning, 21*(6), 718–733. https://doi.org/10.1080/1523908X.2019.1672525.

Pel, B. (2016). Trojan horses in transitions: A dialectical perspective on innovation 'capture.' *Journal of Environmental Policy and Planning, 18*(5), 673–691. https://doi.org/10.1080/1523908X.2015.1090903.

Quitzau, M. B., Jensen, J. S., Elle, M., & Hoffmann, B. (2013). Sustainable urban regime adjustments. *Journal of Cleaner Production, 50,* 140–147. https://doi.org/10.1016/j.jclepro.2012.11.042.

OECD (2016). Regional policy profile Finland. https://www.oecd.org/regional/regional-policy/profile--Finland.pdf. Accessed 15 May 2020.

Roberts, C., Geels, F. W., Lockwood, M., Newell, P., Schmitz, H., Turnheim, B., & Jordan, A. (2018). The politics of accelerating low-carbon transitions: Towards a new research agenda. *Energy Research and Social Science, 44,* 304–311. https://doi.org/10.1016/j.erss.2018.06.001.

Rotmans, J., Kemp, R., & Van Asselt, M. (2001). More evolution than revolution: Transition management in public policy. *Foresight, 3*(1), 15–31. https://doi.org/10.1108/14636680110803003.

Sallinen, S., Majoinen, K., & Seppälä, J. (2017). *Toimiva kunta. Hyvinvointia! Sivistystä! Elinvoimaa!* Association of Finnish Municipalities. In Finnish.

Swilling, M., & Hajer, M. (2017). Governance of urban transitions: Towards sustainable resource efficient urban infrastructures. *Environmental Research Letters, 12*(12). https://doi.org/10.1088/1748-9326/aa7d3a.

UN (2019). *Global sustainable development report: The future is now – science for achieving sustainable development.* New York: United Nations. https://sustainabledevelopment.un.org/content/documents/24797GSDR_report_2019.pdf. Accessed 13 April 2020.

Valencia, S. C., Simon, D., Croese, S., Nordqvist, J., Oloko, M., Sharma, T., … Versace, I. (2019). Adapting the Sustainable Development Goals and the New Urban Agenda to the city level: Initial reflections from a comparative research project. *International Journal of Urban Sustainable Development, 11*(1), 4–23. https://doi.org/10.1080/19463138.2019.1573172.

Von Wirth, T., Fuenfschilling, L., Frantzeskaki, N., & Coenen, L. (2019). Impacts of urban living labs on sustainability transitions: mechanisms and strategies for systemic change through experimentation. *European Planning Studies, 27*(2), 229–257. https://doi.org/10.1080/09654313.2018.1504895.

Zeemering, E. S. (2017). Sustainability management, strategy and reform in local government. *Public Management Review, 20*(1), 136–153. https://doi.org/10.1080/14719037.2017.1293148.

Zeni, J. (1998). A guide to ethical issues and action research. *Educational Action Research, 6*(1), 9–19. https://doi.org/10.1080/09650799800200053.

Zinkernagel, R., Evans, J., & Neij, L. (2018). Applying the SDGs to cities: Business as usual or a new dawn? *Sustainability (Switzerland), 10*(9). https://doi.org/10.3390/su10093201.

15 Opportunities and Challenges for Local Government Institutions in Localising Sustainable Development Goals in Zimbabwe

Cowen Dziva and Itai Kabonga

Abstract

The post-2015 development Agenda is premised on 'leaving no one behind' in its quest to realise sustainable development. Consequently, there have been increasing clarion calls for the localisation of Sustainable Development Goals (SDGs). This qualitative study interrogates the potential of key governance and development institutions in localising SDGs in Zimbabwe. Data were gathered through literature review and key informant interviews with 25 purposively and conveniently sampled leaders and experts in local development. With a focus on traditional leadership and the Rural District Councils (RDCs), the chapter unearths a wide range of opportunities for infusing SDG targets into by-laws, policy and development plans; forging of partnerships for community and resource mobilisation; and funding of SDGs, conflict management, service provision and ecological resource management in rural communities. In practice, however, the potential of local government institutions to localise SDGs is limited by institutional incapacity, resource constraints, limited autonomy and imperfect flow of information and data on SDGs, political patronage and corrupt tendencies. Resultantly, the implementation of SDGs by a majority of local institutions has been piecemeal and insignificant. The chapter vouches for capacity building and resourcing of local institutions and their leaders to understand and effectively localise SDGs.

Keywords

Traditional leadership · Local institutions · Localisation · Development

1 Introduction

The expiry of the Millennium Development Goals in 2015 saw the United Nations General Assembly (2016) adopting the 2030 Agenda for Sustainable Development as the new global development policy. Unlike the MDGs which had 8 goals, the adopted framework is guided by 17 Sustainable Development Goals (SDGs), 168 targets and 247 indicators meant to eradicate poverty and guarantee a secure and sustainable future for the world by 2030. The Agenda for Sustainable Development has been hailed for taking into cognisance the core

C. Dziva (✉)
Nehanda Centre for Gender and Cultural Studies, Great Zimbabwe University, Masvingo, Zimbabwe

I. Kabonga
Department of Humanities and Social Sciences, Zimbabwe Ezekiel Guti University, Bindura, Zimbabwe

G. Nhamo et al. (eds.), *Sustainable Development Goals for Society Vol. 1*, Sustainable Development Goals Series, https://doi.org/10.1007/978-3-030-70948-8_15

dimensions of sustainability, namely, social progress, economic growth and environmental protection (Nabarro 2016). The 17 SDGs are grouped into five categories, namely people, planet, prosperity, peace and partnership (Vaggi 2016). This clearly shows the determination of drafters to deal with real challenges affecting humanity on the globe. Thus, many regard Agenda 2030 to be transformative in nature as it seeks to 'end poverty and hunger, while safeguarding the planet' (Nabarro 2016, p. 23). The former UN Secretary-General, Ban Ki-moon, further touted the Agenda as 'a to-do list for people and planet, and a blueprint for success' (Wills 2016, p. 3).

Of the 17 goals, many UN member states, including Zimbabwe, have prioritised certain goals for implementation. In Zimbabwe, the government prioritised the following ten goals: 2–9, 13 and 17, as italicised in Box 15.1. The ten SDGs prioritised by Zimbabwe impact the jurisdiction of local institutions in relation to the provision of basic services including health, education, water and sanitation, food, proper management and harnessing of ecological resources for sustainable community development.

As good as the vision of the SDG framework seems, it takes strong and committed local institutions to implement and work towards the realisation of sustainable development at grassroots levels. Indeed, the persistence of development challenges in the world is not only a result of poor policies and plans, but more too a result of the failure to translate plans into firm action, results and reality for communities. Of late, the failure of global policies to change lives has been a result of limited buy-in and support at grassroots level. Resultantly, the post-MDG framework has increasingly called for the localisation of SDGs, which denotes the 'process of defining, implementing, and monitoring strategies at the local level for achieving global, national and subnational sustainable goals and targets' (UNDP 2014, p. 3). Localisation relies on varied local stakeholders harnessing their resources for local development. For these reasons, RDCs and traditional leadership institutions' partnerships with communities and private and public actors remain at the heart of SDG localisation processes in Zimbabwe. By their nature, RDCs and the institution of traditional leadership are well positioned to engage and involve more stakeholders to work towards the realisation of Agenda 2030's aim to 'leave no one behind' in society.

Many studies that support Agenda 2030's localisation at grassroots level where poverty remains pervasive have been documented (Reddy 2016; UNDP 2014). Yet, limited studies exist after the Agenda's launch to untangle how local government institutions have fared with SDG localisation in Africa. This study unearths the opportunities and limitations of RDCs and traditional leaders in localising SDGs in Zimbabwe. It is now 5 years after the adoption of the framework, and many local institutions' experiences remain undocumented even though these experiences can be useful in informing policymakers on the course of action to take towards the realisation of SDGs (Fenton and Gustafsson 2017; Reddy 2016; Gustafsson and Ivner 2017). Fenton and Gustafsson (2017) aver that the knowledge gap related to localisation of SDGs into municipalities' existing strategies, policies and practice needs to be elucidated by research to enhance effective and rapid implementation. Therefore, the chapter builds a case for strengthening the localisation of the Agenda 2030 discourse in Zimbabwe and other global South nations.

Box 15.1 : The Sustainable Development Goals Prioritised by Zimbabwe

- SDG 2: Zero hunger
- SDG 3: Good health and well-being
- SDG 4: Quality education
- SDG 5: Gender equality
- SDG 6: Clean water and sanitation
- SDG 7: Affordable and clean energy
- SDG 8: Decent work and growth
- SGD 9: Industry, innovation and infrastructure
- SDG 13 Climate action
- SDG 16: Peace and justice

Source: Authors

The chapter has five sections. It starts with this introduction and background which provide a synopsis of SDG framework and the localisation mantra. This section is followed by a review of literature on the composition and structure of traditional leadership and RDCs in Zimbabwe. Thereafter, the chapter explains the research method used which is followed by the presentation and discussion of findings regarding the opportunities and challenges of local government institutions in localising SDGs. The chapter ends with a conclusion and policy options for effective localisation and realisation of SDGs.

2 Literature Review

There has been growing consensus that the implementation of global development goals is largely hampered by limited grassroots support in the global South (Reddy 2016; Jime´nez-Aceituno et al. 2020; Chirisa 2012; Fenton and Gustafsson 2017; Moallemi et al. 2020). Thus, studies have been calling for strengthening of bottom-up initiatives to improve local knowledge, ownership and, ultimately, localisation of global development goals (Reddy 2016; Moallemi et al. 2020; UNDP 2014). By its nature, the localisation process involves concrete tools, mechanisms, innovations, platforms and processes for effective translation of development agenda into results at grassroots level (Reddy 2016). The localisation process thus places local governments at the centre of all efforts through which global development ideals and targets can be fulfilled (United Cities and Local Governments 2016).

In Zimbabwe, the key local government institutions in rural areas include the RDC and traditional leadership. This chapter deliberately excludes the DA from the focus of the study as this office directly works with RDCs and traditional leaders in varied ways. There exists over 60 RDCs established by the Constitution of Zimbabwe, and operationalised by the RDCA to represent and manage the affairs of rural communities (Chigwata 2018). The specific roles of RDCs include making by-laws on numerous issues that impact the realisation of SDGs, including bush fires, fisheries, fences, agricultural and related services, animal diseases and obstruction of water flow (RDCA 1988). RDCs implement their mandate through the elected Council and its administrative arm comprised of experts in socio-economic development fields. As the highest decision-making body, the Council is made of councillors, each democratically elected by voters to represent, foster accountability and, above all, enhance the performance of the local authority. The councillors sit and participate in five mandatory committees (Finance Committee, Town Board, the Roads Committee, Ward Development Committees (WADCOs) and the Rural District Development Committee (RDDC)) and other committees and subcommittees as may be created by the RDC. Some of these committees have oversight and policy development roles, while others have implementation roles which they undertake with supervision from the Council (Chigwata 2018). While councillors participate in many of these committees as a group, each of them has the chance to chair the WADCO in their respective wards.

The traditional leadership institution remains another key local government institution for SDG localisation in Zimbabwe. It is composed of 271 paramount chiefs, 452 headman and 2500 village heads who are the custodians of religion, culture and agents of development in their respective areas (Chigwata 2018; Musekiwa 2012). The village head chairs the village assembly, which brings together all inhabitants of over 18 years in the village to partake in development planning and implementation (Chigwata 2018). The village head reports to the headman—a sub-chief—who then reports to the paramount chief of an area (Chakaipa 2010; Musekiwa 2012). The headman, together with the local councillor and village heads, also oversees the work of the village assemblies and reviews their plans before their incorporation in rural district development plans. By their nature, the traditional leadership institution and the RDC are well positioned and capable of spearheading local development in Zimbabwe (Chigwata 2018; Musarandega et al. 2018; Ringson 2017).

Although various studies have examined RDCs and the institution of traditional leadership in relation to spearheading development from below (Chirisa 2012; Chigwata 2018; Musarandega *et al.* 2018; Ringson 2017), their potential in localising global development ideals has received less attention. A study by Chirisa (2012) focused on the potential of African local governments to localise MDGs. While the study's results remain key in ensuring a better understanding of the opportunities and pitfalls in localisation of global goals in the global South, the study was concerned with MDG framework, and was only focused on looking at urban and rural councils at the exception of traditional leadership institution. Thus, there exists a paucity of literature on the efforts of local government institutions to implement SDGs in Africa, and Zimbabwe in particular.

The scarcity of studies on SDG localisation persists despite a burgeoning global literature that largely speaks of local government institutions' cutting-edge advantages for fulfilment of this cause (Reddy 2016; Bowen et al. 2017; Fenton and Gustafsson 2017; Florida 2015; Gustafsson and Ivner 2017). Many of these studies unravel the broader spectrum of responsibilities and institutional opportunities, conditions necessary and challenges for SDG localisation (Florida 2015; Morita et al. 2020). One cutting-edge advantage of local governments remains the coincidence that many of the SDG content and targets speak directly to local institutions' mandate in community development (Kemp, Parto & Gibson 2005). For Gustafsson and Ivner (2017, p. 302), this alone is enough evidence that 'many of the goals will have to be dealt with locally'. As they are domiciled in local contexts, local governments are strategically positioned to identify the poor and direct resources towards poverty alleviation and enhancing local people's access to food, training and education, health and other necessities for improved well-being of local communities (Reddy 2016; Kemp et al. 2005).

Reddy (2016) further speaks of local governments' abilities to effectively manage collective resources, and to reform land tenure in ways that protect the rights of the poor (United Cities and Local Governments 2016). They are also strategically positioned to combat climate change, protect the environment and support agricultural production for greater food security of their inhabitants (Morita et al. 2020; Florida 2015). Local governments' improvement of infrastructure attracts investment and development of industries that result in local production, economic growth, access to markets and ultimately employment creation, and local food chains (Morita et al. 2020).

In many studies for localisation of global goals, local governments are increasingly viewed as models for gender equality and women empowerment through non-discriminatory service provision and fair employment practices (Hendriks 2018; United Cities and Local Governments 2015; Reddy 2016). As the major service providers, 'local governments can have a significant impact on women's lives by using fair employment practices and ensuring non-discriminatory service provision to citizens' (United Cities and Local Governments 2015, p. 6). Gender-balanced local government structures are poised to prioritise and mainstream gender equality, and above all challenge gender stereotypes, thereby becoming role models to young women and girls (Hendriks 2015). With their proximity and wielding power in local communities, many local government institutions are strategically positioned to combat violence and other harmful practices that are detrimental to the realisation of equality and sustainable local development ideals (United Cities and Local Governments 2016).

Local governments also remain key in resource mobilisation through public-private partnerships and revenue collection for pro-poor service delivery (UNDP 2014; Reddy 2016; Hendriks 2018). According to Gustafsson and Ivner (2017, p. 301), local institutions are 'constant local actors close to citizens and they can also influence other actors through their strategies' to fund local development efforts. In many developing countries, Zimbabwe included, however, funds generated by local institutions have been marred with accountability challenges (Chirisa 2012; Chigwata 2018). These corrupt and accountability challenges can heavily affect prioritisation of SDG goals and targets (Hendriks 2018). Coupled with underfunding from central government, many municipalities cannot function properly and look after the needs of the poor and special interest communities (Hendriks 2018; Chirisa 2012).

Other identified challenges for localisation of global development ideals include weak governance systems, policy and institutional inconsistency, competing local interests, resource constraints, limited access to information about development strategies and models, and technology (Moallemi et al. 2020; Chirisa 2012; Staford-Smith et al. 2017; Reddy 2016; Bowen et al. 2017; Gustafsson and Ivner 2017). In a study by Morita et al. 2020 in Japan and Indonesia, a weak governance and institutional system was found inhibitive to the implementation, monitoring and evaluation of SDGs at the local level. Similar sentiments are deduced from Weymouth and Hartz-Karp (2018), who found strong governance and accountability to be key in localisation of SDG targets. With regard to localisation of MDGs in Africa, and Zimbabwe, Chirisa (2012) identified gaps to include human and financial resource challenges.

While there is growing consensus that local governments are potentially well positioned to localise SDGs, it is less clear how they are faring in practice. Many emerging global experiences on the subject are generally symptomatic of local governments' potential (Hendriks 2018; Reddy 2016; Gustafsson and Ivner 2017) without a deeper interrogation of their localisation of SDGs in practice. It therefore remains important that this study rigorously analyses local governance systems with a view to understand their performance and effectiveness in implementing SDGs from below.

3 Materials and Methods

The study is premised on a qualitative research approach to understand the opportunities and challenges of the localisation of SDGs in rural Zimbabwe. Data for the study were gathered through a review of extant literature and key informant interviews with 25 respondents working in local governance and development. Of the 25 respondents, 18 were purposively and conveniently recruited from local institutions and non-state actors from Mwenezi and Chiredzi districts of Masvingo province while the remainder 7 were from Mberengwa district in the Midlands province of Zimbabwe. In many instances, the study relied on the contacts of CEOs of RDCs and DAs to reach out to local leaders, and experts in their respective districts for interviews. For state institutions, the study reached out to five traditional leaders (mainly chiefs), three councillors, two chief executive officers and three DAs of the selected districts. The study further interviewed nine officers for government departments including for rural agricultural extension officers, environmentalists and general healthcare providers. Their diversity and the varied socio-economic sectors they represent made them strategic informants with insights on SDG localisation in Zimbabwe. The study further reached out to three local development experts from disadvantaged groups and civil society organisations working to advance local socio-economic development in Zimbabwe. It is from these experts that the researchers gained insights on the mandate, practices and prevailing challenges of RDCs and traditional leadership institutions in localising SDGs.

Due to the COVID-19 restrictions, the study utilised phone and skype interviews. The researchers reached out to the participants to seek consent and make arrangements for the interviews, which were normally conducted a few days from the notification date. With consent from participants, some interviews were recorded for later transcription.

Primary data were triangulated with extant literature on local governance and development. Documents that were reviewed with the aim of understanding the opportunities and gaps in localising SDGs included laws and policies regulating local government institutions, as well as the Constitution of Zimbabwe, the Traditional Leaders Act and the Rural District Councils Act (RDCA). The study also utilised white and grey literature from books, journals and newspaper reports on the efforts and challenges of local government institutions in Zimbabwe.

Data were analysed in accordance with the opportunities and challenges of local government institutions in localising SDGs. With the thematic analysis technique, the study managed to extract the sentiments of participants and documented views in the extant literature on prospects for the localisation of SDGs in Zimbabwe.

4 Presentation and Discussion of Findings

4.1 Opportunities and Practices of Local Government Institutions

Local government institutions are presented with various opportunities to translate SDG targets into firm action and concrete results for local communities in Zimbabwe. By their nature, local institutions have a broad community development mandate, possess legal power, command respect and are closer to the grassroots, something that makes them indispensable in the localisation of SDGs. This section untangles the wider opportunities available for RDCs and traditional leadership to localise SDGs through policy formulation, coordination, resource and community mobilisation, service provision and, above all, peace building.

4.2 By-Laws, Policy and Development Plan Formulation

Local institutions are involved in legal reform, policy formulation and implementation for community governance and development. While many of the laws and policies are made by central government, by-laws and financial and development plans are made by local institutions. Besides, local leaders such as councillors, headmen and village heads sit in various committees and, together with locals, draw financial and development plans for implementation while taking cognisance of the prevailing geopolitical environment in their respective communities. These processes present opportunities for local leadership to infuse and mainstream gender and other critical issues as advocated by the SDG framework. Cognisant of the need to localise SDGs, local leaders are poised to come up with development policies and plans infused with SDG targets. Upon adoption, the SDG infused policy and development plans are relayed and cascaded to grassroots by local leaders, mainly councillors and their respective communities, for onward implementation. The inclusive nature of the process allows for community buy-in and understanding of SDG infused policies and development plans for effective implementation and realisation.

Another opportunity to integrate SDGs presents itself when traditional leaders and their advisors draft guidelines for governing their communities in line with national laws and policies. The guidelines include those for determining fines for diverse crimes committed by community members. Accordingly, leaders come up with fines for varied issues that, if kept unchecked, affect community development and realisation of SDGs. It emerged from this study that for wanton cutting down of trees, streambank cultivation and other practices that damage the environment or affect community harmony, one is liable to pay a fine in the form of cattle, goats, chickens or monetary equivalent, depending on the gravity of the matter. In some instances, the guidelines stipulate that the offender pay a fine to the traditional leader as well as restitution to the complainant.

The institution of traditional leadership is also strategically positioned to lobby and advocate for modification and abolishment of inimical and harmful practices which violate human rights and limit the realisation of SDGs. With the 2013 Constitution having outlawed harmful traditional and cultural practices such as child marriages, child pledging and domestic violence, traditional leaders are also amending their guidelines to stipulate fines for these anti-SDG practices. While some leaders have been accused of fanning inimical practices such as virginity testing and child marriages (Sithole and Dziva 2019), many traditional leaders interviewed for this study are reliable actors in ending many of these harmful practices. In their explanations, traditional leaders confirmed using varied forums to discourage societies from harmful practices. The actions by local leaders are commendable considering the prevalence and implications of harmful practices that militate against gender equality, educational attainment and ultimately social development. If left unresolved, practices such as domestic violence often result in family disintegration and destitution on the part of children, injuries and even death on the part of the victim (Dziva et al. 2020). When this happens, the realisation of many SDGs will largely remain a pipe dream.

4.3 Coordination of Local Development

The local government structures have the opportunity to localise SDGs as they coordinate planning, implementation and monitoring of community development projects in each district. Critical in this coordination is the fact that technical experts (DA, the Council and its secretariat) are involved most in the process at district level. For instance, the legal requirement that the DA coordinates the RDDC and subsequent community development in a district allows for this office to direct development opportunities to where they are needed most. Discussion with some DAs also confirmed how they utilise this opportunity to direct local investors and NGOs to areas where they have comparative advantages to leave a mark in the realisation of SDGs and wider community development. Because they possess knowledge of their districts and development issues, the DAs can critically assist in the localisation of SDGs.

The decentralised structures of RDCs and traditional leadership make such institutions integral players in the endeavour to complete the localisation of SDGs in Zimbabwe. Owing to their decentralised structures and platforms, local institutions can effectively relay SDG targets to communities, and make them participate in identifying local opportunities and strategies for the realisation of SDGs at the lower levels of the society. With so much local participation and flow of information, there can be mutual agenda setting and a shared vision in SDG implementation. This is likely to result in the smooth implementation of SDGs, whereby leaders understand contextual and first-hand challenges and opportunities to tap into for the realisation of SDGs in their communities.

4.4 Enforcement of By-Laws and Policies

Local institutions use various means to enforce laws and policies in their respective communities. The power that local institutions wield in enforcing mechanisms remains an opportunity for the effective implementation of SDGs. With their local courts, traditional leaders preside over cases and fine community members who are found guilty of anti-development actions and practices. For grave criminal cases, traditional leaders directly refer to police and magistrates' courts. In relation to environmental protection, the leaders confirmed to be fining guilty community members for causing veld fires, deforestation and stream-bank cultivation and triggering erosion through the use of sledges.

It also emerged from discussions with environmental health technicians that traditional leaders are playing a key role in enforcing COVID-19 health regulations. All those who disregard social distancing by holding their cultural and traditional functions in the wake of the COVID-19 pandemic are fined by local chiefs. The EHT further explained how traditional leaders' enforcement of rules can help reduce infant and maternal mortality as families of women who give birth at home and not at clinic are made to pay a goat as fine. This resembles the Malawi scenario in which mothers were fined 500 to 3000 Malawian kwacha (US$1–$5) for giving birth at home instead of going to the clinic (Walsh et al. 2018). It also emerged from the study that failure to have one's children immunised during a national immunisation programme attracts a fine for the household. Due to the respect, power and coercive means of traditional leaders, many communities choose to comply with set guidelines, thereby working towards the realisation of SDGs in communities.

Three of the five traditional leaders interviewed for the study, however, complained of resistance from locals and of limited means to enforce their court decisions. In most instances, the failure to pay fines and comply with local court orders will see the chiefs' court instituting other measures including threatening to forcibly evict the offender from the community. In one case scenario, a chief in Matabeleland region of Zimbabwe presided over a case where a villager allegedly caught his wife having sex with another man in their matrimonial home. The chief banished the woman from his community. When the woman refused to leave the area, the chief in question asked his subjects to evict the couple by getting rid of the fence surrounding the homestead and cattle kraal (Mashundu 2019).

It is such radical actions by traditional leaders in compelling non-complying members that result in leaders violating human rights and defeating the SDG spirit. Even though the chief was practising his culture, his actions largely violated the family's wider rights, including their right to property. The case confirms how some traditional leaders, if unmonitored, may violate the rights of their subjects. Relatedly, some communities accuse traditional leaders of using their powers to discriminate and punish their perceived enemies and to extort people's livestock in frivolous and trumped-up charges. As Rukuni et al. (2015) aptly captured, the partiality of traditional leaders when adjudicating cases is most evident in politically sensitive matters, or where these leaders have interests which involve their relatives. The plight of aggrieved persons is made worse by the fact that justice by most traditional leaders lacks proper appeals and review mechanisms. When this happens, the SDG framework's spirit of fairness, inclusivity and justice is compromised.

Nonetheless, traditional leaders make use of traditional and cultural values and norms to ensure community development, and the subsequent realisation of SDGs. As the custodians of culture and tradition, local leaders control, oversee and ensure the sustainable utilisation of ecological resources by their communities. Using cultural and traditional norms and values, communities are often encouraged to practise sound utilisation of 'sacred places' within their areas including ancestral lands, heritage, mountains, rivers and forests (Musarandega et al. 2018). Traditional leaders often instil fear in their communities by telling them that vandalising 'sacred places' comes with severe punishment from ancestors or the gods. In most instances, the punishment manifests itself in different ways as the accused can, for instance, get lost, confused or rather stranded for hours, if not days, while in such forests, mountains and other sacred places.

The need for sustainable utilisation of sacred places also applies to private investors coming into the community. Accordingly, investors are forced to follow proper guidelines and produce clear plans for the sustainable usage of the resources in line with local cultures. It is upon fulfilment of these requirements that leaders can conduct rituals to cleanse and clear such sites for development investments to commence (Musarandega et al. 2018). Courtesy of these cultural and traditional norms, rural communities have seen less anti-SDG ecological damages (Marango 2017). Progressive chiefs also take advantage of such platforms to inculcate environmental sustainability and, above all, demand for their communities to accrue greater benefits and employment from the investment in their area. During the Dande Dam construction in Mashonaland Central Province of Zimbabwe, however, traditional leaders objected to the project and this stalled progress (Dodo 2013). While some blame this for stalling investment, it also results in greater local inclusion and participation in planning and utilisation of their locally found resources for poverty reduction.

4.5 Community and Resource Mobilisation

The localisation of SDGs depends on the availability of financial, natural and human resources. Local institutions are integral in mobilising communities and resources towards the implementation of SDGs. Rural based local institutions often source the greater part of their income from central government to supplement what they get from levies (Chigwata 2018). In addition to these sources, local institutions engage in private, public and local partnerships to fund community development and ultimately the realisation of SDGs. Such partnerships are echoed in SDG 17: 9, specifically targeted to 'enhance international support for implementing effective and targeted capacity building in developing countries to support national plans to implement all sustainable development goals, including through NorthSouth, South-South and triangular cooperation'. Courtesy of these partnerships, local government institutions have improved their financial muscle to execute life-long projects, including the drilling of boreholes and wells, and construction of small dams, bridges, schools, health facilities and communication networks in their communities. In areas where projects of this nature were implemented, they are improving local people's lives through advances in waste management, infrastructural development and access to basic services.

Local leaders are marketing investment opportunities that are found in their areas to small and medium enterprises. This helps in the development of local economies with the potential to provide services, create employment and increase the tax base for RDCs to effectively respond to community problems. Local leaders' proximity and respect in communities make them appeal to wider social, religious, economic and political groupings for increased participation in development initiatives. As diverse groupings partake in community development, they bring with them varied and key skills, interests and experiences for effective and efficient implementation of cutting-edge projects for sustainable development.

Local institutions raise awareness on the strategies needed to combat community development challenges including climate change, environmental degradation and communicable diseases. In this study, traditional leaders and councillors were found to be using their community gatherings to raise awareness on anti-SDG practices such as gender-based violence, climate change and ecological damage. In the wake of COVID-19 disease, the President of Zimbabwe Emmerson Mnangagwa pleaded with traditional leaders to sensitise their communities and combat the spread of COVID-19 in rural Zimbabwe (Machivenyika and Mugabe 2020). As a way of complying with the president's call, some local leaders utilised their community platforms to raise awareness on the novel virus' preventative measures by emphasising and encouraging communities to practise social distancing and good hygienic measures. All these efforts remain critical in the localisation of SDGs, particularly SDG 3.

It also emerged from the study that many of the external organisations operating in rural areas rely on traditional leaders as link persons to reach out to local communities. It is a norm in Zimbabwe for organisations coming to work in rural areas to notify the local leaders of their envisaged intervention. Consequently, local leaders congregate relevant committees and inform the wider community of the intended intervention. As shown by Walsh et al. (2018) in Malawi, health experts and organisations heavily relied on traditional leaders for the mobilisation of communities for campaigns and activities that discourage overreliance on traditional birth attendants. Likewise, this study found local leaders to be working hand in glove with outside organisations operating in rural areas. The outsiders often leave their flyers and pamphlets with local leaders for onward distribution and information dissemination to all corners of their communities. With the assistance of headmen, village heads and messengers, chiefs easily cascade information, distribute IEC materials and place posters at strategic places such as community water sources and service centres to easily catch the attention of and reach out to wider society. It is under these circumstances that the study found local leaders to be reliable partners in the localisation of SDGs in Zimbabwe.

4.6 Maintenance of Peace and Justice

Chiefs and councillors implement SDG 16 through endogenous measures in managing local development affairs, preserving peace and ensuring access to justice in communities. Chigwata (2018) revealed that the dispute resolution role takes up about 55% of the official time of chiefs in their districts. In their conflict management practices, traditional leaders mediate and seek to resolve family and community disputes, including gender-based violence, land conflicts and other disagreements with potential to stall community development.

In relation to domestic violence, traditional leaders provide counselling services, reconcile conflicting couples and discourage the use of violence to resolve disputes. Traditional leaders have local courts that also act as alternative dispute resolution platforms that amicably and economically bring justice closer to the poor. One cutting-edge advantage of traditional courts remains their emphasis on 'reconciliation rather than retribution to ensure harmony among neighbours, relatives and communities in rural areas' (Keulder 1998, p. 173). In many instances, these courts' verdicts are respected and have ensured peace and tranquillity, which this study found to be vital conditions for the attainment of sustainable community development. In their negative ways, local disputes and conflicts degenerate into full-fledged intractable violence that results in internally displaced persons and refugees, deaths and wider human rights violations with potential to stall the realisation of SDGs.

This important role of local leaders has not been without challenges in rural Zimbabwe. In the heat of political competition, traditional leaders are given motor vehicles, fuel subsidies, allowances and improvements of their homes, arguably to buy their allegiance by incumbent parties and leaders (Chigwata 2018). As a consequence of such generosity, many local leaders have been found on the wrong side of the law, instigating political violence for incumbent parties. Unsurprisingly, election reports accuse traditional leaders of participating in electoral politics by playing a central role in promoting hate speech, frog marching villagers to polling stations, vote buying, intimidating and fanning political violence against opposition supporters in contravention of Section 133A of the Electoral Act of Zimbabwe (ZHRC 2018). This disturbs peace in society, and largely defeats the spirit contained in the SDG framework.

4.7 Provision of Services

Local institutions implement SDGs through the provision of essential services including food, health, education, water and sanitation for community functioning. In providing these services, local institutions work towards the realisation of varied SDGs, including goals 1 (no poverty), 2 (zero hunger), 3 (good health and wellbeing), 4 (quality education), 5 (gender equality), 6 (clean water and sanitation), 11 (sustainable cities and communities), 14 (life below water) and 15 (life on land). In relation to ensuring food security and ending hunger, local institutions periodically source agricultural inputs and food aid from the central government and non-state actors for distribution to their communities.

The provision of food by local leaders often takes two forms: unconditional aid to vulnerable groups (chronically ill persons, orphans and vulnerable children, elderly and persons with disabilities), and 'food for assets' or 'food for work' to able-bodied, but poor families. For the latter, the support is advanced after partaking in community-rebuilding projects such as gulley reclamation, rehabilitation of roads and bridges, and brick moulding for schools, halls or health facilities.

Another initiative by traditional leaders includes the *Zunde RaMambo* (the chief's granary), which is largely a measure for ensuring food self-sustenance of indigent families in a given chiefdom. In this initiative, the chief avails a piece of land and inputs for communities to grow crops on the leaders' behalf. In this project, communities come together and collectively render services in ploughing, weeding and harvesting these fields for the leader (Ringson 2017). Community participation in *Zunde RaMambo* is also an expression of oneness, belonging, reinforced collaboration, solidarity, relationships and loyalty to the leader (Mararike 1999). Thereafter, the chief keeps the proceeds and distributes to indigent and food-insecure families in times of need. A similar initiative was also evident in Malawi where chiefs mobilised subjects to cultivate crops to aid poorer and needy families (Walsh *et al.* 2018). In doing so, traditional leaders provide safety nets for the poor families, thereby building resilience against hunger and food insecurity as advocated for by SDG 2.

While challenges of inadequacy, politicisation and erratic supplies often hamper the utility of local institutions' initiatives to fight hunger, the 'real' poor beneficiaries have found the initiatives to be supportive. From a broader sense, the provision of food and other necessities to the poor remains crucial in reducing the impetus for conflicts in rural communities. Indeed, the failure by local institutions to provide essential services has degenerated into intractable and destructive conflicts that largely hamper the realisation of SDGs in many developing nations.

In some instances, local leaders utilise their indigenous knowledge systems (IKS) to ensure the sustainable utilisation of local resources by villagers to ensure food security and mitigate the vagaries of weather (Musarandega et al. 2018). Defined as the totality of information, skills and practices acquired by local people through past experiences, observations, informal experiments and intimate understanding of their environment, IKS is largely used by community leaders to mitigate communities' vulnerabilities to climate change and diseases (Marango 2017; Musarandega et al. 2018). With little access to modern weather forecast technologies and information, local leaders invoke IKS to predict seasons and climate and adequately inform their communities on the definitive course of action to take to alleviate adverse weather conditions. This confirms Nakashima et al.'s (2012) study that IKS have for long been used to interpret incidences of meteorological phenomena such as drought, storms and floods, thereby guiding local communities through offering early warning, risk reduction and management of disasters. Similarly, traditional leaders invoke their IKS to deal with allergies and diseases in communities. In offering these solutions, local leaders have contributed a great deal to the localisation of SDGs.

5 Challenges of Local Institutions in Localising SDGs

Notwithstanding the noted opportunities and practices, local institutions have not fared well in the localisation of SDGs in Zimbabwe. In practice, many local institutions are under-capacitated to infuse, fund, implement and report on SDGs and the set targets. In addition to the already discussed challenges, this section unearths the main limitations that hinder the effectiveness of RDCs and traditional leaders in localising SDGs in Zimbabwe.

5.1 Limited Capacity

The capacity of local institutions to coordinate and implement SDGs is affected by their limited knowledge and understanding of the SDG framework. This, therefore, affects its effective infusion into local by-laws, policies and plans, and its ultimate implementation in respective communities. Within a district, a few office bearers have know-how of SDGs, and these include the DA, local authority secretariat and few councillors, whereas the majority of leaders are not well versed in SDGs. For those with limited knowledge, the localisation of SDGs remains minimal and some happens coincidentally as they are responding to community challenges in their areas of jurisdiction.

Limited education remains another challenge that limits the capacity of local leaders to effectively implement SDGs at local levels. This emanates from the fact that educational qualifications are not a requirement for one to be elected a councillor and be appointed as a traditional leader in Zimbabwe. As such, many councillors and traditional leaders have not acquired formal education (Chigwata 2018). This emanates from the fact that this then brings into question their ability to effectively comprehend and deliver their mandate, especially when it comes to technical issues, including budgeting and infusing SDGs into strategic plans. More often than not, local development plans gather dust in drawers and shelves of local authorities owing to the limited capacity of councillors to implement them (Chakaipa 2010). This negatively impacts the ability of local leaders to understand SDGs, infuse them into local plans, constantly implement, track and monitor progress towards realisation.

5.2 Limited Resources

Resource constraints cause an unfathomable strain on local government institutions' capacity to effectively respond to local people's service needs. Due to many factors, including their inability to engage in vigorous revenue generation, many local authorities are not able to raise enough resources for their planned activities. RDCs in Zimbabwe are owed huge sums of money by local ratepayers (Chigwata 2018; Chakaipa 2010). While some businesses fail to honour their levy obligations because of poor business returns, others boycott as they protest poor service delivery in their areas. RDCs have also been blamed for their ineffective debt recovery mechanisms (Chigwata 2018). Other cited explanations for non-payment of taxes relate to limited awareness and arbitrary, regressive and sometimes forceful tax collection methods. All these actions make rural communities reluctant to honour their tax obligations, resulting in limited revenue for local authorities.

Saddled with these challenges, many RDCs rely on central government and external partnerships to bankroll their community development. Yet, government disbursements have been erratic and far below their requests, especially considering the recurring socio-economic crisis and political polarisation that Zimbabwe has found itself in the first two decades of the twenty-first century. Despite its utility in propelling community development, donor support has its own sustainability challenge. The financial challenges of RDCs further undermine their credit worthiness to moneylenders. Thus, resource constraints largely make the SDG framework a mantra for the DA, CEO and Council chair's opening meeting statements without much action on the ground.

5.3 Political Patronage and Corruption

Local institutions' efforts to localise SDGs are also hampered by corruption and increased political interference by central government into local affairs. As noted in previous sections, the local government laws in Zimbabwe increasingly cede power to the appointing authority and responsible minister to dictate affairs of RDCs and traditional leaders. For instance, the appointments of the CEOs and officers of the RDCs and those of traditional leaders are made with consent from politicians including the president and the minister responsible for local government (RDCA 1988; Constitution of Zimbabwe 2013). The interference is even worse for elected councillors who belong to political parties. The institution of traditional leadership also receives more directives from the DA (Dodo 2013). Consequently, local leaders have become conduits through which governments and politicians of the day gain political mileage than development in rural areas.

As finances, agricultural inputs and food aid for onward distribution to their respective communities often come from and through politicians and central government officials, local leaders are often coerced to politicise such assistance. Several complaints have been made against the partisan distribution of Cyclone Idai floods and COVID-19 food aid by ZANU PF councillors and traditional leaders in uttermost disregard of the local laws and policies requiring such leaders to be non-partisan in their operations and response to disasters (Mavhinga 2019). An investigation by the Zimbabwe Human Rights Commission (ZHRC) (2018) implicated Chief Makuni in partisan distribution of presidential inputs in the Rushinga Constituency. The partisan resource distribution by leaders has been done without considering gender equality and vulnerability levels of beneficiaries. Partisan aid distribution defeats the spirit enunciated in the SDG framework to 'leave no one behind'.

This study also found rampant corruption and bad governance to be significant adversaries against the localisation of SDGs. A study by Chakaipa (2010) documented the arrest and appearance of Chief Negomo of Mashonaland Central Province before the court for defrauding Mvurwi Grain Marketing Board of farming inputs (fertilisers) meant to benefit peasant farmers under the Presidential Summer Crop Season Programme. Corrupt and bad governance tendencies are also rampant within local authorities. Many local authorities in Zimbabwe have been accused of spending around 75% of their incomes on unjustifiably hefty salaries and allowances for senior management instead of service provision (Chigwata 2018). Furthermore, local authority leaders also practise corruption in allocation of resources, residential stands, handling of public funds and tenders. This explains why a considerable number of local government officials have been prosecuted for abusing their powers and council property (Chakaipa 2010). These corrupt tendencies defraud local authorities of revenue meant to fund community development and the ultimate realisation of SDGs.

The challenges of local institutions are compounded by the adversarial relationship that develops when RDCs and local chiefs compete for power in rural communities. In some instances, the adversarial relationship exists within RDCs, between experts (secretariat) and politicians (the council). The net result of these adversarial power relations and corruption becomes loss of focus concerning local government institutions' core service provision mandate and ultimately the achievement of SDGs.

5.4 Lack of Data

Equally challenging for local institutions is limited access to national data on SDG priorities, targets and progress. This study established that many local authorities find it difficult to access national SDG data and plans. Even with government prioritisation of the SDG framework, there remains limited data that has been shared with local authorities. Thus, many RDCs find it difficult to report on SDGs as the reporting templates hardly include such information. The need to create a national database of SDG localisation targets, progress and challenges remains vital for a shared national vision. This will go a long way in sharing notes on the practices and challenges of RDCs in implementing SDGs across the country. Without this data to support learning, planning and monitoring, the capacities of RDCs to localise SDGs remain limited and disjointed.

6 Conclusions

The localisation of Agenda 2030's clarion call to 'leave no one behind' hinges on concerted efforts by local government institutions, including RDCs and institution of traditional leadership, to mobilise and harness resources for their effective realisation in the global South. In rural Zimbabwe, RDCs and the institution of traditional leadership are constitutionally mandated to oversee local governance and development. This broader mandate placed upon local institutions and their proximity to local communities strategically positions them to mobilise local people and resources around them for the cause. By ensuring increased local people participation, RDCs and traditional leadership institutions improve on the capacities of locals to tap into wider community resources for the realisation of SDGs. Furthermore, local institutions are well placed to forge partnerships with outsiders to manage conflicts; reduce inequality, vulnerability to hunger and poverty; and work towards environmental sustainability. It is because of local institutions' lack of participatory approaches that ultimate efficiency, accountability, ownership and sustainable community development become unachievable.

Despite these opportunities, the chapter noted limited localisation of SDGs in Zimbabwe. The capacity of traditional leadership institutions and RDCs to localise SDGs is largely undermined by limited resources, limited education of leaders and their corrupt tendencies, political interference and patronage. Burdened with these challenges, the localisation process has only been a preserve of the DA and local authority secretariat, which have knowledge of SDGs and the set targets. For many local leaders, the implementation of SDGs has been accidental and piecemeal for they are not conversant with SDGs let alone the targets.

Overcoming these underlying impediments creates well-functioning and empowered local institutions which remain key in localising SDGs. The chapter, thus, vouches for resourcing and capacitation of local government institutions and their leaders. Capacity building poised for effective localisation of SDGs should therefore emphasise on improving institutional governance, leaders' knowledge and their abilities to spearhead positive and visionary management in all facets of life. Equally important is the need to inculcate good governance practices among players in local institutions.

References

Bowen, K. J., Cradock-Henry, N. A., Koch, F., Petterson, J., Hä'yha, T., Vogt, J., Barbi, F. (2017). Implementing the "Sustainable Development Goals": towards addressing three key governance challenges—collective action, trade-offs, and accountability. *Current Opinion in Environmental Sustainability, 26–27*, 90–96.

Chakaipa, S. (2010). Local government institutions and elections. In J. de Visser., N. Steytler & N. Machingauta (Eds.), *Local government reform in Zimbabwe: A policy dialogue* (pp. 31-68). Western Cape: Community Law Centre.

Chigwata, T. C. (2018). *Provincial and Local Government Reform in Zimbabwe: An analysis of the law, Policy and Practice*. Cape Town: Juta and Company.

Chirisa, I. (2012). African local governments and the MDGs: Contraints and policy alternatives. *Global Advanaced Journal of Peace, Gender and Development Studies*, 1(1), 001-011.

Constitution of Zimbabwe. (2013). *Amendment, No. 20 Act of 2013*. Harare: Fidelity Printers and Refiners.

Dodo, O. (2013). Traditional leadership systems and gender recognition: Zimbabwe. *International Journal of Gender and Women's Studies, 1*(10), 29-44.

Dziva, C., Dewa, D., Khumalo, P. (2020). Recurrent male-to-female partner violence in Mkoba high density suburb of Gweru town in Zimbabwe, *Eastern Africa Social Science Research Review, 37*(2), 35-52.

Fenton, P., Gustafsson, S. (2017). Moving from high-level words to local action—governance for urban sustainability in municipalities. *Current Opinion in Environmental Sustainability, 26–27*, 129–133. https://doi. org/10.1016/j.cosust.2017.07.009.

Florida, R. (2015). 11 Reasons the UN Should Make Cities the Focus of Its Forthcoming Sustainable Development Goals. http://www.citylab.com/work/2014/04/11-reasonsun-should-make-cities-focus-its-new-sustainable-development-goals/8896/. Accessed 10 September 2020.

Gustafsson, S., Ivner, J. (2017). Implementing the global sustainable goals (SDGs) into municipal strategies applying an integrated approach. In: L. Filho (Eds), *Handbook of sustainability science and research* (pp. 301-316). Berlin: Springer International Publishing.

Hendriks, C. J. (2018). Municipal financing for sustainable development: A case of South Africa, *Local Economy*, 33 (7), 757-774.

Jime´nez-Aceituno, A., Peterson, G. D., Norstro¨m, A.V., Wong, G.Y., Downing, A.S. (2020). Local lens for SDG implementation: lessons from bottom-up approaches in Africa. *Sustainability Science*, 15, 729–743.

Kemp, R., Parto, S., Gibson, R. (2005). Governance for sustainable development: Moving from theory to practice. *International Journal of Sustainable Development*, 8, 12–30.

Keulder, C. (1998). *Traditional leaders and local government in Africa: lessons for South Africa*. Pretoria: Human Sciences Research Council.

Machivenyika, F., Mugabe, T. (2020, April 5). President briefs traditional, religious leaders on Covid-19. *The Herald*. https://www.herald.co.zw/president-briefs-traditional-religious-leaders-on-covid-19/. Accessed 15 April 2020.

Mashundu, N. (2019, August 16). Chief Ndiweni, 23 others convicted of destroying property. *Chronicle*. https://www.chronicle.co.zw/chief-ndiweni-23-others-convicted-of-destroying-property/. Accessed 25 March 2020.

Marango, T. (2017). Potential strategies for harnessing indigenous rainmaking practices to combat the negative effects of climate change in Chimanimani District of Zimbabwe, PhD thesis, Institute for Rural Development, University of Venda, Thohoyandou, South Africa.

Mararike, C. G. (1999). *Survival strategies in Rural Zimbabwe: The role of Assets, Indigenous Knowledge and Organization*. Harare: Mond Press.

Mavhinga, D. (2019). Zimbabwe Should Investigate Allegations of Aid Abuse: Cyclone Victims

Reportedly Denied Food on Political Grounds, Human Rights Watch, March 29. https://www.hrw.org/news/2019/03/29/zimbabwe-should-investigate-allegations-aid-abuse. Accessed 20 March 2020.

Moallemi, E. A., Malekpour, S., Hadjikakou, M., Raven, R., Szetey, K., Ningrum, D. D., Dhiaulhaq, A., Bryan, B. A. (2020). Achieving the Sustainable Development Goals Requires Transdisciplinary Innovation at the Local Scale. *One Earth*, 3(3), 300-313.

Morita, K., Okitasari, M., Masuda, H. (2020). Analysis of national and local governance systems to achieve the sustainable development goals: Case studies of Japan and Indonesia, *Current Opinion in Environmental Sustainability*, 15, 179-202.

Musarandega, H., Chingombe, W., Pillay, R. (2018). Harnessing local traditional authorities as a potential strategy to combat the vagaries of climate change in Zimbabwe. *Jàmbá: Journal of Disaster Risk Studies*. https://doi.org/10.4102/jamba.v10i1.651

Musekiwa, N. (2012). The role of local authorities in democratic transition. In E. Masunungure & J. Shumba (Eds.), *Democratic Transition* (pp. 230-251). Harare: Weaver Press & IDAZIM.

Nabarro, D. (2016). Transforming our world: How sustainable development will help us achieve zero hunger. In Welthungerhilfe, International Food Policy Research Institute, and Concern Worldwide. *2016 Global hunger index: Getting to zero hunger* (22-27). Bonn, Washington DC and Dublin: Welthungerhilfe, International Food Policy Research Institute, and Concern Worldwide. https://doi.org/10.2499/9780896292260_03. Accessed 20 February 2020.

Nakashima, D. J., McLean, G. K., Thulstrup, H. D., Ramos, C. A., Rubis, J. T. (2012). *Weathering uncertainty: Traditional knowledge for climate change assessment and adaptation.* Paris and UN, Darwin: UNESCO.

Staford-Smith, M, Griggs, D, Gafney O et al (2017). Integration: the key to implementing the sustainable development goals. *Sustainability Science*, 12, 911–919. https://doi.org/10.1007/s11625-016-0383-3

Sithole, L., Dziva, C. (2019). Eliminating harmful practices against women in Zimbabwe: Implementing article 5 of the African Women's Protocol. *African Human Rights Law Journal*. https://doi.org/10.17159/1996-2096/2019/v19n2a1.

Reddy, P. S. (2016). Localising the sustainable development goals (SDGs): The role of Local Government in context. *African Journal of Public Affairs*, 9(2), 1-15.

Ringson, J. (2017). Zunde RaMambo as a traditional coping mechanism for the care of Orphans and Vulnerable Children: Evidence from Gutu District, Zimbabwe. *African Journal of Social Work*, 7(2), 52-59.

Rukuni, T., Machingambi, Z., Musingafi, M., Kaseke, K. (2015). The role of traditional leadership in conflict resolution and peace building in Zimbabwean rural communities: The case of Bikita District. *Public Policy and Administration Research*, *5*(3), 75-79.

Rural District Councils Act (RDCA) (1988). [Chapter 29:13].

United Cities and Local Governments (2015). The Role of Local Governments in Promoting Gender Equality for Sustainability. Paris: UCLG. www.uclg.org/sites/default/files/the_role_of_local_governments_in_promoting_gender_equality_for_sustainability.pdf. (Accessed 12 September 2020).

United Cities and Local Governments (2016). The Sustainable Development Goals: What Local Governments Need to Know. Paris: UCLG. Available at: www.uclg.org/sites/default/files/the_sdgs_what_localgov_need_ to_know_0.pdf. (Accessed 21 September 2020).

United Nations Development Programme (UNDP). (2014). *Localising the Post-2015 Development Agenda: Global Taskforce of Local and Regional Governments for the Post-2015 Development Agenda Towards Habitat 111*. New York: United Nations.

United Nations General Assembly (UNGA). (2016). Final list of proposed Sustainable Development Goal indicators: Report of the Inter-Agency and Expert Group on Sustainable Development Goal Indicators. https://sustainabledevelopment.un.org/content/documents/11803Official-List-of-Proposed-SDG-Indicators.pdf. Accessed 18 February 2020).

Vaggi, G. (2016). Making the Sustainable Development Goals work. *International Journal of Cooperation and Development*, 3(2), 34-58.

Walsh, A., Matthews, A., Manda-Taylor, L., Brugha, R., Mwale, D., Phiri, T., Byrne, E. (2018). The role of the traditional leader in implementing maternal, newborn and child health policy in Malawi. *Health Policy and Planning*, https://doi.org/10.1093/heapol/czy059.

Weymouth, R and Hartz-Karp, J. (2018). Principles for Integrating the Implementation of the Sustainable Development Goals in Cities *Urban Science,* 2 (77): 1-22.

Willis, K. (2016). Viewpoint: International development planning and the Sustainable Development Goals (SDGs). *International Development Planning Review*, 38 (2), 105-111.

Zimbabwe Human Rights Commission (ZHRC). (2018). *2018 Harmonized Elections Report*. ZHRC, Harare.

16 Water, Energy, Health and Sanitation Challenges in Masvingo's Low-Income Urban Communities in the Context of SDGs in Zimbabwe

Mavis Thokozile Macheka and Donald Chikoto

Abstract

To keep pace with the demand for urban housing, the Government of Zimbabwe (GoZ) instituted a low-income urban housing policy in 1980. This resulted in the construction of many low-income houses in most urban areas. This chapter seeks to assess the efficiency of service delivery provisions to a selected marginal society in Zimbabwe's oldest city, Masvingo. It draws on the experiences of Victoria Ranch residents. This research study used a qualitative research approach. Data were gathered from ten ($n = 10$) key informant interviews, ten ($n = 10$) in-depth interviews and three ($n = 3$) focus group discussions. Purposive sampling was used to select key informants while participants for in-depth interviews and focus group discussions were conveniently selected. Findings point to inadequate service delivery provision in Victoria Ranch. The government systems in place created parallel structures in the management of the Victoria Ranch low-income residential area, which was affecting service delivery negatively. As such, there were challenges of water scarcity, no sanitation and health facilities, and no electricity supply. The study recommends that Zimbabwe's local government authorities must ensure that each urban environment is functional, with an efficient service delivery system to enhance the health and well-being of residents.

Keywords

Low-income houses · Marginal societies · SDGs · Service delivery · Urban housing

1 Introduction and Background

Achieving sustainable development has become a global policy priority. In 2015, the United Nations identified 17 Sustainable Development Goals (SDGs) (UN 2015a). Since the global commitment to accelerate the achievement of the SDGs was declared, several strategies and policies were instituted and implemented by different countries globally. Zimbabwe is no exception. Since independence from British rule in 1980, the Government of Zimbabwe has focused on the provision of housing for the urban poor (Moyo 2014). However, the scale of the service delivery provision problems has remained unabated in Zimbabwe. Many scholars have largely focused

M. T. Macheka (✉) · D. Chikoto
Great Zimbabwe University, Masvingo, Zimbabwe
e-mail: mtmacheka@gzu.ac.zw;
chikotodonald@gmail.com

G. Nhamo et al. (eds.), *Sustainable Development Goals for Society Vol. 1*, Sustainable Development Goals Series, https://doi.org/10.1007/978-3-030-70948-8_16

on the provision of housing for the urban poor, where government, non-governmental organisations, multilateral agencies and communities can play a critical role (Gibb et al. 2008; Muchadenyika 2015; Mutekede and Sigauke 2007). However, the research focus is limited to the living standards of the urban poor residing in low-income housing areas. This chapter analyses the service delivery to urban dwellers in low-income housing areas in marginal societies of Masvingo. The study focuses on marginal areas which are residential areas that were developed for low-income earners in Zimbabwe. These low-income houses were mainly provided by housing cooperatives in Zimbabwe. The study seeks to answer the following research questions: What is the nature of challenges of water, energy, health and sanitation in Victoria Ranch and how do Victoria Ranch residents account for the challenges they face in accessing these basic services? What measures have been put in place to ensure the availability of water and energy as well as ensure health and sanitation for all in Victoria Ranch? Through this research question, the chapter seeks to examine the progress local authorities have made towards achieving SDGs 6 and 7. SDG 6 aims to ensure availability and sustainable management of water and sanitation, and SDG 7 seeks to guarantee people's access to affordable, reliable, sustainable and modern energy (UN 2015b). These questions were raised to help us assess whether the Government of Zimbabwe is on the right path to achieving the SDGs.

The agenda to ensure availability of water, energy, health and sanitation predates the interventions of the SDGs. The Millennium Development Goals (MDGs) make reference to access to safe drinking water under MDG7 that aimed at ensuring environmental sustainability. The issue was mentioned under target 7c, which aimed at halving, by 2015, the proportion of the population without sustainable access to safe drinking water and basic sanitation (UN 2000). The significant role of water in achieving MDGs led the United Nations to declare 2005–2015 as Water for Life Decade (Sanusi 2010). During this decade, nations were expected to maintain the gains made in water access, push ahead quickly to provide drinking water and sanitation services to millions of people living in rural areas, and accelerate the successful efforts in urban areas to keep pace with the rising urban population, particularly by focusing on low-income and disadvantaged groups (WHO and UNICEF 2006). Though the importance of water and sanitation was recognised, the goal did not get attention in the MDGs.

Despite the availability of water being crucial in achieving sustainable development, by 2015, the target was not achieved. Since 1990, the proportion of the population without access to improved drinking water in most parts of the world has been halved. However, sub-Saharan Africa fell short of achieving the MDG target on water (UN 2015a). Further, the report highlighted that in 2015, it was estimated that 663 million people worldwide still used unimproved drinking water sources, including unprotected wells and springs and surface water, and nearly half of all people using unimproved sources lived in sub-Saharan Africa, while one-fifth lived in South Asia. The literature further revealed that most problems of water and sanitation challenges were experienced in sub-Saharan Africa. Angoua et al. (2018) reported that despite the progress made during the period 2005–2015, achieving the target on access to safe water and sanitation still remains a major challenge in urban areas of sub-Saharan African countries. As earlier noted by Cohre (2008) the problem of access to water and sanitation mainly affects people who are vulnerable and marginalised, for instance those people residing in precarious settlements in deprived urban areas (Hawkins et al. 2013) and those in rural areas or disadvantaged urban fringes (Perez et al. 2012). The situation was not unique to sub-Saharan Africa. In West Africa, for instance, the target 7c of MDGs was not achieved (UN 2015b).

With regard to the availability of energy, the MDGs, as originally formulated, did not prioritise energy as a key goal. However, scholars agree that although there is no MDG that directly addresses energy, it was clear that without energy services of adequate quality and quantity, countries cannot meet the MDGs (Takada and Fracchia 2007; Modi et al. 2006). The nexus between energy and MDGs was identified 5 years after the development of the MDGs. In 2005, at a World Summit in Sustainable Development, governments of the world met and agreed to take joint actions and improve efforts to work together at all levels to improve access to reliable and affordable energy services for sustainable development sufficient to facilitate the achievement of the MDGs (ESMAP 2002). Energy can unlock sustainable economic growth, improve human health and well-being and enable women and children to lead more productive lives (UN 2018a, 2018b; NCE 2018). By so doing, they acknowledged that the availability of energy is key in achieving all the MDGs.

By the end of the lifespan of the MDGs in 2015, nations realised the importance of achieving universal access to water, health, sanitation and energy as key in achieving sustainable development goals. As a result, when the United Nations formulated the SDGs in 2015, they incorporated water and energy within the sustainable development agenda as goals 6 and 7.

In Zimbabwe, the immediate solution to address the issue of water, health, energy and sanitation was in promulgating the Urban Councils Act in 1996 and charging urban councils with a service delivery mandate (Urban Councils Act 1996b, Chapter 29:15). The councils were required to ensure adequate provision of services such as water, wastewater, solid waste, energy, transport and other related services (Bachmann and MacCleery 2006). Though urban councils are mandated to deal with service provision, they are often overwhelmed by numbers of people in urban areas. Makwara (2012) adds that most urban local authorities in Zimbabwe continue to experience numerous challenges in maintaining an efficient service delivery, in key areas such as water, sanitation, health, shelter, road infrastructure and sewage reticulation. Scholars identify an increase in urban population, proliferation of housing cooperatives and local governance issues, among others, as being barriers to effective service delivery in Zimbabwe. Challenges such as inconsistent water supply, uncollected refuse, derelict roads, uncoordinated housing developments and load-shedding all point to catastrophic service delivery (Muchadenyika 2017a).

The process of urbanisation has led to challenges of urban sprawl, urban poverty, inadequate housing for the urban poor, inadequate infrastructure and poor service provision (Munzwa and Jonga 2010). Resultantly, there is immense pressure on local governance departments in most cities to deliver more services to an increasingly urban population (Muchadenyika 2017b). Dewa et al. (2014) cite corruption, incompetence and negation of residence to be the challenges causing poor service delivery in local authorities in Zimbabwe. The municipalities have also been blamed for spending finances on hefty salaries at the expense of service delivery (Sifile et al. 2015). Earlier, Chirisa et al. (2014) had averred that the cities in Zimbabwe were robbed of treasures that must accrue to them to finance infrastructure and services. Muchadenyika and Williams 2016 argues that local authorities in Zimbabwe are failing to provide water, sanitation, health, education, transport and housing services to citizens. Similarly, Murimoga and Musingafi (2014) argued that there is a general decline of municipal service delivery and poor governance systems in Masvingo and Harare municipalities. As a result, the failure of formal service delivery systems has given rise to the resurgence of societies and cooperatives (Chirisa et al. 2014).

Again, an urban governance crisis has also led to the failure of service delivery (Muchadenyika and Williams 2016). For example, the formation of housing cooperatives has also led to the manifestation of the ineffectiveness of these state-sponsored, formal service delivery mechanisms (Chirisa et al. 2014). Further, by the 2000s, local cooperative organisations re-emerged as an instrumental way of dealing with the problem of housing and land tenure, particularly among the residents of peri-urban areas. Such challenges, coupled with the economic meltdown from the year 2000, resulted in galloping inflation and subsequent collapse of service delivery (Makuruva 2012). Resultantly, service provision among local authorities in Zimbabwe is a matter of trying to make something out of a crisis situation (RTI and IDAZIM 2010). These have implications on the service delivery because residents in housing cooperatives must pool their resources to deal with service delivery challenges. We argue that these, among other problems, have affected the provision of adequate service delivery in low-cost housing residential areas. This chapter draws on the experiences of residents from Victoria Ranch to assess service delivery provision (water, sanitation and energy, among others) in the marginal communities.

2 Materials and Methods

This study was conducted in Victoria Ranch residential area in Masvingo urban. Masvingo is the oldest city in Zimbabwe. It is in the Southern part of Zimbabwe, about 300 km from the capital city, Harare. Victoria Ranch is located 8 km to the southwest of Masvingo city centre (Musingafi et al. 2015). Victoria Ranch is physically located in Masvingo city but under the jurisdiction of MRDC. The land used to be peri-urban and was previously owned by a white farmer (Chikomwe 2014). Prior to land resettlement in 2002, the land was managed as idle land by MRDC. Later, the land was expropriated and declared as state land (Masvingo Department of National Housing 2007). The land was allocated to a housing cooperative which later signed a memorandum of understating (MOU) with MRDC, but after the development of stands, all the residents were handed over to MCC. Through the National Housing Delivery programme of 2006, it was converted to residential use. The residential area has a total of approximately 10,000 stands, and that makes up 30% of the city's housing stock (Chikomwe 2014). The area has residential stands, recreation stands designated as parks, church stands, stands for primary and secondary schools, clinics, commercial stands and light industrial stands (Masvingo Department of National Housing 2007, 2015). Surprisingly the residential area was built without basic services such as electricity, water and sewerage.

This study adopted a qualitative research approach. The approach allowed for the study of participants in their natural setting. In other words, it allowed us to study the experiences of Victoria Ranch residents. Qualitative research was adopted because it uses any methods that rely upon primary source information, where very often the 'data' is not numerical (Yates 2004). The approach is subjective and holistic and thus allows for data analysis during the process of collection. Thus the approach gives the researchers an insider perspective.

Interviews and focus group discussions were conducted with the residents of Victoria Ranch residential area, responsible local authorities and representatives of the housing cooperative in the area. Interviews and focus group discussions were conducted with 50 (n=50) participants. In selecting the participants, the emphasis was on representational diversity of key informants in service delivery systems. These methods allowed us to appreciate the extent to which the urban councils are offering basic services to its citizens in low-income housing areas.

The study purposively sampled 10 ($n = 10$) key informants and these are officials from the local municipality, officials from the housing cooperative, local leadership, community subcommittees and residents association. The key informants were asked to explain the measures they are taking to ensure regular service delivery to the people. They were further interrogated to explain the challenges that affect efficient service delivery. The municipality was chosen because they are mandated through the Urban Councils Act, Chapter 29:15, to provide services to the people in urban areas. The Vashandi Housing cooperative, local leadership and residents association are seized by ensuring that residents in Victoria Ranch access water and electricity. In-depth interviews were conducted with 10 ($n = 3$) residents from various sections of the residential areas. These participants were conveniently selected because they experience everyday life in Victoria Ranch; thus, they have a better understanding of the challenges related to service delivery. The researchers engaged a total of 30 ($n = 30$) Victoria Ranch residents who were divided into three focus group discussions. The participants were conveniently selected to highlight the challenges they face with regard to service delivery such as water, energy, health and sanitation. The groups met at different places and meetings to ensure that the interview involved people from the different sections of Victoria Ranch.

During the study, the researchers followed ethical guidelines expected when conducting social research. The researcher has a moral and professional obligation to be ethical when carrying out research. Ethics include debriefing, voluntary participation, informed consent, confidentiality and anonymity of the right to and withdrawal of participation. The researchers debriefed the participants so as to allay any fears that participants might have. Participants were granted the opportunity to ask questions. Participants willingly participated in the study. The research participants were not coerced into participating in the research. The researchers assured privacy to participants, and private information was to remain relatively confidential. Creswell (2003) asserts that researchers should guarantee that collected data remains confidential, and sources of information should be kept anonymous. Hence participant's names will remain anonymous. Participant names remained anonymous throughout the study.

The data were analysed using thematic analysis. As Braun and Clarke (2006) note, thematic analysis is a qualitative analytic method for identifying, analysing and reporting patterns (themes) within data and it minimally organises and describes the data set in detail and interprets various aspects of the study. The major themes that were identified during data analysis were (i) water scarcity crisis, (ii) sanitation and health facility challenges and (iii) electricity challenges. The analysis and presentation of findings are in the form of detailed descriptions, using narrative vignettes and direct quotes from interviews and focus group discussions.

3 Results and Discussion

Interviews and focus group discussions with residents of Victoria Ranch point to some challenges in service delivery. These challenges are the complexity of service delivery in Victoria Ranch, water scarcity challenges, sanitation and health facility challenges and electricity challenges. The challenges are detailed below:

3.1 Water Scarcity Crisis in Zimbabwe: Victoria Ranch Experiences

Interviews with Victoria Ranch residents revealed several gaps in the way that the local government is seeking to improve residents' access to basic services. User fees such as administration fees, plan approval, building inspection, water connection and sewer connection were paid to MCC (Interview with MCC official, Masvingo urban, 10 February 2020). This is so because MCC is responsible for providing services to residents. However, based on the findings of the study, Vashandi Housing cooperative that developed Victoria Ranch is responsible for providing the services (Interview with Housing Cooperative member, Masvingo urban, 10 March 2020). Thus there is passive involvement of the urban and rural councils in service delivery, and that makes the issue of service delivery both challenging and cumbersome. These systems impact negatively on service delivery. In view of the foregoing, there is a disempowerment of Masvingo local municipalities in providing services directly to urban residents. In terms of the provision of water to Victoria Ranch residents, the payment of user fees to MCC did not result in the effective provision of water. To begin with, there are few residents who have paid for the water connection, and the majority did not connect water at their own residential houses. Few residents who connected water managed to access water services from MCC for a short time. The following vignette from a resident sheds more light:

> Since 2016, we used to receive water two to three times per week, but the water would only be accessed between 01:00-05:00. Those who did not connect water at their houses must queue at a public tap during those odd hours. From August 2019 to date, we have not received even a drop of water from the municipality (Interview with resident Mrs Shchie, Victoria Ranch, 27 February 2020).

Water connection and scarcity challenges are surprising, given that MCC has collected administrative service charges from residents. As noted earlier, Victoria Ranch residents rely on developers for water provision. Since the inception of the project in 2005, Vashandi Housing cooperative drilled two boreholes. In one focus group, it emerged that 'when the water supply situation got worse at the end of 2019, Vashandi Housing co-operative drilled two more boreholes. However, the additional boreholes are not even functional, and one of the boreholes is just 15 metres deep' (Focus group discussion with Victoria Ranch residents, 02 April 2020). Although the effort was meant to address water challenges, residents complained that these boreholes were not enough for the households in the area. One resident explained 'although there are boreholes in the area, they are not enough to serve +/-10 000 households' stands in the area. And we pay to get water from the boreholes to co-operatives' (Focus group discussion with Victoria Ranch residents, 06 April 2020). Further, the persistent water shortage in Victoria Ranch has forced some residents to take advantage of the situation and start selling water to other residents. Water selling in Masvingo urban has become a viable enterprise that generates income and employment, and contributes to urban livelihood diversity. The following interview excerpt sheds more light:

> Some vendors who are clever have bought water tank and bowsers. They fill these with water and sell it to residents. We just buy despite the fact that we do not know where they get the water from. It is either from the boreholes, dams or other suburbs, I do not know. I just need water for drinking and other household use (Interview with resident Dr Stenns, Victoria Ranch, 27 January 2020).

The residents indicated that due to water scarcity, they were compelled to using water from small streams for laundry. This situation is typical of rural areas in Zimbabwe, yet Victoria Ranch is an urban setting. However, using water from the streams has health implications since the streams are located in areas where there is no sewer system. The streams are used for waste disposal, including open defecation. Thus, such practices constitute a health risk for residents. Other residents used water from the same streams for bathing, and most of them were complaining of irritating effects on their skins. In a study conducted in Harare and Masvingo, Murimoga and Musingafi (2014) found that water scarcity usually had adverse effects on consumers. The scholars cited an example of the 2009 cholera outbreak in Harare, which spread throughout the country, that was linked to water scarcity and polluted water sources. The cholera outbreak killed more than 5000 people and sickened more than 83,631 as a result of the water crisis (Murimoga and Musingafi 2014).

Political problems were also identified as one aspect affecting the effectiveness of service delivery in Masvingo urban. In some instances, MCC officials argue that the pumping capacity affects water provision in the city. However, the situation on ground nevertheless points to the issue that the politically driven land grabs left the city council incapacitated to service the area. The suburbs under the jurisdiction of land developers are the worst affected because of the controversies surrounding their inclusion into the city council's conventional water supply system. One participant said, 'the ruling government is not providing resources to the opposition dominant council for fear that the opposition will gain credit for ameliorating the condition of poor urban residents' (Interview with MCC official Johns, Masvingo urban, 28 March 2020). Dissatisfied by the local authority's response, one of the members of the Vashandi cooperative highlighted:

> The council is claiming that their pumping capacity is affecting water provision to Victoria Ranch but that is not the correct position. This is partly due to the attitude of council leadership. It is mostly political-oriented. The council is run by councillors from the opposition, and as such they are sabotaging programmes of housing provision from the ruling party (Interview with a member of the Housing Cooperative, Victoria Ranch, Masvingo urban, 28 February 2020).

Political squabbles affect ordinary citizens who are neither politicians nor local authorities. Undeniably, in urban Zimbabwe, the urban governance crisis has led to service delivery failure (Muchadenyika and Williams 2016). Thus, the politicisation of municipalities in Zimbabwe has negative implications for basic service delivery.

While municipalities are regarded as the custodians of public funds, tasked with utilising these resources to address the basic needs of local communities (Reddy 2016), service delivery in Victoria Ranch is currently managed by housing cooperatives. There are parallel structures between local councils and housing cooperatives with regard to service delivery in Victoria Ranch residential area. Chikomwe (2014) explained that the parallel development at Victoria Ranch emanated from the unwillingness of the MCC to connect Victoria Ranch to the water and sewerage mains. Further, MCC operated under an overload since the mains had the capacity for use by 10,000 residential stands and were never upgraded despite an increase in population to 30,000 residential stands for the whole city. Chirisa et al. (2014) argue that the issue is more political where housing cooperatives, commonly known as land developers, have access to state land. They argue that desperate people are allocated and given stands or houses and guaranteed permanent residence in unplanned areas. The local authorities are being directed by the government to distribute and allocate land for development of new properties to developers, employers, housing cooperatives and development banks (Aghimien et al. 2018). In the end, these developers cannot provide the basic services expected by urban dwellers. The housing cooperatives are manipulated by politicians seeking to gunner support from the urban poor, albeit with very unpredictable consequences. Resultantly, the scarcity of freshwater is rapidly becoming a nightmare amongst urban residents in Zimbabwe (Chigwenya 2010). The overall impression suggests that such a scenario would compromise the attainment of goal 6 of the UN's SDGs. This points to the local municipality's failure to address the basic service needs of residents.

This situation prevalent in Victoria Ranch typifies the scenario in most of Southern African countries. A study of Lusaka urban council in Zambia illustrated how inter-party competition thwarts the delivery of public services to, and harassment of, the urban poor who live in cities controlled by an opposition party (Resnick 2010). Further, the provision of goods and services for the urban poor can be used as a political tool when political decentralisation is not accompanied by sufficient administrative and fiscal decentralisation. In South Africa, Reddy (2016) demonstrated that political infighting and related clashes between the political and management components in local government in South Africa have also adversely affected municipal service delivery. These studies demonstrate that it is very difficult for Southern Africa as a region to meet sustainable development goals because of such challenges. The governments are more concerned about power than delivering basic services to its poor citizens. Historically, the public sector in an African context has always been deemed to be political (Booysen 2012; Cameron 2003, 2010; De Visser 2010). Sadly, it is difficult to ensure the availability of water, sanitation and access to affordable, reliable, sustainable and modern energy for all in a context where political issues outweigh any concerns for service delivery.

In addition, the Sustainable Development Goals Center for Africa (SDGC/A) report (2019) revealed that the data collected in Africa on SDG 6 from 2014 to 2019 reflected that access to improved drinking water within a 30-min round trip is below the world average and off-target. However, the experience of water delivery challenges is not only unique to Southern Africa's urban residents, but also experienced in most parts of the world. This is evidenced by the report of UN (2018a) that reviews the opportunities and challenges for progress towards achievement of the SDGs. It indicates that in contrast with the target goals of the SDGs, the global world is not on track to achieve the global SDG 6 target by 2030 at the current rate of progress (UN 2018b). The report specifically outlines that by July 2018, extending access to safe drinking water presents a huge challenge, and billions of people still need access to basic toilet and handwashing facilities.

3.2 Sanitation and Health Facility Challenges in Victoria Ranch

The challenge of water in Victoria Ranch has further affected sanitation and health facilities. Despite the fact that Victoria Ranch residents have paid for sewer connection, there are no sewer systems in the residential area. Very few residents had erected their own septic tanks; some had EcoSan and Blair toilets while some residents did not have toilets at all. The situation is unhealthy as one resident explained:

> I feel that I wasted my money by constructing a septic tank, what is its use if there is no water in the area. A septic system cannot operate without running water. I only get a few buckets for household use and drinking. Thus, I cannot have extra for the toilets, we just have to use our neighbour's eco-toilet or the bush toilet (Interview with resident Mrs Jojo, Victoria Ranch, 30 January 2020).

Upon further enquiry on the use of ecological sanitation (EcoSan) toilets referred to, we were informed that these toilets were not user friendly to residents and they posed a health hazard. EcoSan toilets were an idea developed in Zambia (Interview with Resident Association representative, Masvingo urban, 02 February 2020). As a measure to address the non-existence of sewerage systems, the authorities compelled all stand owners to have the eco-toilets before they could reside in their houses. However, there are a number of houses with neither sewer system nor EcoSan toilets.

Residents without the eco-toilets were using either bush toilets or unfinished houses as toilets. Some unscrupulous residents were defecating in other residents' houses that were under construction. In view of this, these urban residents were just urban dwellers with rural identities, and their situation was worse off, constituting a health hazard. During transect walk, we observed filthy conditions in some parts of the residential area. This activity exposed residents of Victoria Ranch to serious health risks. Further, there is a greater likelihood of contaminating underground water because the water table was very high and Blair toilets were flooding during the rainy season, thereby posing a serious environmental problem. Therefore, there is a need for local authorities to ensure adequate provision of all services to the people to mitigate such challenges.

The regulatory framework for urban house development in Zimbabwe requires the establishment of sewerage system before any development is done in an area (Urban Councils Act 1996b, Chapter 29.15; Regional Town and Country Planning Act 1996a, Chapter 29.12). However, because of the high demand for housing in urban areas in Zimbabwe, local authorities are compelled to approve the development of houses in areas with no such basic services. This is evident in most residential areas developed by cooperatives. Most settlements being developed by social movements are served by wells, Blair toilets, pit latrines, Skyloos and subgrade gravel roads (Muchadenyika 2017a). Further, it resembles the 'villagisation' of cities in progress, where cities are being ruralised. The lack of a sewerage system and the existence of Blair toilets have become a common feature in urban residential areas where most low-income people reside (Muchadenyika 2015). This also mirrored the experiences in some parts of Zimbabwe. For example in Bulawayo urban, in low-income houses, sanitation issues were a big-time bomb because households are still using EcoSan and Blair toilets in a community where wells and boreholes were the primary sources of water for daily use (Chigwenya 2019).

More worrisome is how waste is being managed in Victoria Ranch. The MCC, which is usually the responsible authority, is not managing waste in Victoria Ranch. The increase in the number of houses, general dealers and retail shops has resulted in the generation of lots of waste, and unfortunately there is no refuse collection in the area. The existence of many people in the area has seen a number of shops stocking a variety of foodstuff and other items packed in various packages of different materials, ranging from plastics and cans to bottles. This has increased littering due to lack of refuse collection in the area. One resident has this to say, 'people dump waste anywhere, some dump in the nearby forest, those with cars collect their refuse to the dumping site at their own expense. The majority is just dumping waste in unfinished houses under construction' (Interview with resident X, Victoria Ranch, 30 January 2020). The waste simply finds its way to the dumpsite because the responsible authorities have failed to adequately manage waste due to financial constraints. This was explained by one of the key informants, who explained that the local municipality was experiencing financial challenges. Because of its limited capacity, the municipality transferred the responsibility to manage the neighbourhood to housing cooperatives.

The housing cooperatives were also inadequately resourced. Therefore, they neglected the issue of waste management due to incapacitation. The rate of litter collection in Victoria Ranch was being surpassed by the rate of dumping, and this has seen lots of empty cans, disposable nappies, food containers and other various types of waste accumulating. This confirms findings by Mohee and Simelani (2015), who argued that many cities and towns in developing countries are not adequately meeting their refuse collection obligations due to financial constraints.

The current challenges in Victoria Ranch have the potential to undermine all the efforts by the Government of Zimbabwe to achieve sanitation and health for all by 2030. Mutetwa et al. (2016) argue that although there is a comprehensive legal framework that guides waste management in both urban and rural areas in Zimbabwe, it does not have an overall waste act that provides the essential legal basis for a consolidated waste management strategy. This is demonstrated by the MCC's unresponsive behaviour towards waste management in the area, yet the volume of waste generated continues to increase at a faster rate than the ability of the residents to improve on the financial and technical resources needed. There is no refuse control system in place, and that results in indiscriminate illegal dumping. We established that waste was crudely dumped at open dumpsites that do not meet basic environmental standards. This results in leachate from the dumpsites polluting underground and surface water sources. Loose papers and plastics are blown away by wind resulting in the aesthetic intrusion of the surrounding environment. This is a threat to human health. The presence of houseflies and a strong stench that pervades the area is a constant reminder that pollution is a common problem in Victoria Ranch. These realities point to the weaknesses of the state in handling service delivery challenges.

3.3 The Fight to End Energy Poverty in Zimbabwe: Victoria Ranch Experience

Apart from the shortage of water and sewer facilities in Victoria Ranch, the Government of Zimbabwe has also failed to provide energy for all in the form of electricity. Victoria Ranch residential area, Masvingo urban, does not have electricity and residents expressed that they had no hope that electricity will be made available to them in the foreseeable future. The Director for the Housing Scheme reiterated:

> We have signed an agreement with the Zimbabwe Electricity Supply Authority (ZESA), the power utility service provider in the country that each household should pay US$510 for electricity supply. However, the economic quagmire in the country coupled with financial regulations in the country that the United States dollar is no longer an official currency; the issue of having electricity remains a pipe dream. The electricity situation can only be addressed when the country's economy stabilises (Interview with Director Housing Cooperative, Victoria Ranch, 10 March 2020).

In terms of access to electricity, the current unstable economic environment presents a bleak future for Victoria Ranch residents. 'The majority are already poor residents who cannot raise such a figure because of the economic meltdown in the country. For a few who can afford the figure, they are also not allowed to use foreign currency when doing business in the country' (Interview with resident Mr. Mhoi, Victoria Ranch, Masvingo urban, 15 February 2020). In view of the above, it was possible that it would take a long time for the area to have electricity as there was no evidence of the preliminary work such as erected electricity poles and ZESA had not scheduled electricity provision. One of the residents sarcastically said:

> There is no hope to have electricity in the near future. We know it will be made available but obviously after a very long time. Currently, it is the least of our worries; we are now used to the darkness. Look at those who have electricity connected; they rarely have it because of electricity blackouts (Interview with resident Mr Johns, Victoria Ranch, 29 February 2020).

Having a community that is used to accepting the abnormality is a testimony of a country that is very far from attaining the sustainable development goals. While provision of energy, including electricity, is key, especially in urban areas, we are presented with an urban area that is not expecting such services in the foreseeable future. To address the challenge of electricity, the residents of Victoria Ranch in particular, and Zimbabwe, in general, were resorting to the use of solar power. In Zimbabwe, solar power was traditionally used as a source of energy in rural areas. Therefore, its adoption in urban areas reflected that the Government of Zimbabwe is incapacitated to deliver basic services to its citizens. The use of solar energy does not go without criticism because it has resulted in several incidences of theft. This was evidenced by reports from residents that a number of solar panels were stolen from rooftops at night.

As an alternative source of energy, some residents were using gas. Many women interviewed revealed that they were using gas for stoves and refrigerators. However, because of the high rate of poverty in Zimbabwe, the use of gas was common among the better-off residents who could afford to buy gas stoves and to refill the gas. Worse still, many residents argued that the use of gas indoors was very dangerous and they pointed out a number of cases where houses were gutted by fire as a result of gas explosions. Those who could not afford gas were using firewood as a source of energy. There was increased demand for firewood for both domestic use and commercial use in Victoria Ranch. As a result, the poor have no alternative except to benefit from the ecological resources. For poor families to meet short-term needs, they mine the natural capital by excessively cutting down trees for firewood and fail to replace soil nutrients (World Bank 1992). The lack of electricity supply in Victoria Ranch compelled residents to look for fresh firewood in the nearby Maparanyanga forest. Discussions with participants revealed that due to a large number of residents depending on firewood, the rate of vegetation depletion was too high. We acknowledge that residents had found a cheaper way of solving the challenges caused by lack of energy and electricity services in the area. However, this impacted heavily on the environment. The consequences would make it difficult for the government to mitigate. For example, the government may not be able to address all the challenges linked to deforestation by 2030.

Studies have shown that Africa will account for 75% of the world population without access to electricity and the continent will be the last one to be 'left behind' in the global electrification process (International Energy Agency 2017). Zimbabwe's urban areas are not an exception. The study argues that the failure to address issues of service delivery further incapacitates government because it all leads to environmental degradation, which should be addressed by 2030. Similarly, the Sustainable Development Goals Center for Africa, which was mandated by African leaders to champion the implementation of SDGs, produced an SDG monitoring report in 2019 that covered the period from 2015. Generally, only three goals are likely to meet the 2030 target (SDGC/A 2019). The three include SDG 5 gender equality, SDG 13 climate action and SDG 15 life on land. The report further presents that with regard to goal 7, data collected from 2016 to 2019 showed that half of the continent had electrification rates of less than 40%. North Africa is on track to achieve 100% electrification by 2030, and East Africa could be on track for universal electrification if investment increases.

4 Conclusion

This chapter aimed at assessing the efficiency of service delivery systems in Zimbabwe's marginal societies in the form of water, energy and sanitation. Data analysis shows that Victoria Ranch in Masvingo urban is being ruralised. It points to the very limits of the Government of Zimbabwe to achieve sustainable development goals 6 and 7 by 2030. Although low-income house owners had devised strategies in coming up with service delivery provision, the scale of the service provision problems had remained unabated in Zimbabwe. Zimbabwe urbanites were taken back to the rural areas through their way of living vis-à-vis service delivery. As Muchadenyika and Williams (2016) argues, Zimbabwe recorded unprecedented regression in key human development indicators as the state failed to provide public goods and services to its citizens. Like other urban areas in Zimbabwe, the local authorities in Masvingo failed to provide basic services to Victoria Ranch residents. The case of Victoria Ranch is not worth to generalise. However, what the discussion illustrates are the limits of Zimbabwe's local government authorities to ensure an urban environment that is functional, with an efficient service delivery system that accords residents all the requisite necessities.

References

Aghimien, D., Aigbavboa, C., Ngwari, T. (2018). Challenges of Sustainable Low-Income Housing Delivery in Harare, Zimbabwe Proceedings of the International Conference on Industrial Engineering and Operations Management Washington DC, USA, September 27-29, 2018

Angoua, E. L. E., Dongo, K., Templeton, M. R., Zinsstag, J., Bonfoh, B. (2018). Barriers to access improved water and sanitation in poor peri-urban settlements of Abidjan, Cote d'Ivoire. *PLoS ONE 13 (8)*

Bachmann, J., & MacCleery, R. (2006). *Managing municipal services delivery*. New York: USAID.

Booysen, S. (2012). Sideshow or heart of the matter? Local politics and South Africa's 2011 local government elections. In S. Booysen, (ed.), *Local elections in South Africa: Parties, people and politics*, (pp. 1-10), Bloemfontein, Sun Press.

Braun, V., & Clarke, V. (2006). Using thematic analysis in psychology. *Qualitative Research in Psychology, 3, 2,* 77-101

Cameron, R. (2003). Politics–administrative interface: The case of the city of Cape Town. *International Review of Administrative Sciences 69*, 51-66.

Cameron, R. (2010). Redefining political–administrative relationships in South Africa. *International Review of Administrative Sciences 76,* 676-711.

Chigwenya, A. (2010). Decentralization without devolution and its impacts on service delivery: the case of Masvingo municipality in Zimbabwe. *Journal of Sustainable Development in Africa 12, 1,*

Chigwenya, A. (2019). Low income housing problems and low-income housing solutions: opportunities and challenges in Bulawayo. *Journal of Housing and the Built Environment 34*, 927-938

Chikomwe, S. (2014). An Analysis of Public–Private Partnerships in Housing in the Zimbabwe National Housing Delivery Programme: A Case of Masvingo City. Unpublished Research Report, MBE Housing, University of Witwatersrand, Johannesburg.

Chirisa, I., Gaza, M., Bandauko, E. (2014). Housing Cooperatives and the Politics of Local Organization and Representation in Peri-Urban Harare, Zimbabwe. *African Studies Quarterly, 15 (1)* 37-53

Cohre, W. (2008). SDC and UN-HABITAT. Sanitation: A human rights imperative. Geneva: 78-92-95004-47-4.

Creswell, J. W. (2003). *Research design: Qualitative, quantitative and mixed methods approach (2nd ed.).* Thousand Oaks, CA: SAGE Publications.

De Visser, J. (2010). The political–administrative interface in South African municipalities: Assessing the quality of local democracies, *Commonwealth Journal of Local Governance, Special Issue, March,* 86-101.

Dewa, D., Dziva, C. Mukwashi, K. (2014). Exploring Local Governance Challenges in Zimbabwe under the Government of National Unity era and beyond, *Journal of Political Science Development, 2, 8*

ESMAP (UNDP/World Bank Energy Sector Management Assistance Programme). (2002). ESMAP Business Plan, 2002-2004. World Bank: Washington, DC, USA.

Gibb, K., O'Sullivian, T. and Glossop, C. (2008). *Home Economics: How Housing Shapes City Economies.* London, Centre for Cities.

Government of the Republic of Zimbabwe. (1996a). Regional, Town and Country Planning Act, Chapter 29:12 (Revised Edition 1996). Harare: Government Printers.

Government of the Republic of Zimbabwe. (1996b). Urban Councils Act (Chapter 29:15). Harare: Government Printers

Hawkins, P., Blackett, I., Heymans, C. (2013). Poor-inclusive urban sanitation: An overview. http://hdl.handle.net/10986/17385.

International Energy Agency, (2017). Energy access outlook (World Energy Outlook Special Report)

Makuruva, C. T. (2012). Community Based Emergency Management: A case Study on a Cholera Outbreak in Zimbabwe. School of Public Health and Psychological Sciences, Aukland University of Technology

Makwara, E. C. (2012). Water Woes in Zimbabwe's Urban Areas in the Middist Of Plenty: 2000 -Present. *European Journal of Sustainable Development, 1,2,* 151.

Masvingo Department of National Housing, (2007). NHDP: Operation Garikai/Hlalani Kuhle. Department of National Housing, Masvingo.

Masvingo Department of National Housing, (2015). Housing Pay Schemes. Department of National Housing, Masvingo.

Modi, V. S., McDade, D., Lallement, and Saghir, J. (2006). Energy and the Millennium Development

Goals. New York: Energy Sector Management Assistance Programme, United Nations Development Programme, UN Millennium Project, and World Bank.

Mohee, R., & Simelani, T. (2015). Future directions of municipal solid waste management in Africa. African Institute of South Africa, Pretoria, South Africa.

Moyo, W. (2014). Urban Housing Policy and Its Implications On The Low-Income Earners Of A Harare Municipality, Zimbabwe. *International Journal of Asian Social Science, 4(3).* 356-365

Muchadenyika, D., & Williams, J. J. (2016). Social change: Urban governance and urbanization in Zimbabwe. *Urban Forum, 27*, 253-274.

Muchadenyika, D. (2015). Land for Housing: A Political Resource-Reflections from Zimbabwe's Urban Areas. *Journal of Southern African Studies, 41*(6), 1-20.

Muchadenyika, D. (2017a). Civil society, social accountability and service delivery in Zimbabwe. *Development Policy Review, 35*,178-195

Muchadenyika, D. (2017b). Social Movements and Planning Institutions in Urban transformation: Housing in Metropolitan Harare, Zimbabwe (2000-2015), A thesis submitted to Institute for Social Development, Faculty of Economic and Management Sciences, University of the Western Cape, South Africa

Munzwa, K. M., & Jonga, W. (2010). Urban development in Zimbabwe: A human settlement perspective. *Theoretical and Empirical Researches in Urban Management, 4*, 120–146.

Murimoga, R., & Musingafi, M. C. C. (2014). Local Governance and Service Delivery in Zimbabwean Local Authorities: The Case of Harare and Masvingo Urban Municipalities. *International Journal of Public Policy and Administration Research 1, 3,* 94-107

Musingafi, T. M., Musingafi M.C.C, Kaseke, K. E. (2015). Ecological Sanitation Systems in Urban Areas: The Case of Victoria Ranch Residential Area, City of Masvingo. *Public Policy and Administration Research, 5, 6,* 1-12.

Mutekede, L., & Sigauke, N. (2007). Low Income Housing Finance – Post-Independence Experiences, Prospects and Policy Implications for Zimbabwe, Paper prepared for the ENHR 2007 International Conference: Sustainable Urban Areas.

Mutetwa, M., Mafume, P. N., Zendera, W., Musimbo, N. (2016). Challenges of Solid Waste Management in Zimbabwe: A case study of Sakubva high density suburb. *Journal of Environment and Waste Management, 3, 2*, 142-155.

NCE, (2018). Unlocking the Inclusive Growth Story of the 21st Century: Accelerating Climate Action in Urgent Times. A Report of the Global Commission on the Economy and Climate

Perez, E., Cardosi, J., Coombes, Y., Devine, J., Grossman, A., Kullmann, C. et al. (2012).What does it take to scale up rural sanitation? Washington, DC. World Bank: Water and sanitation program: Working paper 70944.

Reddy, P. S. (2016). The politics of service delivery in South Africa: The local government sphere in context. *The Journal for Transdisciplinary Research in Southern Africa 12(1),* a337

Resnick, D. (2010). Failing to Capitalize? Urban Service Delivery in Opposition-Controlled African Cities. Paper presented at the World Urban Forum V in Rio de Janeiro on March 22, 2010 and at the Places We Live: Slums and Urban Poverty in the Developing World policy workshop in Washington, DC on April 30, 2010.

RTI, & IDAZIM. (2010). *Local governance in transition: Zimbabwe's local authorities during the inclusive government.* Retrieved from Research Triangle Institute website: https://www.rti.org/publication/ local-governance-transition-zimbabwes-local-authorities-during-inclusive-government

Sanusi, Y. A. (2010). Water, Sanitation and Human Development in Urban Fringe Settlements in Nigeria. *Theoretical and Empirical Researches in Urban Management 8 (17),* 14-29

Sifile, O., Madzorera, S., Chavunduka, M. D. (2015). Towards Improving Service Delivery in Local Authorities. A Case of Chegutu Municipality in Zimbabwe. *Journal of Humanities and Social Science 20 (11),* 55-63

Sustainable Development Goals Center for Africa (SDGC/A) (2019). Africa 2030 Sustainable Development Goals Three-Year Reality Check, The Sustainable Development Goals Center for Africa

Takada, M., & Fracchia, S. (2007). A review of energy in national MDG reports. New York, NY: United Nations Development Programme.

United Nations. (2000). The Millennium Development Goals. New York, United Nations.

United Nations. (2015a). The Millennium Development Goals Report. New York, United Nations.

United Nations. (2015b). Transforming our World: the 2030 Agenda for Sustainable Development. Resolution adopted by the General Assembly 25 September 2015. Available at: http://www.un.org/ga/search/view_doc.asp?symbol=A/RES/70/1&lANG=e (accessed 15 August 2020).

United Nations. (2018a). Sustainable Development Goal 6: Synthesis Report 2018 on Water and Sanitation, New York, United Nations Publications

United Nations. (2018b). Sustainable Energy 'Golden Thread' Linking 2030 Agenda with Pledge to Leave No One Behind, Especially Rural Women, Deputy Secretary-General Tells Side Event, Press Release, 16 July

WHO and UNICEF. (2006). Meeting the Millennium Development Goals: Drinking water and sanitation target; the urban and rural challenge of the decade. WHO and UNICEF, Washington, D. C. WHO Resource office for the future Sanitation

World Bank. (1992), *World Development Report 1992,* Oxford University Press, New York

Yates, J. S. (2004). *Doing Social Science Research,* London, Sage Publications in association with the Open University Press.

Trends in Research Around the Sustainable Development Objectives: A Bibliometric Analysis

17

Daniela Morales-Zapata, Alejandro Valencia-Arias, Luis Fernando Garcés-Giraldo, Esaú Toro-Vanegas, and Jefferson Quiroz-Fabra

Abstract

The Paris Agreement on climate change determined a strong disposition to generate actions to meet the urgency of sustainable development for the world, and consolidated the Sustainable Development Goals (SDGs) as the most important global course in history, in terms of poverty and hunger reduction, sustainability in production and consumption patterns, and mitigation of dangers of climate change. Considering this interest, this work is aimed at exploring the evolution and trends in research work around SDGs, as of a bibliometric analysis between 1990 and 2019. The methodology used is a bibliometric analysis of statistical information from 1990 scientific publications, recorded in the Scopus database. The compiled records were used to calculate the bibliometric indicators of quantity, quality, and structure for the academic output found. The main findings are around a great deal of interest in research on this type of topic, mainly driven by the political dynamics of the UN. Evidence shows that in the published journal article works significant emphasis has been placed on SDGs focusing on gender equity, climate action, and global health. Another area of greater focus includes the opportunity to create synergies between public and private entities to address SDGs.

Keywords

Sustainable · Bibliometric analysis · Gender equity · Climate action · Global health

1 Introduction and Background

The countries that make up the United Nations (UN) have advanced in the generation of alliances and agreements in favor of sustainable development, eradication of poverty, promotion of educa-

D. Morales-Zapata
Institución Universitaria Escolme, University of Essex, Essex, UK
e-mail: dm19409@essex.ac.uk

A. Valencia-Arias (✉)
Fundación Universitaria Católica del Norte, Antioquia, Colombia

Universidad Católica Los Ángeles de Chimbote, Instituto de Investigación, Chimbote, Peru

L. F. Garcés-Giraldo
Corporación Universitaria Remington, Medellín, Colombia
e-mail: luis.garces@uniremington.edu.co

E. Toro-Vanegas
Fundación Universitaria Católica del Norte, Antioquia, Colombia
e-mail: etorov@ucn.edu.co

J. Quiroz-Fabra
Institución Universitaria Escolme, Antioquia, Colombia
e-mail: ciesadministracion@escolme.edu.co

G. Nhamo et al. (eds.), *Sustainable Development Goals for Society Vol. 1*, Sustainable Development Goals Series, https://doi.org/10.1007/978-3-030-70948-8_17

tion, access to water and basic sanitation, as well as strengthening of health systems, among other objectives. In 1990 the United Nations Conference on Environment and Development (UNCED) adopted Agenda 21, and then in 2000, the member states led to the establishment of the Millennium Goals (MDGs). The year 2002 saw the declaration of Sustainable Development in Johannesburg. Later in 2012, at the Rio+20 Conference, world leaders decided to establish the Sustainable Development Goals (SDGs) as a transition from the MDGs (UN Department of Economic and Social Affairs 2019). There was the need to continue working to reduce child mortality rates and economic poverty, reduce educational gaps, and join efforts against HIV/AIDS, malaria, and tuberculosis (Sachs 2019). Other critical issues that needed global attention include public policy issues and investment to improve the quality of life of millions in the world. The shaping of the SDGs marked a route for UN member countries to frame their agendas and policies between 2015 and 2030 (Hák et al. 2016).

SDGs have the vision of the planet's environmental sustainability and reduce the impact of human-induced climate change (Gómez-Lee 2018). The SDGs were adopted for the period between 2015 and 2030, considered the most critical planetary courses in history (Gómez-Lee 2018). The SDGs have, in turn, become the guide for all governments, some with a mainly socioeconomic focus, while others with a focus on the biophysical environment. In both scenarios, human existence is an essential factor, so there is a dependency relationship, either due to the socioeconomic or ecosystem approach and the objectives involved. In this way, it is important to consider the behavior associated with the dynamics even in the economic field or environmentally speaking, and how humans interact in these aspects (Keesstra et al. 2016).

The idea of SDGs has rapidly gained importance due to the growing urgency of sustainable development for everyone. Although specific definitions vary, sustainable development encompasses the so-called triple bottom line approach to human well-being, a combination of economic development, environmental sustainability, and social inclusion. However, specific objectives differ globally, between and within societies. Certainly, so far, no consensus has been agreed regarding trade-offs and synergies between economic, environmental, and social objectives. Still, a shared focus on economic, environmental, and social goals is a hallmark of sustainable development and represents a broad consensus on which the world can build (Sachs 2012).

The Sustainable Development Goals (SDGs) that replaced the Millennium Development Goals (MDGs) came into effect in September 2000 at the United Nations Millennium Summit, where commitments to human rights, good governance, and democracy were defined. The MDGs were primarily focused on eradicating extreme poverty and hunger, while the SDGs have a vision of the environmental sustainability of the planet and seek to contain the dangers of human-induced climate change (Gómez-Lee 2018). In this way, the SDGs are a continuation of the Millennium Development Agenda, which for 15 years has driven progress in several important areas: reducing income poverty, providing access to water and sanitation, reducing child mortality, and significantly improving maternal health. MDGs succeeded in reducing the rate of extreme poverty from 40% in 1990 to 20% today and with the potential to reach 0% by 2030, as part of the inspiration for SDGs (Sachs 2019).

Given the growing interest in developing synergies between entities and understanding the impact of SDGs in areas highlighted before as poverty, gender, health, and environment, it is necessary to review the studies that have been investigated around sustainable development and its implications. This document structures the content of the analysis, starting with the methodology followed for the search of information; then the results are presented under the indicators of quantity, quality, and structure, where the different aspects and impacts of the publications are analyzed; and then emphasis is made on trends and emerging topics of interest, ending with the conclusions.

2 Materials and Methods

To develop an analysis of research trends on the research topic in detail, a useful tool used is bibliometric analysis. This can be defined as the science that studies the nature and course of discipline, as long as it leads to publications (Sancho 2002). In this way, bibliometric analysis is useful for measuring the scientific activity and impact through the quantification of publications and citations subject to an individual, research group, institution, or country, in addition to allowing monitoring of trends and associated changes, contributing in this way to obtain a clearer panorama and work route (Arenas and Santillán-Rivero 2002).

For this case, a review is carried out for the subsequent bibliometric analysis, with the help of the indexed database SCOPUS, which is the largest database of citations and summaries of peer-reviewed literature which is characterized for its quality and great access on the Web, thus giving a great compendium of documents in the different areas of science (Cañedo et al. 2010). A search of the currently published literature was then carried out on the Scopus platform using the keywords "sustainable development goals" and "sustainable development goal."

The keywords were organized in the following search equation: TITLE ({sustainable development goals} OR {sustainable development goal}) OR KEY ({sustainable development goals} OR {sustainable development goal})

Based on the 1913 results obtained when using the search string in the SCOPUS database, the respective analysis of the different bibliometric indicators is carried out, which allows studying science under three aspects: activity, productivity, and scientific progress; these aspects of the study are classified as indicators of quantity, quality, and structure, in order to obtain more information on the behavior of this important subject of study. The search results were processed using Microsoft Excel®.

From these data, a descriptive analysis of the publications that were found was carried out taking into account the characteristics of the publications such as the year of publication, a number of publications per journal, participating institutions, areas of knowledge covered, type of document published, and participating countries. Additionally, relevant keywords were identified in the search results, as well as their variation by year. Likewise, the impact factor per journal and the outstanding authors in the area with their respective impact factor were considered in the analysis. The previous information was represented in indicators of quantity, quality, and structure.

3 Presentation and Discussion of Findings

3.1 Quantity Indicators

The calculated indicators that refer to scientific production in terms of quantity are presented below.

3.1.1 Publications Per Year

The results obtained from the search regarding the behavior that the different advances have had over the years show that this indicator began to move from the year 1990 because it was the year in which publications of the topic coincided with the establishment of Agenda 21 at UNCED. It is observed how after 2011 the scientific production in the SDGs grew rapidly, going from 7 publications in 2012 to 120 in 2015, rising to 421 in 2017 and 642 publications as of March 2019. These data then show the great interest registered by the SDGs, having an annual increase of 255%, which shows the upward trend that it has in the academic community.

In addition, it is of great importance to know the value of the half-life that this consulted information has since it indicates the useful life of the publications and with it the use of this in future research. In this way, the validity of this information has had an increasing behavior according to the consigned data, giving an idea of mobility in the development of publications on the subject. This value increased greatly since 2017, amounting to more than 1 year as it had been managed in previous years. In this way, the average life of this information consulted is 1.3 years, a short life span due to the number of publications that are continuously developed.

3.1.2 Publications Per Journal

Figure 17.1 shows the top ten journals with the largest number of associated publications on the SDGs. First, with a considerable number of publications compared to the others, there is the Swiss Sustainability journal with approximately 120 publications; this international journal specializes in environmental, cultural, economic, and social sustainability of human beings. In second place is Elsevier's Journal of Cleaner Production, with more than 40 publications; this is a journal focused on research and practice on topics such as cleaner production, environment, and sustainability. Third, with around 30 posts, is Springer's Sustainable Science, which studies interactions between global, social, and human systems. In general, the other journals listed focus their study on sustainability from the environmental, social, and economic fields, as expected since the SDGs involve various areas of knowledge. For example, this is the case of The Lancet and The Lancet Global Health, which are specialized journals in medical and health matters, as well as Marine Policy that focuses on ocean studies.

In the distribution of publications among the journals that were consulted, it is found that 50% of academic production is done by 12.7% of the journals, which means that a large part of the studies carried out focus on a few of these journals. Furthermore, according to the transience index, it shows that 65.6% of the journals have published a single work on the subject of interest.

3.1.3 Publications by Country

This indicator highlights the countries that have the largest number of publications associated with the SDGs, as can be seen in Fig. 17.2. As is notable, the countries that head the list are the United States and the United Kingdom with 342 and 249 publications, respectively. Following the list is Australia with 126 publications and Germany with 97 associated publications; in general, the countries that follow in the list have less than 100 publications each, highlighting the participation of Asian countries such as India and China that have an important role within this field to be the countries with the largest world population, in addition to South Africa as the only African country.

The distribution of the publications associated with the different countries shows that only 6.1% of the countries consulted are responsible for 50% of academic production, in addition to having a transience index that reflects that 30.4% of countries have only published a single paper around the SDGs.

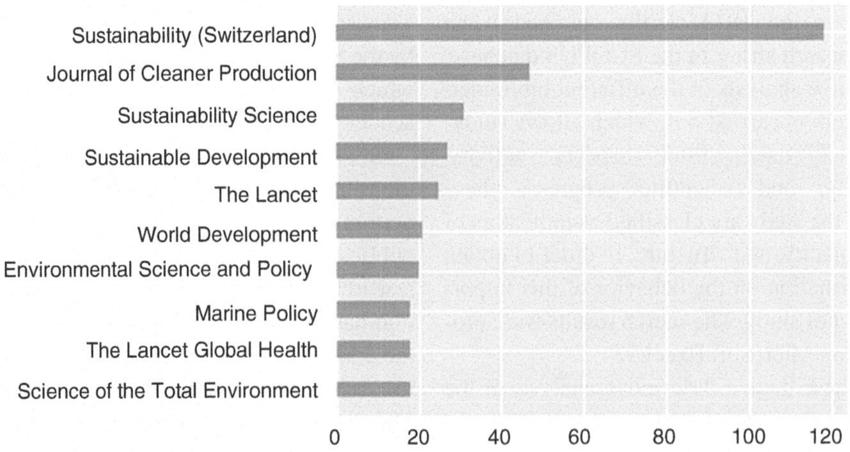

Fig. 17.1 Number of publications per journal. Source: Own elaboration

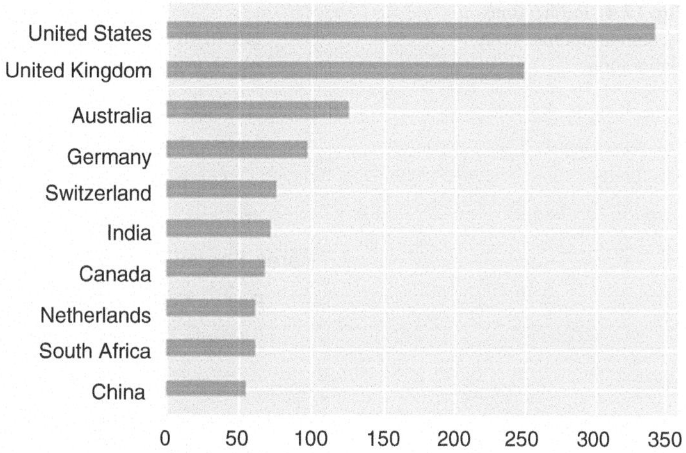

Fig. 17.2 Publications by country. Source: Own elaboration

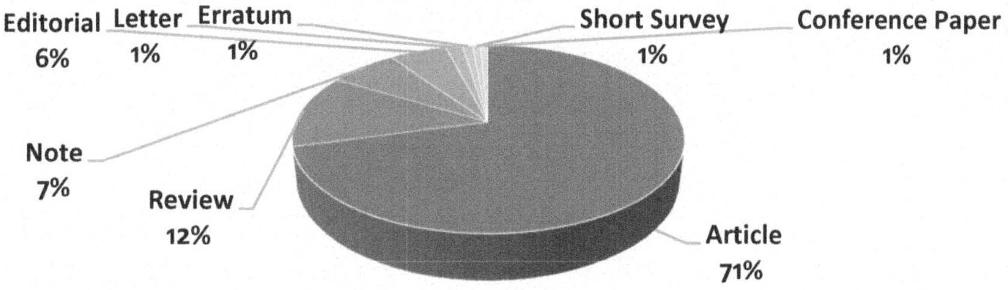

Fig. 17.3 Publications by type of document. Source: Own elaboration

3.1.4 Type of Document

This indicator highlights the formats in which knowledge is disseminated and therefore the channels through which this information circulates (see Fig. 17.3). In this indicator, it was found that the participation of the different formats is distributed mainly under one, having the publication with articles as the largest segment, covering 71.7%, indicating that most of the studies were developed using a scientific method and they are published in journals. With 12.2% there are reviews or revisions and with 5.8% the notes. The remaining 10.2% is part of the other categories, in which book chapters, monographs, or conference papers can be considered.

3.1.5 Publications by Author

This indicator reflects the productivity of the different authors who have the greatest relevance in the field of study, which is why Fig. 17.4 shows the ten authors with the greatest number of publications. The list is headed by Bartram, J. with 11 publications, followed by Bhutta, Z. A. and Allen, C. who have 10 publications each. The authors Agrawal, A., Hill, P.S., and Li, Y. have nine publications each. The rest of the authors on the list have eight publications related to the SDGs. It is noticeable that this participation is distributed in a similar way among these authors, where 50% of the academic production is developed by 29.8% of the authors consulted.

3.2 Quality Indicators

Next, the analysis will be addressed from the quality indicators that refer to the citations that are associated with the publications and that allow measuring the impact of these as appropriate.

3.2.1 Impact Per Journal

This indicator has the top ten journals with the highest number of publications with citations as can be seen in Fig. 17.5. The Lancet journal is the one that heads the list with more than 1000

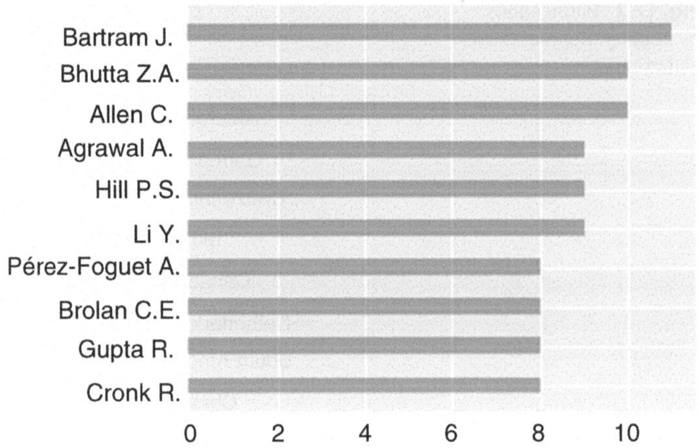

Fig. 17.4 Publications by author. Source: Own elaboration

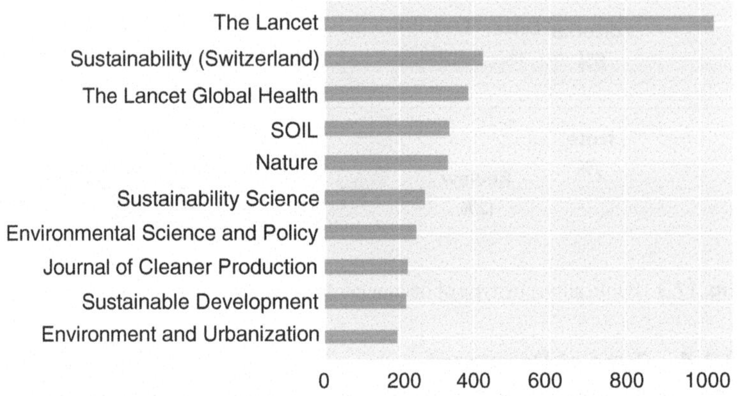

Fig. 17.5 Number of citations per journal (ten most cited). Source: Own elaboration

associated citations to its publications, which represents dominant participation over the rest of the journals, such as Sustainability, the Swiss journal that gathers just over 400 citations. Third, there is The Lancet Global Health journal, which is part of the family of journals associated with The Lancet, the British journal with great prestige in the medical sector, which is the top of the list. Next in the list are SOIL and Nature journals. Although the remaining journals have less citations, it is noteworthy that in the last position there is a journal with around 200 citations; in the same way, it can be seen that the 3 journals that head this list together with the Journal of Cleaner Production, Sustainable Science, and Sustainable Development can also be found among the journals with the largest number of publications.

Furthermore, according to the distribution of citations in the subject of interest, only 3.2% of the consultation journals are responsible for 50% of the citations in the field of study and with a transience index that results in which 33% of the journals do not have any associated citations in the subject of this research.

3.2.2 Impact by Author

This indicator highlights the quality of the top ten authors according to the number of citations associated with their studies related to the topic of interest. The distribution of citations is very even with very low differences between authors in terms of number of citations; for example, first, there is the author Jun Zhu with 495 citations, followed by the author Gupta, R. with 481. The authors Robert Black, Perin, J., Liu, L., and Chu, Y. that follow in the list have the same number of citations with a total of 468, which may be due to having articles that they developed together (see Fig. 17.6).

On the other hand, it should be noted that with the distribution of citations, only 6.5% of authors have 50% of citations in the field of interest, according to search results, while 22.8% of the authors do not have associated citations.

3.3 Structure Indicators

Figure 17.7 represents the network of authors on the subject who have the highest number of citations in their work, specifically those who have more than 200 citations. To begin, we have the authors Colin Mathers, Dan Hogan, Yuen Chu, Simon Cousen, and Robert Black, who have collaborated in various works that relate SDGs and the global agenda on sustainability to public health issues, mortality rates of children, as well as fatal diseases in this population.

Professor Jun Zhu is an author associated with the National Office of Motherhood and Child Health Affairs and has conducted some studies based on a systematic analysis of child health and mortality issues and their implications for compliance to the SDGs in this area of science. On the other hand, the authors of the block, such as Joshua Salomon, carried out studies on its impact on health, economic well-being, and incorporation of the SDGs. In this way, both authors of the first red, along with authors such as R. Gupta, JM. Haro, and O. F. Norheim, are related through collaborations that try to measure the progress made in the different countries with the SDGs related to the health sector that emphasizes children's health.

On the other hand, the authors Johan Bouma and Gerben Mol focused on another topic of importance related to the intervention and participation of soil sciences as a relevant input within the fulfilment of the SDGs, as a helpful factor in subtopics such as agriculture, land use, and their properties, which can be used as a point to consider within the pertinent applications within the development of the SDGs in the specific situations of each territory. This is a separate segment that has quite a few citations as can be seen in the figure.

3.4 Emerging and Growing Fields

According to the bibliometric analysis, the emerging and growing fields of research that result from the analysis of keywords, built from the standardization of the search criteria and association of concepts with the same meaning, show that the topics with the greatest number of publications in the last decade are associated with gender equality (SDG 5), global health (SDG 3), and climate action (SDG 13); therefore, the revision of trends will be mainly framed around these SDGs. Nevertheless, there has been a constant call to address actions from an intersectoral approach (Caroline Mair 2014).

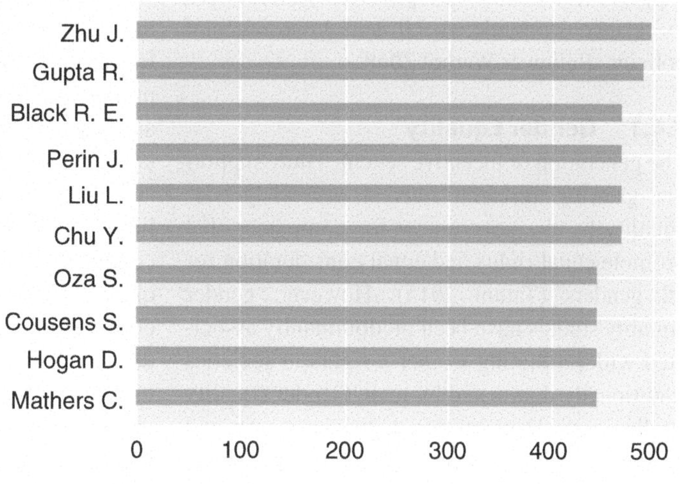

Fig. 17.6 Number of citations by authors. Source: Own elaboration

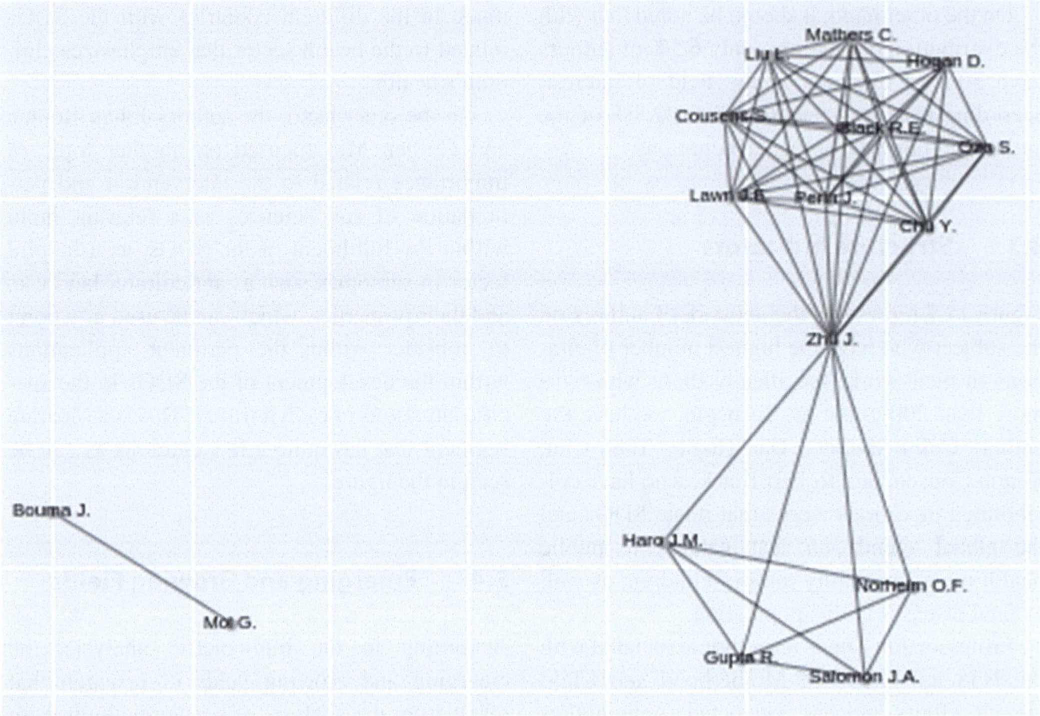

Fig. 17.7 Network of most cited authors. Source: Own elaboration

"Sustainable development" is related to the process of implementing advanced technology, increasing capital, and improving the standard of living without causing land degradation or depleting natural resources, and it is a concept that has evolved from economic development, and likewise from the adoption of agreements and commitments to act responsibly by corporations to support local and national government initiatives to balance economic, social, and environmental policies (Bailey & Warhol 2018).

3.4.1 Gender Equality

The generation of incentive schemes may support the growing research interest around gender equality by the government for companies that promote equal rights and equal consideration for all genders (Tatum 2015). However, gender equality studies have been predominantly associated with promoting women's rights in societies traditionally dominated by men. Gender equality studies have brought into focus issues around violence against women such as rape and sexual assault, but it has also fostered the need to consider that a woman of color may face different obstacles and experiences of sexism from a White woman (Rich and Walker 2019). Likewise, gender equality seeks to reduce the disparities around access to resources, services, power, and decision-making processes and structures, as well as to education and job market (Nesti 2019).

Although international organizations such as the United Nations, the OECD, and the European Union have been strongly engaged in promoting equal opportunities, gender issues are still not perceived as important in some countries due to cultural resistance and gender stereotypes and in some cases since some public officials do not have enough resources and competences to manage gender issues adequately (Nesti 2019). However, different governments around the world have played a decisive role and have accelerated the progress towards parity, implementing legislation, fiscal incentives, and partnerships with the private sector (Gupta et al. 2019).

At the local communities' level, women have established relations such as the Associations for the Empowerment of Women—AEW—to promote entrepreneurship and self-employment. Women can raise awareness of the opportunity to become financially and socially independent in patriarchal societies. For example, in Serbia, female refugees are educated to raise awareness on self-employment as part of supporting international institutions' programs. Those programs have had a gradual effect in the last decade, increasing the number of women's enterprises and women holding positions in other fields, such as services, bookkeeping, financial consulting, education, rural tourism, and production of agricultural products (Maksimovic et al. 2019). In this line, some research institutes in India have contributed to sustaining women's livelihood and contributed to the economy. They are promoting floriculture as an activity for generating gainful self-employment among farmers by conducting entrepreneurship programs and courses to provide women with skills needed for floriculture business, raise awareness about the industry's potential, and generate income (Janakiram et al. 2018).

As an initiative of empowerment, the Self-Help Groups (SHGs) was an idea of Grameen Bank of Bangladesh in India to rural credit without asking borrowers to provide collateral or engage in paperwork. SHG is formed by 10–20 volunteer women and works as an association of people for collective goals, with a social and/or economic purpose. The idea is to create a common fund and provide loans to its members to promote entrepreneurship and skill development. This initiative has evolved, and nowadays, there exist three different models with banks: (1) SHGs formed and financed by banks, (2) SHGs formed by NGOs and other agencies as financial intermediaries, and (3) SHGs formed by banks, using NGOs and other agencies as financial intermediaries. Furthermore, as part of the CSR efforts, corporates are promoting women empowerment by the formation of SHGs. They want to impact fewer privileged women, make them self-reliant, increase their self-confidence, and equip them with financial planning, business, and leadership development (Kapoor 2019).

From a different perspective, literature has shown the critical role that *workplace social support* for women, such as role models, formal and informal mentors, sponsors, supportive supervisors, and peer support, plays in closing gender gaps. Women account for around 5.2% of CEO positions, and, according to the WEF, there is a 32% gender gap across indices of economic, opportunity, education, empowerment, and well-being around the world. Part of this gender inequality is due to the different situations that women experience related to stereotypes, through which women feel marginalized and concerned about confirming these stereotypes.

Organizations and professional communities consider investing in efforts to generate recognizable models of success for women, mainly in contexts where women are underrepresented (Cortland and Kinias 2019). In the case of the United Kingdom, after a review about the current state of women on boards, chairpersons, and chief executives, investors and executive search firms backed by the government took action to close the gender gap by increasing the percentage of women on board and senior leadership positions up to 33% of FTSE 100 companies, which is currently 32.1% (Davies 2011; GOV.UK 2019); by disclosing each year the proportion of women on the board, senior executive positions, and female employees in the whole organization; by reporting the gender pay gap every year; and by providing employers with different practices to close the gender gap within organizations (GOV.UK 2017).

3.4.2 Global Health

Scientific knowledge around health issues has been increasingly spread due to the use of technologies such as the Internet, helping to promote more discoveries. Although healthcare has become a political issue for all the governments around the world, it has a broad scope that goes from infections and chronic diseases to environmental pollution, injury prevention, war, and hunger. The poorest countries face unequal healthcare mainly due to the marginalization of large groups of people, injustices in trade, and availability of care. Furthermore, the reality for poorest countries is that healthcare is not affordable, they lack tools and cures, some cures are too expensive, and there are not enough personnel available (Wagner 2019).

The G7 has taken into consideration health issues as part of its key agenda, besides core topics such as downside risks to the global economy, terrorism, and a record-high number of refugees. Global health has become a matter of attention since it has been related to having an effect on security and economic challenges, just as the achievement of sustainable development. Health emergencies are increasing to become threats, similar to war conflicts and natural disasters, where these threats are claiming human lives on a scale equivalent to pandemics, the latter with severe impacts on the lives and livelihoods of human beings, generating large numbers in terms of deaths similar to armed conflicts (1 to 1.5 million deaths per year in the twentieth century). This has demanded to strengthen health systems, especially in vulnerable nations (Hara and Ezoe 2019).

The mitigation of the spread of infectious diseases has promoted the adoption of nonpharmaceutical intervention (NPI) measures and effective strategies for communication with the public to ensure a timely response to this type of crisis. According to a Pandemic Influenza Readiness Assessment made in 2015 in the United States, community-level NPIs consider home isolation, school closure, isolation of sick people, and social distancing as critical measures to mitigate the spread of diseases. However, communication with the public is fundamental to ensure the dissemination of guidance about NPI measures, mainly among the most vulnerable and non-English-speaking populations, during the different stages of a potential outbreak (Naik et al. 2019). Apart from NPI measures, outbreaks such as Ebola promoted the launch of the Pandemic Emergency Facility (PEF) of the World Bank, an emergency funding mechanism complementary to the Contingency Fund for Emergencies (CEF) of the WHO for swift fundraising, and the creation of the Standard Operating Procedure (SOP) for health emergencies, to improve health and humanitarian coordination among WHO and other important UN partners, in order to strengthen health systems across developing countries and finally achieve universal health coverage (UHC) (Hara and Ezoe 2019).

UHC has become a priority for main countries, but this requires efforts to reduce fragmentation in terms of financing and service delivery and increase efficiency in the health system as a whole. The creation of formal coordination mechanisms and the improvement of the capacity at the subnational level are crucial elements to ensure that any resource spent has a great positive impact on the healthcare system. The adoption of budgetary practices aligned with OECD recommendations around good budgetary governance could make planning and budget execution easier at all governmental levels, as making the approved budget closer to what is expected to be executed would increase the usefulness of this one. Peru is a case of improvement in terms of budgetary practices, by implementing *results-based budgeting*, moving the public sector to a results-based management model, which seeks effectiveness of public expenditure by aligning resource allocation with government priorities linked to specific goals. Through this methodology, which starts in the budgeting process, all the information available is used to determine the needs of drugs, human resources, and capital goods to attain particular objectives. Furthermore, unlike previous budgeting practices which assigned resources on a historical basis, this methodology evaluates the needs and the most efficient methods to meet them (Vammalle et al. 2017).

From a different viewpoint, e-health solutions as a combination of technology and health have proven quality in terms of healthcare delivery and well-informed stakeholders. They have been associated as a reengineering of the healthcare sector, allowing to reduce expenditure, solve the limited capacity of the healthcare problems, and increase productivity (Banna & Ottesen 2018). However, there are some limitations of the use of e-health, such as the accessibility to technology in some regions. The US federal e-health web page has made efforts to provide tools on how to create websites, educate elderly people who are not active users of technology, and even make strategies to broaden access to cover limited-resource populations (Barnhill 2019). For instance, Kenya implemented Totohealth in 2015, a mobile-based application for tracking maternal health, helping to reduce maternal and child mortality through SMS technology by sending advice and reminders about nutrition, reproductive health, parenting, and development needs to expectant and new mothers (Constable 2016).

Financial mechanisms have had an impact on the acceleration of social policy innovation and promotion of UHC. For instance, social impact bonds (SIBs) are a form of contract with the government, by which it pays for better social outcomes in certain areas and gives part of the savings achieved to investors (Misawa & Sengoku 2019). SIBs work like any other debt securities, offering a rate of return, but are tied to a social or environmental objective (Smink 2017). Nevertheless, SIBs associated with healthcare objectives are few, and SIBs have predominantly covered the social welfare sector. Moreover, the allocation of capital in this type of investment is still low. Main investors are foundations and organizations, such as NGOs and CDFIs (Gruyter et al. 2020). In Hachioji City, Tokyo, Japan, in 2017, a SIB project where Cancer Scan Co Ltd and K-three Inc., as providers, and Japan Social Impact Investment Foundation, DIGISEARCH & ADVERTISING Inc., and Mizuho Bank, Ltd. as investors, participated was successful by increasing the colorectal cancer screening examination rate and Hachioji City rewarded Cancer Scan, which compensated the investors according to the achievement (Misawa and Sengoku 2019).

3.4.3 Climate Action

Climate change has resonated throughout the last several decades, due to natural variability and anthropogenic changes. Global warming as an example of climate change has increased the magnitude and frequency of floods and droughts in many places around the world, becoming a serious threat to human communities and the entire natural environment. However, the contribution of human activity towards climate change remains crucial and controversial (Misawa and Sengoku 2019). Although the 2030 Agenda has proposed different actions to tackle climate change, it shows a significant challenge around the unwillingness in how to approach this issue in line with economic development and health and well-being needs, thus evidencing the need to implement inter-sectoral governance mechanisms (Belesova et al. 2016).

Although multi-stakeholder alliances have been articulated, which have become mechanisms of conventional implementation to achieve the SDGs, and are perceived as a positive contribution to addressing climate change, there is very little evidence of positive performance. Furthermore, it has been proposed that, through coalitions and cooperation between government agents, business actors, and civil society, the probability of remaining in what has been called "safe operating space for humanity" will be increased, with which, in some way, more efficient, effective, and inclusive actions are expected to be taken to address global policy issues. Concerning climate change, multi-stakeholder associations have been established that have been dedicated to the implementation of the SDGs since the 2002 World Summit on Sustainable Development in Johannesburg and that have become an example to address other areas of global governance (Pattberg and Widerberg 2016).

To create awareness of the consequences of climate change as a severe threat, the Climate Action Simulation has been implemented in more than 22 countries in diverse educational scenarios between May 2015 and October 2019. This simulation that works through an interactive computer model simulating the physical climate and energy systems allows people to understand the interaction of both systems and the effect of policies and assumptions around climate dynamics. The simulation works as an educational tool creating awareness around climate change and sustainability and drives engagement around social dynamics that seek to promote the adoption of best practices on energy by allowing the users to take on roles during the decision-making process during the simulation. The Climate Action Simulation seeks to educate people around ongoing climate change, urgency of mitigation to achieve SDGs, potential impacts such as temperature and sea-level rise, and ocean acidification (Rooney-Varga et al. 2019).

On the other hand, local climate governance has been limited probably because local communities tend to focus mainly on strategies of technology-oriented ecological modernization rather than individual behavior change and unsustainable social practices. However, some intentional communities and low-carbon municipalities have adopted *low-carbon initiatives* by which the former focuses on substituting practices and eliminating unsustainable practices which reduce carbon-intensive practices as a consequence and the latter invest considerable effort in changing people's attitudes creating awareness by using marketing campaigns; most practices are optional or informative as they offer a low-carbon alternative and provide incentives (Hausknost et al. 2018). For instance, *bikesharing systems* have been implemented in urban settings as systems open to the public to serve as a form of public transport with very low emission form of mobility, offering a variety of pickup and drop-off locations. The system has evolved throughout the years; in Rennes, France, a third-generation system was incorporated; this new system included the use of technology to have greater control over the bicycle use, such as the use of sophisticated docking stations and automated smartcards (or magnetic stripe cards) electronic bicycle locking and payment, and the use of websites and apps to provide real-time information for users and a portal through which customers can manage their profile (Parkes et al. 2013).

The use of digital technology has become increasingly relevant in the way communication works across the world and also in the way communities cope with different types of issues. Polisdigitocracy is a new form of governance that seeks to engage citizens through digital technology and tackle climate change challenges efficiently (Lawrence et al. 2017). This is the way cities invest in smart technologies such as open data stores, citizen engagement platforms, and intelligent transport systems to help them provide good services at a lower financial and environmental cost. An example of this is citizen communication platforms, a space created for communities to explore local issues, enhance citizen engagement, and have their voice heard by the government. London, England, in 2010 took the initiative of crowdsourced ideas for climate change; this led to Talk London, a place where different communities discuss the city's issues, promoting the open debate and dialogue on public policy challenges, not just between politicians but also citizens, which is a source of empowerment for individuals and communities to take actions that support the vision of the city (Parkes et al. 2013).

4 Conclusions

According to the selected database, the first time in which sustainable development began to be discussed was in 1990 at the United Nations Conference on Environment and Development (UNCED). The first publications were produced only in 2013, the year in which the development of the proposal on the SDGs began; the number of publications began to be significant compared to that of previous years; that year 20 publications were made, and 5 years later, in 2018 the number of publications reached 615. Additionally, during the first semester of 2019, 407 publications were made; these publications are written mainly around the need to create synergies and alignment between public and private sector entities and the SDGs.

It is observed that in terms of scientific production, there has been an increasing trend, in line with the historical milestones around sustainable development. Most studies have been developed under a scientific method and are published in journals; however, the useful life of this information is on average 1.3 years. The journals with the highest scientific production are Sustainability, Journal of Cleaner Production, and Sustainable Science, all of them focused on sustainability in the social, environmental, and economic fields. However, the British Journal of the Lancet records the largest number of publications cited. For its part, the countries with the highest scientific production around the SDGs are the United States, the United Kingdom, and Australia, with the United States being the country with the highest number of publications cited. Additionally, Jamie Bartram should be highlighted, as the author with the highest number of publications, although Zhu, J. is the most impactful, reaching the highest number of citations.

Concerning research topics, the greatest number of publications has been generated in the last decade, with growth in the scientific production of key issues in the promotion of sustainable development, such as gender linked to SDG 5: gender equality, around which the rights of women, the conditions of migrants around the world, and the inclusion, participation, and empowerment of women are becoming increasingly relevant, generating an alert about the need to create more progressive and substantial statements that allow closing gender justice gaps. Global health is framed in SDG 3: health and wellness, whose research calls attention to the need of creating synergies between emerging economies (BRICS), G7, G20, and the 2030 Agenda that respond with greater commitment and quality to the needs of coverage and equity, and the protection of natural resources and climate change are framed in SDG 13: climate action, whose scientific production has led to the conclusion of the need to address its effects from an inter-sectoral approach and strengthen governance around the SDGs.

Given the growing deal of interest in terms of research around gender equality, global health, and climate action, some countries have made efforts and have taken action to address the gaps and accomplish what has been set in the 2030 Agenda, through entrepreneurship and self-employment programs for women, skill development programs, and gender inclusion for city development. The adoption of no pharmaceutical interventions to address health emergencies and the adoption of budgetary practices are aligned with OECD recommendations for the health system, and e-health solutions. Educational tools to raise awareness about climate issues and the implementation of low-carbon practices. However, there are still challenges that have to be overcome around violence against women, discrimination of women of color, and opportunities in the job market such as top management positions. In terms of global health, issues around how to address health emergencies, especially in poor countries where tools and cures are scarce or too expensive, remain. Finally, regarding climate change, finding the balance between environmental issues and economic development is a challenge that also demands cooperation between different actors.

However, a lot of research has been published and different strategies have been proposed and put into action around SDG in developed countries and emerging economies such as the United States, the United Kingdom, India, South Africa, and China. There are very little research and integrated strategies presented in developing countries such as in Latin America and Africa and some countries in Asia. This raises an opportunity to develop further research in those countries around SDG, considering those areas that have contributed to the destruction of the environment by human activity. Topics around food security, responsible consumption and production, agriculture, and migration support the shift to a more sustainable global economy, especially when it comes to facing the setbacks that different crises bring to many countries in terms of inequalities, global health, and economic development.

References

Arenas, J., & Santillán-Rivero, E. (2002) Bibliometría ¿Para qué? Biblioteca Universitaria Nueva Época, 5(1), 3-10. https://rdu-demo.unc.edu.ar/handle/123456789/715?locale-attribute=es

Banna & Ottesen, A. (2018). Health Solutions in Developing Countries: Case of Kuwait. 2018 IEEE International Conference on Innovative Research and Development, ICIRD. 1–6.

Barnhill, J H. (2019) E-Health. Salem Press Encyclopedia.

Belesova, K., Kelman, I., & Boyd, R. (2016). Governance through Economic Paradigms: Addressing Climate Change by Accounting for Health. *Disaster Policies and Governance: Promoting Community Resilience 4.*

Bailey & Warhol, T. (2018). Sustainable Development Debate. *Salem Press Encyclopedia.*

Cañedo, R., Rodriguez, R., & Montejo, M. (2010). Scopus: la mayor base de datos de literatura científica arbitrada al alcance de los países subdesarrollados. *Revista Cubana de Información en Ciencias de la Salud*, *21*(3). DOI: https://doi.org/10.36512/rcics.v21i3.14.g45

Constable, C. (2016). The Rise of the Social Enterprise in Nairobi. *African Business*. 20–22.

Cortland, C. I., & Kinias, Z. (2019). Stereotype Threat and Women's Work Satisfaction: The Importance of Role Models. *Archives of Scientific Psychology*, *7*(1), 81–89.

Davies, Lord of Abersoch. (2011). Women on Boards. 1–40. www.Gov.Uk/Government/Upload Accessed February 2011.

Gómez-Lee, M. I. (2018). Agenda 2030 de desarrollo sostenible: comunidad epistémica de los límites planetarios y cambio climático. *OPERA*, 69-93. https://doi.org/10.18601/16578651.n24.05

GOV.UK. (2017). Gender Pay Gap Reporting: Overview. FTSE 350 Urged to Keep up the Pace to Meet Women on Boards Target. https://www.gov.uk/government/news/ftse-350-urged-to-keep-up-the-pace-to-meet-women-on-boards-target.

Gruyter, E., Petrie, D., Black, N., & Gharghori, P. (2020). Attracting Investors for Public Health Programmes with Social Impact Bonds. *Public Money and Management*, *40*(3). 225–36.

Gupta, V., Solveigh, H., Krishnan, M. & Madgavka, A. (2019). Accelerating Gender Parity: What Can Governments Do? *McKinsey & Compay*. https://www.mckinsey.com/industries/public-sector/our-insights/accelerating-gender-parity-what-can-governments-do

Hák, T., Janoušková, S., & Bedřich, M. (2016). Sustainable Development Goals: A need for relevant indicators. *Ecological Indicators*, 565-573.

Hara & Ezoe & Ezoe, S. (2019). Why does global health matter to diplomacy? global health as a security and economic challenge and as an opportunity for world leaders, with a special focus on the G7 Ise-Shima Summit. *Global health governance IX* (1&2).

Hausknost, D. et al. (2018). Investigating Patterns of Local Climate Governance: How Low-Carbon Municipalities and Intentional Communities Intervene in Social Practices. *Environmental Policy and Governance*, *28*(6), 371–82.

Janakiram, T., Jain, R., Vyas, A.K, & Srilekha, R. (2018). Florienterpreneurship in India: New Avenue of Economic Empowerment of Women. *ASCI Journal of Management*, *47*(1), 72–80.

Keesstra, S., Bouma, J., Wallinga, J., Tittonell, P., Smith, P., Cerdà, A., Fresco, L. (2016). The significance of soils and soil science towards realization of the United Nations Sustainable Development Goals. *Soil Journal*, 111-128. DOI: https://doi.org/10.5194/soil-2-111-2016

Kapoor, S. (2019). Entrepreneurship for Economic and Social Empowerment of Women: A Case Study of Self-Help Credit Program in Nithari Village, Noida, India. *Australasian Accounting, Business and Finance Journal*, *13*(2). 123–42.

Lawrence, S., Doody, L., López, J., & Peracio, P. (2017). Polisdigitocracy: Citizen Engagement for Climate Action through Digital Technologies. *Field Actions Science Report 2017*, (Special Issue 16), 58–65.

Mair, C. (2014). Climate Change: The Greatest Challenge for the Future and a Major Cross-Sectoral Area of Intervention. *International Community Law Review*, *16*(2). https://brill.com/view/journals/iclr/16/2/article-p177_3.xml#d10132948e1758.

Maksimovic Ivanović, T., & Vujko, A. (2019). Self-Employment of Women through Associations in the Rural Areas of Sirinicka Zupa. *Ekonomika poljoprivrede*, *66*(1), 251–63.

Misawa & Sengoku, S. (2019). Social Impact Bonds: Current Context and Implementation Model in the Healthcare Industry. *Technology Management in the World of Intelligent Systems*.

Naik et al. (2019). Influenza-Related Communication and Community Mitigation Strategies: Results From the 2015 Pandemic Influenza Readiness Assessment. Health Promotion Practice, *20*(3), 338–43.

Nesti, G. (2019). Mainstreaming Gender Equality in Smart Cities: Theoretical, Methodological and Empirical Challenges. *Information Polity*, *24*(3), 289–304.

Parkes, S., Marsden, G., Shaheen, S., & Cohen, A. (2013). Understanding the Diffusion of Public Bikesharing Systems: Evidence from Europe and North America. Institute of Transportation Studies, UC Berkeley.

Pattberg, P., & Widerberg, O. (2016). Transnational Multistakeholder Partnerships for Sustainable Development: Conditions for Success. *Ambio A Journal of the Human Environment*, *45*(1), 42–51.

Rich, AK., & Walker, K. (2019). Gender Equality: Overview. *Salem Press Encyclopedia*.

Rooney-Varga, J N. et al. (2019). The Climate Action Simulation. Simulation and Gaming.

Sachs, J. (2012). From Millennium Development Goals to Sustainable Development Goals. *The Lancet*. https://doi.org/10.1016/S0140-6736(12)60685-0

Sachs, J. (2019). Transforming Our World: Achieving the Sustainable Development Goals [Material del Aula]. https://courses.edx.org/courses/course-v1:SDGAcademyX+TOW001+3T2018/course/

Sancho, R. (2002). Indicadores bibliométricos utilizados en la evaluación de la ciencia y la tecnología. Revisión Bibliográfica. http://www.temarium.com/serlibre/recursos/pdf/79059.Inteligencia%20Competitiva.Lecturas.pdf#page=77

Smink, V. (2017). Qué Son Los Bonos de Impacto Social, La Novedosa Solución Que Aplican Algunos Países Para Resolver El Tema de Los Ni-Ni y Otros Problemas Sociales. BBC.

Tatum, M. (2015). Sustainability: How FMCG Can Help Save the World in 17 Steps. Grocer: p28-33. 3p.

UN Department of Economic and Social Affairs. (2019). Sustainable development goals: sustainable development knowledge platform. https://sustainabledevelopment.un.org/

Vammalle, C., Rivadeneira, C., James, AM., Prieto, L., & Montañez, V. (2017). Health Financing and Budgeting Practices for Health in Peru. *OECD Journal on Budgeting*, *2*(1), 93–149.

Wagner, G. (2019). Global Health Issues and Policies. *Salem Press Encyclopedia*.

18

Harnessing the Potential of Information and Communication Technologies (ICTs) in Agribusiness for Youth Employment: Lessons from Bikita, Zimbabwe

Joseph Tinarwo and Dominique E. Uwizeyimana

Abstract

Agribusiness has become a key strategy to achieve socio-economic transformation and has been promoted by both governments and development partners in recent years. Yet, there is little evidence on the use of information and communication technologies (ICTs) in agribusiness in order to create employment for the youth, especially in rural areas. Conventional employment creation strategies by governments for the youth are not yielding the much-anticipated results, and some innovative ways of determined actions are imperative to create youth employment, achieve socio-economic transformation and ultimately realise the Sustainable Development Goals (SDGs). This research primarily follows a qualitative approach to collect data to understand the potential use of ICTs in agribusiness for youth employment in Bikita. The results reveal that, notwithstanding the efforts by the Government of Zimbabwe in promoting the use of ICTs in agribusiness, rural youths are confronted by a horde of challenges including lack of ICT skills and knowledge, lack of financial support for business start-ups and poor infrastructure to support the use of ICTs. The chapter recommends a cohesive approach that factors capacity development and skill upgrading for the youths, ensures strong legal and vibrant institutional frameworks that support use of ICTs, provides financial support for business start-ups for the youths and expands ICT infrastructure to remote areas of the country.

Keywords

ICTs · Agribusiness · Youth employment · Bikita Zimbabwe

1 Introduction

The future of development in Africa depends on young people (AfDB 2016; World Bank 2018). Nevertheless, youths are facing a myriad of challenges, and greatest amongst them is unemployment or underemployment (AfDB 2016; FAO and ECA 2018). Most of the youths in Africa

J. Tinarwo (✉)
Julius Nyerere School of Social Sciences, Great Zimbabwe University, Masvingo, Zimbabwe

D. E. Uwizeyimana
School of Public Management, Governance and Public Policy, University of Johannesburg, Johannesburg, South Africa
e-mail: dominiqueu@uj.ac.za

G. Nhamo et al. (eds.), *Sustainable Development Goals for Society Vol. 1*, Sustainable Development Goals Series, https://doi.org/10.1007/978-3-030-70948-8_18

reside in rural areas and are currently facing minimum chances of gainful employment (Allen et al. 2016). Considering that agriculture is a key enabler for economic growth and employment creation, tapping into the potential of youth has proved challenging. Agriculture faces an imminent problem of generational succession, but researchers and policymakers are failing to recognise it (Elias et al. 2018). Proctor and Lucchesi (2012) argue that the majority of people involved in agriculture in most countries are old and seem not to have a successor and young people seem not to be interested in farming. Zimbabwe is home to the continent's youngest population, with over 60% of people below the age of 25 (UNFPA 2019). However, this sub-Saharan nation is amongst the countries with the highest youth unemployment statistics in the region with unemployed young people between the ages 15 and 34 years constituting 84% of total unemployed people in Zimbabwe (ILO 2018; Gukurume 2018).

In Zimbabwe, agriculture is regarded as an essential driver for economic growth and policymakers have the opportunity to harness the dividends of this youthful population. It is imperative for the policymakers to rebrand agriculture as the new untapped frontier for youth employment and ultimately achieve socio-economic transformation (Yami et al. 2019). This youthful labour force has the ability to use ICTs and innovate in agribusiness, which can result in employment creation (Frost and Sawa 2017). Koira (2014) claims that agribusiness hosts vast opportunities for youth employment if supported with investments and a strong policy framework. Youths have the ability to harness the untapped potential of agriculture through innovation and entrepreneurship (Betcherman and Khan 2015). Inherently, tapping prospects in ICT innovations in agro-entrepreneurship and value chains result in enhanced agriculture sector's image to the young people and can ultimately create employment for them.

In this research, the author uses the case study of Bikita district in Zimbabwe to understand the potential role of ICTs in agribusiness in order to create youth employment. The objective is to articulate through a case study of Bikita that using ICTs in agribusiness may contribute to the creation of jobs for young people, especially in rural areas. The author argues that the use of ICTs in agribusiness if supported with capacity development and empowerment can eventually lead to youth employment.

The remaining parts of the paper are organised as follows: The following reviews the literature on the use of ICTs in agribusiness for youth employment. This section also reviews the legal frameworks and key institutions that promote the use of ICTs in agribusiness. The third section outlines the methodological approach used in this research. The fourth section presents the results from the case study of Bikita, while the fifth section discusses the results and proposes a conceptual framework to classify the components for attaining youth employment and ultimately achieving the SDGs. The final section of the chapter gives the conclusion.

2 Literature Review

2.1 Conceptualising the Term Agribusiness in Achieving Socio-Economic Transformation

While agribusiness is a common term in international development, there are several interpretations of what it means. For instance, Davis and Goldberg in Zylbersztajn (2017) defined agribusiness as the entire operations in the production and supply of farm inputs, agriculture activities on the farm, and the keeping, manufacturing and distribution of farm produce. In fact, agribusiness refers to all agricultural activities, industries and facilities that form the supply chain from farming through manufacturing and finally selling to final consumers (Allen et al. 2016; Filmer and Fox 2014). Furthermore, Koira (2014) argues that agribusiness involves the entire value chain from food production, storage, transport and processing up to marketing. Although the above definitions appear neutral, they are indirectly founded on normative assumptions that are all commercial undertakings done from the farm to the fork. Fundamentally, definitions of "agribusiness" clearly have normative content, recognising what the institution or author offering the definition considers as correct. The complete value chain, comprising the distribution of farming inputs, manufacturing and processing of agricultural goods, and their supply to ultimate users, is a shared characteristic in various definitions of agribusiness.

Globally, agribusiness is considered as an engine for economic growth and is increasingly recognised in the global, regional and national policies and strategies that strive to promote sustainable development (Abraham and Pingali 2020). The agribusiness sector remains the main global source for the employment creation and socio-economic transformation and contributes to the achievement of several SDGs including food and nutrition security (UN 2015; Nhemachena et al. 2018). FAO (2017) points out that the importance of agribusinesses is underscored in the point that they:

- Are regarded as the main source of off-farm jobs in developing countries, particularly their rural areas
- Have a beneficial effect on poverty alleviation and contribute to youth empowerment, especially in countries that produce high-value agriculture products for exports
- Generate off-farm job prospects in agro-industrial companies situated in remote areas, enhance income generation of rural people including youths and increase the ability to use ICTs
- Aid to create the linkages between agriculture and manufacturing industries, which can result in the growth of the production industries by supplying inputs for value addition from agriculture

2.2 The Relationship Between ICTs, Agribusiness, Youth Employment and SDGs

Notwithstanding the fact that most food is produced by old people, young people are likely to adopt and use ICTs in agribusiness (Njenga et al. 2012). In fact, the use of ICTs has the potential to stimulate and attract the majority of young people into agribusiness, especially those staying in rural areas (Irungu et al. 2015). White (2012) notes that increasing access to ICTs as stipulated in SDG 9, target 9.C, can motivate the youths to engage in agribusiness (SDG 2) and create employment for them. Inherently, SDG 2 aims to end hunger, achieve food security and improved nutrition and promote sustainable agriculture (UN 2015).

In many developing countries like Kenya and Nigeria, young farmers use ICTs more frequently than old farmers, and this has made it possible to attract youth in agribusiness as well as create employment as stipulated in SDG 8 (Rono et al. 2012). Goal 8 of the SDGs seeks to promote sustained, inclusive and sustainable economic growth, full and productive employment and decent work for all (UN 2015). Irungu et al. (2015) argue that young people can use ICT skills to get agribusiness entrepreneurship opportunities and hence address the problem of unemployment. The use of ICTs in agribusiness can enhance information access and connect rural young people to business networks (Rono et al. 2012). IFAD (2014) state that the integration of ICTs in agribusiness has the potential to help rural young people access agriculture extension services and facilitates the access of financial services and markets. Therefore, the growing use of ICTs in agribusiness entrepreneurship, particularly in developing countries, can be used as a strategy to attract the majority of rural young people to gainful employment and ultimately reduce rural to urban migration (Njenga et al. 2012).

It is worthwhile to look at countries that have integrated ICTs in agribusiness in order to create youth employment. Over 90% of the youth in Kenya have mainstreamed ICTs in their farming practices (Frost and Sawa 2017). For example, youths employed at iHub innovation in Kenya, a platform for generating and disseminating farming ideas in rural communities, are advocating the use of technologies and innovation into agribusiness. As a result, two main initiatives were incubated by iHub, namely M-Farm, which is a platform that provides the latest market prices to young farmers (via WhatsApp or SMS) and links them to customers, and iCow, which gives young farmers into livestock production specific information, including livestock market prices and veterinary services (Frost and Sawa 2017).

UjuziKilimo also uses mobile SMS to link young farmers with each other, including devices to collect soil and agriculture data to send young farmers actual guidance on farm tips, pest management, market prices and other relevant agriculture information. This business enterprise only employs youths that are below 30 years. Frost and Sawa (2017) argue that the Technical Centre for Agricultural and Rural Cooperation (CTA) also embraced ICTs in agriculture for the youths by initiating the AgriHack Talent Programme in 2013. The project has targeted more than 600 youthful innovators and entrepreneurs, by developing agriculture ICT innovations and start-ups, like FarmDrive. A new initiative of the project is Pitch AgriHack—a training boot camp, pitching competition and a chance for young farmers to win grants and investments to upscale their services.

Similarly, the Government of Rwanda has attempted to encourage the use of ICTs in agribusiness since the late 1980s (IFAD 2014). For example, a High-Tech Lab is assisting youths to use ICT farming initiatives and turn them into business undertakings. In 2012, kLab was established to attract tenants (the term is given to youths with entrepreneurial concepts) and gurus (mentors devoted to assist the youthful innovators turn their ideas into commercial ventures) (IFAD 2014). The mentoring process gives the youthful businesspeople access to ICTs and capacity building to advance their abilities and market their ideals as well as access capital for business start-ups (IFAD 2014). Training, seminars, workshops and competitions assist the youths to become entrepreneurial, mentoring how to develop business proposals, networking and where to market their products. The majority of the youths in the kLab are university graduates and secondary school-leavers, and they developed two key business innovations (IFAD 2014). OSCA Connect, a start-up youth entrepreneurial group, created a mobile phone application called Sarura, which assists agricultural farmers to enter the crop they want to grow. This mechanism then verifies the meteorological information to decide if the proposed crop is ideal, based on the season and region. Results have shown that Sarura increases agricultural production of farmers and saves them both time and financial resources. The other agricultural innovation that was developed is the Farm In Bytes Application (FIBA). This platform links farmers to agronomists, sellers to buyers, and also other actors in the agribusiness, and assists farmers in daily activities, assisting them to update and manage their agricultural records so that farming advisors and specialists can give them relevant and targeted help (IFAD 2014). In doing this, kLab has created employment for these young people in Rwanda.

Section 2.3 below reviews the legal and institutional frameworks that underpin the use for ICTs in agribusiness.

2.3 Legal and Institutional Frameworks Guiding ICT Use in Agribusiness in Zimbabwe

Effective use of ICTs in agribusiness is hinged on the robust legal and institutional frameworks, including clear policies together with strong coordinating organisations (Tinarwo and Uwizeyimana 2019). The Government of Zimbabwe has made a political commitment to a number of international and regional policy frameworks that support ICT innovations in agribusiness and other sectors. However, the legal and institutional frameworks informing the use of ICTs are highly fragmented and lack harmonisation and coordination (ACBF 2017). This is also worsened by the fact that the Zimbabwean Government has for the past decade failed to provide adequate budgetary support to the institutions that are responsible for the use of ICTs (Tinarwo and Uwizeyimana 2019). In view of the foregoing, ACBF (2017) notes that nations with robust and effective ICTs allocate up to 3.5% of their gross domestic product (GDP), a rate that is far higher for Zimbabwe and other developing countries. The following seeks to present the legal and policy initiatives that are some of the commitments adopted by the Zimbabwean Government in order to promote the use of ICTs in agribusiness and other sectors.

The Sustainable Development Goals are the 17 worldwide initiatives adopted by the world governments to achieve a better life and sustainable future for everyone by 2030. Goal 9 of the SDGs states that investment in ICTs is crucial to achieving sustainable development and empowering communities in many countries (Tjoa and Tjoa 2016). Specifically, target 9.C of the SDGs aims to increase access to ICTs in order to deliver worldwide and inexpensive access to the internet in the poor and developing countries by 2020. In relation to the use of ICTs in agribusiness, the SDGs generally call governments and stakeholders to end hunger, achieve food and nutrition security and promote sustainable agriculture by increasing investment in rural infrastructure, agricultural research and extension services and technology development in order to enhance agricultural productive capacity and create opportunities for value addition and non-farm employment (UN 2015).

In an effort to realise Africa's vision, Agenda 2063 emphasised ICTs as an essential pillar to achieve economic growth and transformation. Agricultural development, along with agribusiness, is at the heart of Africa's developmental plan as a way to help in developing rural areas and create employment (Juma 2015). Zimbabwe, along with other African leaders, has also embraced the Science, Technology and Innovation Strategy for Africa (STISA-2024) in 2014. In fact, STISA-2024 is a 10-year strategic framework for Agenda 2063 that strives to assist African governments to an innovation-driven, knowledge-led economy. This 10-year strategic framework for African countries attempts to provide an effort to improve Africa's Science, Technology and Innovation (STI) status in human resources, technical skills and development of infrastructure and create an enabling atmosphere for modernisation and entrepreneurial culture (ACBF 2017). According to Agenda 2063, embracing STI will result in economic development as well as assists a quarter of companies originating from technological innovations and breakthrough produced by African citizens.

The Government of Zimbabwe launched the country's Information and Communication Technology Policy in 2016 to guide national economic development through systematic and coordinated use of ICTs in all sectors, including agribusiness (ACBF 2017). In fact, the national ICT policy in Zimbabwe is a result of the updated ICT Policy of 2005. The ICT policy in Zimbabwe is regarded as an economic enabler for the growth and creation of sector-appropriate solutions to achieve transformation through technology and information revolution. The goals underpinning the Zimbabwe ICT policy are to assist the establishment and upkeep of infrastructural services essential for ICT growth. Fundamentally, the ICT policy aims to do widespread capacity building and training programmes in order to give enough supply of qualified ICT human resources in all segments of the economy (Tinarwo and Uwizeyimana 2019). The ICT policy strives to create institutional instruments and measures for defining sectoral and also encourage, assist and enhance the expansion and use of ICTs. Finally, the Zimbabwean ICT policy endeavours to safeguard fair access to benefits obtainable by ICTs in all areas of society as well as encourage research and development of home-grown domestic ICTs. Along the ICT policy is the Innovation Drive which is a funding initiative to support young innovators in Zimbabwe, particularly those engaged in agribusiness which is regarded as the backbone of the Zimbabwean economy (ACBF 2017).

The Transitional Stabilisation Programme is the current blueprint guiding Zimbabwe's developmental agenda for the period 2018–2020 (Chitongo et al. 2020). Fundamentally, TSP outlines the policies, strategies and projects that strive to stimulate Zimbabwe's economic growth and employment creation for a prosperous and empowered upper-middle-income society (Chitongo et al. 2020). The TSP states that embracing ICTs contributes significantly to economic growth and has the potential of creating jobs for the youth (Chitongo et al. 2020). The Government of Zimbabwe through the TSP targets the promotion of innovation and infusion of technology in agriculture in order to boost production, improve food and nutrition security and regain the breadbasket for Southern Africa status. The TSP notes that mainstreaming ICTs in agribusiness offers immediate scope for job opportunities, especially for the majority of the unemployed youths in Zimbabwe.

The ICT Ministry in Zimbabwe aims to develop a favourable environment for the development of ICT-based society that cuts across all levels of the society, including agribusiness (Tinarwo and Uwizeyimana 2019). Fundamentally, the ICT Ministry's vision is to ensure an ICT-based republic with universal internet connectivity by 2020 and it also has a mission to exploit the potential of ICT for socio-economic transformation in Zimbabwe (Tinarwo and Uwizeyimana 2019). Currently, the ICT Ministry is undertaking the following programmes in order to create an ICT-based society:

- Creating community information centres aimed at promoting the use of ICT solutions, especially in rural areas
- Commination infrastructure which is the mainstay for universal connectivity in Zimbabwe
- Analysis of national ICT policy agenda and the growth multi-sector policies
- e-Government for service delivery

The Ministry of Agriculture has established a Youth Desk in order to mainstream the young farmers in the agriculture sector. The Ministry is motivated by the notion that the agriculture sector is yet to fully exploit the potential of the youth, and it remains unattractive to the majority. The Youth Desk is guided by the need to rebrand agriculture as a career option for the young people and also introduce mentorship programmes that guide the youth on making a living from agriculture. Moreover, the Youth Desk is aimed at supporting the country's economic reform through creating an enabling environment for youth participation at all levels of the agricultural value chain alongside addressing the compelling evidence of ageing farmer population and its implications in the country.

The Zimbabwe Government formed the SIRDC in 1993, under the Research Act of 1986. SIRDC is a centre under the Office of the President and Cabinet and provides ICT expertise in the diverse areas throughout its institutes (ACBF 2017). The SIRDC's duties include expanding the use of ICTs to all the country's sectors and also regional countries in assorted disciplines and sectors (Tinarwo and Uwizeyimana 2019). In an effort to increase ICT innovations, SIRDC is mandated to do planned research and development for all sectors of the economy, including agribusiness. The SIRDC works together with other domestic and foreign institutions and institution of higher education in doing research and development as well as for capacity building and its use in industrial processes. The SIRDC also serves as a source and communicator of STI, and this is achieved through domesticating imported technology and innovation into the country's needs (Tinarwo and Uwizeyimana 2019). There are 11 institutes within the SIRDC and all deal with areas related to science, technology and innovation to enhance employment opportunities and ultimately national development.

Boxes 18.1 and 18.2 below provide a detailed description of the success stories on the use of ICTs in agribusiness in Zimbabwe. The Eco Farmer ICT project in Box 18.1 has been implemented by one of the biggest mobile network operators (MNOs) in Zimbabwe while Box 18.2 gives a detailed description of the eMkambo Project, implemented by Knowledge Transfer Africa in partnership with Afrosoft Private Limited Company to provide farming and market solutions to farmers in major food markets like *Mbare Musika* in Harare.

Box 18.1: The EcoFarmer ICT Innovation

EcoFarmer is an innovative way of doing agribusiness using cell phones in Zimbabwe and was developed by Econet Wireless Private Limited Company. Farmers that include youths use the cell phones to obtain regular meteorological conditions and information, agriculture tips and advice on market prices for their crops. By using EcoFarmer, the results have demonstrated that the yields have been improved and jobs are created, especially for the youths who can use technology better. Hence, the use of mobile technologies is facilitating the transfer of agriculture extension advisory services to rural areas. The use of mobile technologies to get agricultural advisory services and market tips has improved farming, especially for rural farmers and other actors in the entire agricultural value chain in Zimbabwe. Ecocash, a platform for paying and receiving money, is changing the lives of many people and helping farmers in rural areas to engage in agribusiness. Farmers who include youths get the following information on a daily basis when their cell phone lines are registered and subscribed:

Everyday weather information

Agricultural and market information

Weekly data for the crops

Monthly market information and pricing requests

Credit assessments

Marketing advice and links

Financial connections

Source: Tinarwo and Uwizeyimana (2019)

Box 18.2: The eMkambo Project

In 2012, Knowledge Transfer Africa partnered with Afrosoft, a private limited company into software development to launch a market-driven agricultural solution platform called eMkambo (Ndebele name that means the market). The platform seeks to integrate ICTs into agribusiness and strives to link farmers (of which the majority are youth), traders, financial institutions, input suppliers and NGOs that are into farming. In fact, eMkambo has the database of farmer's mobile numbers and shares SMS messages with them to update them of noble agricultural practices, market information and what is being sought after by traders at major agricultural produce markets like Mbare Musika. With primary financial support from Hivos and SNV (international non-governmental organisations working in Zimbabwe), eMkambo is getting data of all agricultural produce that is being sold at the market from city council officials and from this they can determine the price and demand of all types of crops including their farmers. The data generated is also monetised as banks, and other financing organisations can use it to facilitate short-term loans to buyers. The eMkambo platform also has a calling centre to help farmers and traders with adequate market information for planning purposes. Through eMkambo the majority of the unemployed youth have managed to venture into profitable business enterprises.

Source: Author's compilation based on field survey

3 Materials and Methods

This study adopted the qualitative research approach and entailed the discovering and understanding of the experiences, perspectives and thoughts of participants (Creswell & Creswell 2017). Data were gathered using a review of studies, key informant interviews and focus group discussions. The author started by reviewing past studies on the use of ICT in agribusiness for employment creation and economic growth. Key informant interviews were done with 15 purposively sampled participants of which 4 were from the government (2 from the Ministry of ICTs and 2 from the Ministry of Agriculture), 3 development partners, 5 from academia, and 3 from the private sector. Two focus group discussions were conducted, one with the youths between the age of 18 and 25, of which six were females and five were males. All youths who participated in the FGD attained either secondary or tertiary education. The second FGD was done with the academia at Great Zimbabwe University, of which four were from the department of agriculture while three were from the ICT department. Responses from both the key informant interviews and focus group discussions were recorded using a cell phone and transcribed to English. Data analysis was based on thematic analysis, where each response was read repeatedly in order for the researcher to be familiar with the contents. When done, the records were repeated to get the emerging themes from the original texts that were transcribed.

4 Results

This section presents the results from both the KIIs and FGDs. The results from KIIs are presented in Sect. 4.1, while the findings from the FGDs are presented in Sect. 4.2.

4.1 Key Informant Interview Results

Results from both KIIs and FGDs reveal that the use of ICTs plays a crucial role in enhancing agro-entrepreneurship, improving rural livelihoods and creating employment for the young people since they have a positive attitude towards technological innovation. The following KII verbal quotes illustrate the above:

> Generally, young people are more business-minded and enthusiastic about embracing innovations and contemporary ICTs in agribusiness than older people. In fact, the youth are more recognised by different stakeholders in agribusiness as agents of change. They are often approached as technological resource persons by older farmers and many other stakeholders (Advisor to the largest Donor Agency in Zimbabwe).

However, a key informant from one of the biggest MNOs revealed that its operations had been severely crippled by prolonged power outages throughout the country lasting at least 16 h per day. The MNO's services such as voice, data and mobile money transfer platform have been severely disrupted especially in rural areas where it is difficult to get an alternative power supply to power the network base stations (Econet, 2019). Therefore, continuous power outages have a negative bearing on ICT use in agribusiness. The following verbal quotes illustrate the above:

> Our system developed to sustain a certain level of electricity outages and, as such, the maximum level has been surpassed because of long loading shedding periods being experienced in the country. The unavailability of both electricity and diesel in the country results in a number of our base stations, especially in rural areas, will not be operational when there is no electrical power (Senior Technical Manager).

Furthermore, key informant interviews with academia revealed that the use of ICTs in agribusiness in Bikita is confronted by a host of challenges. It was noted that the ICT infrastructure in the district is very poor to support agro-entrepreneurship ventures and innovations. The poor ICT infrastructure is worsened by the fact that most ICT tools are complicated to use and unaffordable to some of the youth residing in rural areas. The following KII verbal quotes illustrate the above:

> Though Zimbabwe is advocating for devolution of government, it is important to establish remote agricultural information centres, which countries like India have been implementing successfully for some couple years. Use of ICTs in Indian Agribusiness is very impressive, especially in ensuring agriculture and extension coverage, particularly in remote districts (Professor of Agricultural Economics with one of the state universities in Zimbabwe).

A KII with the government mentioned that expanding rural electrification beyond the growth point is critical to support the use of ICT in agribusiness in remote villages. However, the respondent was quick to point out that most villages and districts, especially in Bikita, are yet to be covered under the rural electrification programme implemented by the Zimbabwean Government.

> … Rural electrification programme ends only at the major business centres. There is no electricity to support the use of ICT innovations in agribusiness in many villages of the district where the majority of the rural youths reside … (Provincial Head, Ministry of Lands, Agriculture, Water and Rural Resettlement).

4.2 Focus Group Discussions

The academia through an FGD noted that young farmers get information about good farming practices through radio and television programmes like *Murimi Wanhasi/Umlimi Wamhla* (a vernacular term that means today's farmer) where they receive farming tips like crop and animal diseases, conservation farming practices, markets and climate change-resilient strategies. The same FGD with academia noted that youth also benefit from SMS messages sent by one of the MNOs called EcoFarmer platform where they can ask questions and get responses about farming practices. However, the academic noted that not all the youths in rural areas have the gadgets to benefit from these radio and television programmes. The following verbal quotes illustrate the above:

> … The general poverty trends in Zimbabwe and Bikita district, in particular, have seen most households failing to acquire the necessary ICT gadgets. In fact, gadgets are a luxury that cannot be prioritised especially when people are struggling to get food to eat. Moreover, poor network and unaffordability of bandwidth data bundles from the majority MNOs make use of ICTs in agribusiness exclusively for the rich.

A FGD with the youths in Bikita revealed that the use of ICTs in agribusiness has the potential to reform the way agribusiness information is produced and disseminated since they (youths) can get information on which crop to grow and where to sell it. The following verbal quotes illustrate the above:

> We use ICTs such as cell phones to get market information, including updates on current prices of our agricultural produce at the market. In this regard, we can make choices on which market to sell our produce at the best prices.

The same FGD with the youths indicated that some of them benefit from Facebook, YouTube and Twitter to discuss agribusiness topics. In fact, the majority of agro-food entrepreneurs said that they use social media frequently to connect with business partners and markets. The youths that said that they do not use social media and internet cited lack of airtime and bandwidth data bundles coupled with poor network connectivity in some villages in the district. The following verbal quotes illustrate the above:

> Network connectivity here in the village is very poor, so we don't use the internet or social media. Maybe those with Wi-Fi at the Nyika growth point have access to the internet, but here in Majoni as well as in Neganda village network is very poor. The poor network is mainly caused by the fact that MNOs are still to ensure full network coverage in some parts of the district, and this is also worsened by continuous electricity power cuts due to load shedding.

5 Discussion

The research found that use of ICTs in agribusiness is an important strategy in promoting agro-entrepreneurship, improving rural livelihoods and creating employment for the youths. This finding from interviews is in line with literature which indicated that agribusiness has substantial prospects for employment and economic fortunes for the rural youth and goes a long way in curbing rural-urban migration (Rono et al. 2012).

Lack of ICT infrastructure and gadgets was cited in both KIIs and FGDs as one of the main constraining factors for the youth to embrace ICTs in agribusiness. The results confirm previous findings by Yami et al. (2019) who observed that lack of ICT infrastructure is one of the main challenges facing the rural youths to embrace the use of ICTs in agribusiness. To that end, it is imperative for the Zimbabwean Government to invest and expand ICT infrastructure to rural areas as a way to promote the use of ICTs in agribusiness that has been lagging behind. Expansion and provision of adequate, sustainable energy are equally critical to ensure reliable business (Dube and Nhamo 2020) and network operations in Zimbabwe if the aspirations to achieve Sustainable Development Goals are to be achieved.

Public-private partnerships in ICT infrastructure development could be promoted, particularly in rural areas as a way of stimulating the development and growth of rural economies through the provision of the necessary supporting rural infrastructure (Irungu et al. 2015). Countries that have been successful in building strong ICT infrastructure encourage infrastructure sharing amongst telecommunication companies, a move that has been advocated by the Zimbabwean policymakers to MNOs in the country in recent years (Chifamba 2019). Infrastructure sharing would ensure cost reduction for mobile operators but most importantly ensure a wider and more reliable connectivity for the consumers as companies can leverage on cost reduction to improve maintenance.

The study found that most youths in rural areas in agribusiness lack ICT knowledge, capacity development initiatives and skill upgrading programmes. According to Irungu et al. (2015), most youths in rural areas face several problems, including lack of information on recent agricultural technologies and agribusiness opportunities. Capacity building, including skill upgrading, makes the youth like agribusiness as a profession, engage in agro-entrepreneurship and reduce rural to urban migration (Njenga et al. 2012). Therefore, it is imperative for policymakers to urgently address the knowledge and capacity challenges that young people in rural areas face in embracing ICTs in agribusiness in order for them to get jobs. In fact, the Zimbabwean Government must make a concerted effort in prioritising the rolling out of e-skills for better utilisation of ICT services in agribusiness.

Agribusiness capacity development, information and education can be mainstreamed in schools, colleges and universities to ensure that youth get the skills for agro-entrepreneurship and farming innovations (ACBF 2017). Tertiary institutions must design agribusiness programmes that integrate ICTs with domestic agribusiness approaches. University and college agribusiness programmes could also be combined with collaborative research with agro-entrepreneurs and ICT companies and can result in a wide variety of capabilities for digital jobs. The focus should go beyond offering basic computer and digital literacy and include a range of ICT innovations in agribusiness.

The study found that a lack of strong legal and institutional frameworks to support the use of ICTs in agribusiness is one of the hindrances to create youth employment. The development of sound policies and vibrant institutions that foster the use of ICTs and innovation in agribusiness is key in creating youth employment and achieve SDGs and other development-related variables (ACBF 2017). There is a need to amend and reform policies in most developing countries to ensure that youth access agricultural land and embrace ICTs in agribusiness as a way to reduce unemployment (IFAD 2014). Strengthening institutional capacities and mechanisms that support the use of ICTs in agribusiness must be given top priority starting from the central government down to subnational levels (Tinarwo and Uwizeyimana 2019). Institutional capacity building, especially at subnational levels, is likely to improve policy implementation and enhances the participation of young people in decision-making. These institutions must be supported with adequate financial and human resources in order to effectively coordinate the use of ICTs in agribusiness for rural youth programmes (ACBF 2017).

Finally, access to agricultural markets is one of the inhibiting factors for rural youths to participate in agribusiness. Most young people, especially in rural areas, find it difficult to access viable markets due to lack of ICTs and innovation (IFAD 2014). This is also worsened by the increasing worldwide dominance of supermarkets and the demanding principles used in their supply chains (IFAD 2014). It is imperative for the Zimbabwean Government to mainstream the use of ICTs in agribusiness, provide resources and

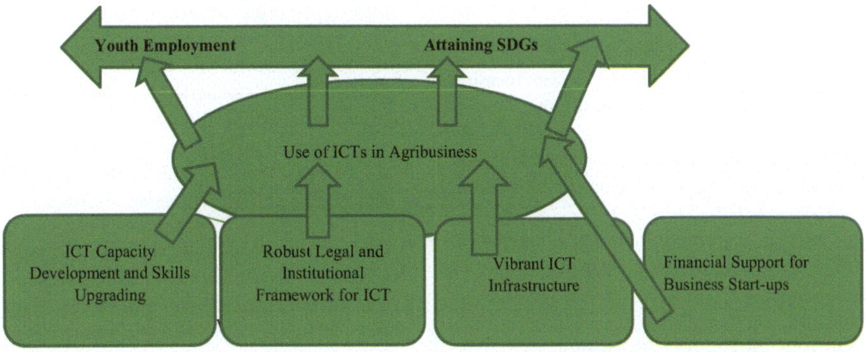

Fig. 18.1 Proposed conceptual framework for harnessing ICTs in agribusiness for youth employment. Source: Authors' own compilation

link rural youths to local and international buyers along the agribusiness value chain and also build up forward and backward business connections to create youth employment, achieve economic growth and ultimately achieve the global developmental targets by 2030. Multi-stakeholder partnerships amongst the key stakeholders like the government, private sector, development partners, civil society, research organisations and academia are critical in order to pool the financial and technical resources needed to harness the use of ICTs in agribusiness in the rural areas (Tinarwo and Uwizeyimana 2019).

Based on the findings and discussions above, the authors propose a conceptual framework for harnessing ICTs in agribusiness for youth employment. Figure 18.1 shows the pathways that are necessary to harness ICTs in agribusiness to create youth employment and ultimately achieve SDGs. The use of ICTs in agribusiness by young people is vital as they still have the energy to work and adapt to technology compared to the ageing population. Few older farmers will be willing to use new technologies required to boost agricultural production in order to feed the ever-increasing population (IFAD 2014).

6 Conclusion

Harnessing ICTs in agribusiness has the potential to address youth unemployment and ultimately achieve several SDGs. However, using ICTs in agribusiness is not a universal remedy to youth unemployment, particularly in developing countries. It is imperative to address the following issues: putting in place strong ICT infrastructure including in remote areas, roll out capacity-building programmes for youths, give financial support for business start-ups to youths and formulate and create robust legal and institutional frameworks that support the use of ICTs in agribusiness, especially at decentralised levels of government. Given the energy deficit being experienced in most developed countries, clean adequate energy supply is an anchor on which to build smart agriculture for the continent and other developing countries which requires urgent attention and investment. Tackling ICT and energy challenges can unlock several industries which can be leveraged to trigger economic development and by default address high unemployment levels, poverty and inequality.

Furthermore, ICT gadgets and bandwidth data bundles need to be affordable while the use of Twitter, WhatsApp and the internet in agribusiness should be promoted amongst the organisations and youths in rural areas. To address the above challenges, solid multi-stakeholder partnerships could be promoted to pool the required technical and financial resources from the government, development partners, civil society, private sector, research organisations and academia. Further studies can be conducted to understand the multiple strategies that policymakers and development partners can use to tap the potential of the youths' involvement in agribusiness.

References

Abraham M., Pingali P. (2020) Transforming Smallholder Agriculture to Achieve the SDGs. In: Gomez y Paloma S., Riesgo L., Louhichi K. (eds) The Role of Smallholder Farms in Food and Nutrition Security. Springer, Cham. https://doi.org/10.1007/978-3-030-42148-9_9

ACBF (The African Capacity Building Foundation). 2017.African Capacity Indicators: Building Capacity in Science, Technology and Innovation for Africa's Transformation. Harare, Zimbabwe.

AfDB, (2016). Jobs for Youth in Africa Strategy (2016–2025). https://www.afdb.org/fileadmin/up. Accessed 12 September 2019.

Allen, A., Howard, J., Jamison, A., Jayne, T., Kondo, M., Snyder, J., & Yeboah, F. (2016). Agri-food Youth Employment and Engagement Study (AGYEES). http://scholar.google.com/citations. 13 August 2019.

Betcherman, G., & Khan, T. (2015). Youth Employment in Sub-Saharan Africa Taking Stock of the Evidence and Knowledge Gaps. https://www.africaportal.org/publications/youth-employment-in-sub-saharan-africa-taking-stock-of-the-evidence-and-knowledge gaps/&ved. Accessed 27 August 2019.

Chifamba, J. (2019, June 4). MNOs Agree to Share Infrastructure. *All Africa*, p.1.

Chitongo, L., Chikunya, P., & Marango, T. (2020). Do Economic Blueprints Work? Evaluating the prospects and Challenges of Zimbabwe's Transitional Stabilisation Programme. *African Journal of Governance and Development*, 9(1), 7-20.

Creswell, J. W., & Creswell, J. D. (2017). *Research design: Qualitative, quantitative, and methods approaches*. Sage publications.

Dube, K., & Nhamo, G. (2020). Vulnerability of nature-based tourism to climate variability and change: Case of Kariba resort town, Zimbabwe. *Journal of Outdoor Recreation and Tourism*, *29*, 100281. https://doi.org/10.1016/j.jort.2020.100281

Econet. (2019, July 22). Econet bemoans power cuts. *The Newsday*, p. 3.

Elias, M., Mudege, N., Lopez, D. E., Najjar, D., Kandiwa, V., Luis, J., & Njuguna-Mungai, E. (2018). Gendered aspirations and occupations among rural youth, in agriculture and beyond: A cross-regional perspective. https://repository.cimmyt.org/bitstream/handle/10883/19632/5975. Accessed 27 October 2019.

FAO and ECA. (2018). Regional Overview of Food Security and Nutrition. Addressing the threat from Climate Variability and Extremes for Food Security and Nutrition. http://www.fao.org/3/CA2710EN/ca2710en.pdf&ved. Accessed 3 September 2019.

FAO (2017). Agribusiness, value chains and the rapid transformation of agrifood systems. http://www.fao.org/3/a-i6811e.pdf. Accessed 24 August 2019.

Filmer, D., & Fox, L. (2014). Youth Employment in Sub-Saharan Africa. World Bank Publications. https://ideas.repec.org/b/wbk/wbpubs/16608.html. Accessed 24 August 2019.

Frost, O., & Sawa, P. (2017). YOUNG PEOPLE: Energising African agriculture. *Spore*, (184), 4-7.

Gukurume, S. (2018). Strengthening small and medium enterprises to address youth unemployment crisis in Zimbabwe. https://opendocs.ids.ac.uk/opendocs/handle/20.500.12413/1. Accessed 16 August 2020.

IFAD, FAO & CTA (International Fund for Agricultural Development, Food and Agriculture Organization of the United Nations& Technical Centre for Agricultural and Rural Cooperation. (2014). Youth and agriculture: Key challenges and concrete solutions. http://www.fao.org/3/a-i3947e.pdf&ved=2ahUKEwi. Accessed 12 August 2019.

ILO (International Labour Organisation). (2018). Labour statistics in Zimbabwe. https://zimbabwe.ilo.org/ilostat/fa. Accessed 1 September 2019.

Irungu, K. R. G., Mbugua, D., & Muia, J. (2015). Information and Communication Technologies (ICTs) attract youth into profitable agriculture in Kenya. *East African Agricultural and Forestry Journal*, *81*(1), 24-33.

Juma, C. (2015). *The new harvest: agricultural innovation in Africa*. Oxford University Press.

Koira, A. K. (2014). Agribusiness in sub-Saharan Africa: Pathways for developing innovative programs for youth and the rural poor. https://www.cabdirect.org/cabdirect/abstract/20198400483. Accessed 14 August 2019.

Nhemachena, C., Matchaya, G., Nhemachena, C. R., Karuaihe, S., Muchara, B., & Nhlengethwa, S. (2018). Measuring baseline agriculture-related sustainable development goals index for southern Africa. *Sustainability*, *10*(3), 849.

Njenga, P., Frida, M. & Opio, R. (2012). Youth and women empowerment through agriculture in Kenya. https://www.semanticscholar.org/paper/Youth-and-Women-Empowerment-through-Agriculture-in-Njenga. Accessed 14 August 2019.

Proctor, F., & Lucchesi, V. (2012). Small-scale farming and youth in an era of rapid rural change. 2012https://africa.ypard.net/sites/ypard.net/files/14617IIED.pdf. Accessed 15 August 2019.

Rono, F., Tonui, C., Nderitu, S., & Kirui, S. (2012). (Rep.) Could ICT be the Solution to Woo Youth to Agriculture?. https://www.africaportal.org/publications/could-ict-be-the-solution-to-woo-youth-to-agriculture/. Accessed 12 September 2019.

Tinarwo, J. & Uwizeyimana, D.E. (2019). Breaking the Traditional Trap: Improving Food and Nutrition Security through Science Technology and Innovation in Zimbabwe. *Bangladesh Sociological Society*, 16(2), p. 178.

Tjoa, A. M., & Tjoa, S. (2016). The role of ICT to achieve the UN Sustainable Development Goals (SDG). https://doi.org/10.1007/978-3-319-44447 Accessed 12 August 2019.

UN 2015. Transforming Our World: The 2030 Agenda for Sustainable Development. https://sustainabledevelopment.un.org/content/documents/21252030%20Agenda%20for%20Sustainable%20Development%20web.pdf. Accessed 8 September 2019

UNFPA (United Nations Population Fund). (2019). *Youth Population in Zimbabwe*. https://zimbabwe.unfpa.org. Accessed 06 August 2019.

White, B. (2012). Agriculture and the generation problem: rural youth, employment and the future of farming. *IDS Bulletin*, *43*(6), 9-19.

World Bank, 2018. Youth Key to Strengthening Africa's Future. https://www.worldbank.org/en/n. Accessed 11 August 2019.

Yami, M., Feleke, S., Abdoulaye, T., Alene, A., Bamba, Z., & Manyong, V. (2019). African Rural Youth Engagement in Agribusiness: Achievements, Limitations, and Lessons. *Sustainability*, *11*(1), 185.

Zylbersztajn, D. (2017). Agribusiness systems analysis: origin, evolution and research perspectives. *Revista de Administração (São Paulo)*, *52*(1), 114-117.

Part V

Conclusion and Policy Recommendations

19 Conclusion and Recommendations: Challenges and Opportunities in SDGs Localisation and Implementation

Kaitano Dube, Godwell Nhamo, and Muchaiteyi Togo

Abstract

The quest for societal development forms the foundation of human aspiration. The path to a prosperous world is fraught with challenges that need urgent attention to improve human livelihoods. The demand for a just, equitable society has never been stronger than now due to the impact of COVID-19 and the increased cost of disasters. This chapter summarises and concludes on critical aspects that are faced by communities to achieve societal sustainability as deliberated in the book. What emerged is that society is faced with numerous challenges at all spatial levels from the international, through national, to local levels. A significant portion of the book looks at various leadership styles that are essential in addressing the global challenges threatening society leading to 2030. Critical to the debate is leadership that is accountable, inclusive, transparent, transformative and driven by the need to see social justice and cohesion in response to the Sustainable Development Goals (SDGs). Several chapters are also dedicated to addressing service delivery and SDGs at the local government level, while others focus on addressing poverty in the context of the SDGs.

Keywords

SDGs · Governance · Accountability · Leadership · Society · Stakeholders · Local government · Poverty

K. Dube (✉)
Department of Hospitality, Tourism and PR, Vaal University of Technology, Vanderbijlpark, South Africa
e-mail: kaitanod@vut.ac.za

G. Nhamo
Exxaro Chair in Business and Climate Change, Institute for Corporate Citizenship, University of South Africa, Pretoria, South Africa
e-mail: nhamog@unisa.ac.za

M. Togo
Department of Environmental Sciences, University of South Africa, Florida, South Africa
e-mail: togom@unisa.ac.za

1 Introduction

Apart from ensuring environmental protection, one of the key aspirations of global leaders was to achieve a prosperous society through the adoption of the Sustainable Development Goals (SDGs). A prosperous society entails communities free of poverty (SDG 1) where there is finan-

G. Nhamo et al. (eds.), *Sustainable Development Goals for Society Vol. 1*, Sustainable Development Goals Series, https://doi.org/10.1007/978-3-030-70948-8_19

cial inclusivity and equality (SDG 10) and gender-responsive development (SDG 5). Given that the bulk of the global population is now living in urban areas (Rogers and Hunt 2019; Zander et al. 2018), global leaders envisaged that under the 2030 Agenda for Sustainable Development there is a need to ensure that cities and other human settlements become more inclusive, safe, resilient and sustainable (SDG 11). Liveable human settlements also need adequate water and sanitation as envisaged under SDG 6. This will not be achievable without robust and capable leadership. It is in this light that the United Nations called for global leadership to promote peaceful and inclusive societies for sustainable development, provide access to justice for all and build effective, accountable and inclusive institutions at all levels under SDG 16 (United Nations 2015).

With only 10 years to go before the lapse of the 2030 deadline, there is increased interest in taking stock of the progress made so far and milestones yet to be achieved. The United Nations (2020) dubbed the next years a decade of action, noting that in as much as progress has been made, in many places, there is evidence that things are not moving at the anticipated speed and pace required to meet all the targets. Calls are now being made to ramp up mobilisation of government, civil society and global citizens to localise and make SDGs their own in an accelerated manner. The decade of action (2020–2030) has three focus areas, as shown in Fig. 19.1.

With the advent of COVID-19, there are fears that the pandemic will derail the progress made thus far and set other countries totally off track with regard to the implementation of the SDG agenda (Nhamo et al., 2020). A combination of the impact of the pandemic and the raging increased cost from the impact of natural disasters and climate change is proving costly to the attainment of the SDGs, particularly in sub-Saharan Africa (SSA) and other developing countries and Small Island Developing States (SIDS). Instead of witnessing reduced poverty, some regions of Africa have been observing an increasing trend in poverty against decreasing global trends in most regions (Fig. 19.2). Over the years the number of people living in extreme poverty, i.e. on less than US$1.90 a day, has increased in the African region with projections showing that Africa will not likely meet its SDG targets on poverty (SDG 1).

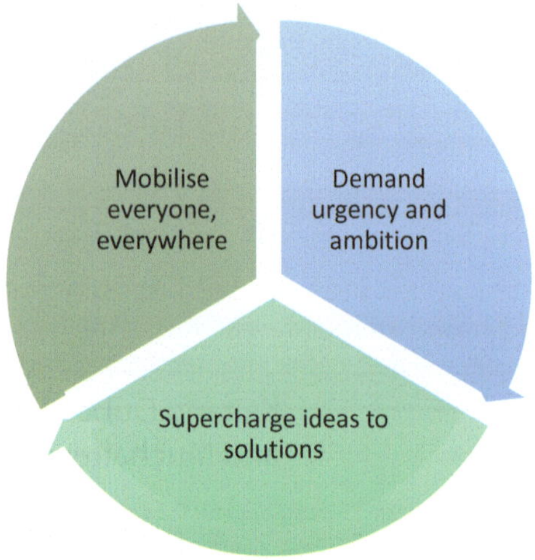

Fig. 19.1 Focus areas for the Decade of Sustainable Development Goals Action. Source: Authors, based on the United Nations (2020)

2 Revisiting the Context

The proportion of people living in poverty in developing countries and SIDS including SSA, South America and the Asia Pacific (Fig. 19.3), which are most vulnerable to weather extremes, presents unique challenges requiring urgent attention if global inequality is to be addressed fully. The Intergovernmental Panel on Climate Change—IPCC (2018)—cited SSA as one of the most vulnerable regions to the impact of extreme weather events ushered in by the continued rise in temperature. On the other hand, in its Special Report on Ocean and Cryosphere, the IPCC (2019) paints a bleak picture on the fate of SIDS as they stand to experience significant losses due to rising sea levels and other climate extremes and natural disasters. Such a scenario presents governance and leadership challenges for communities and demands capable and robust leadership.

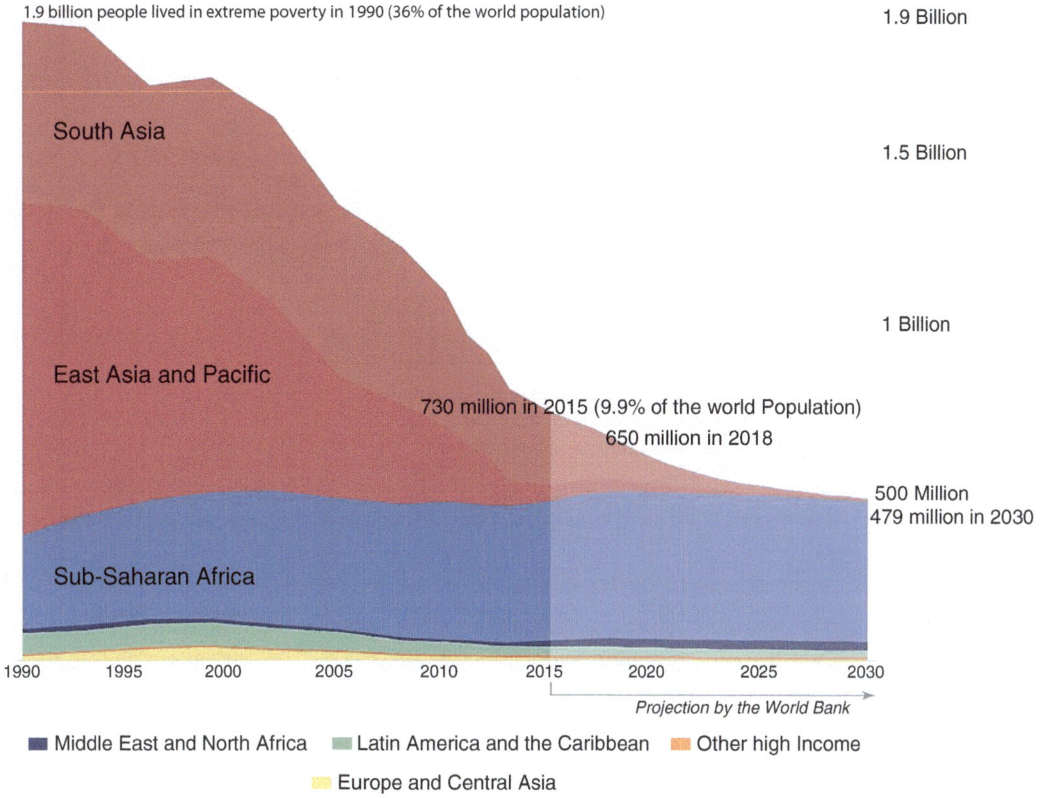

Fig. 19.2 Statistics of people living in extreme poverty projected to 2030. Source: Our World in Data (2019)

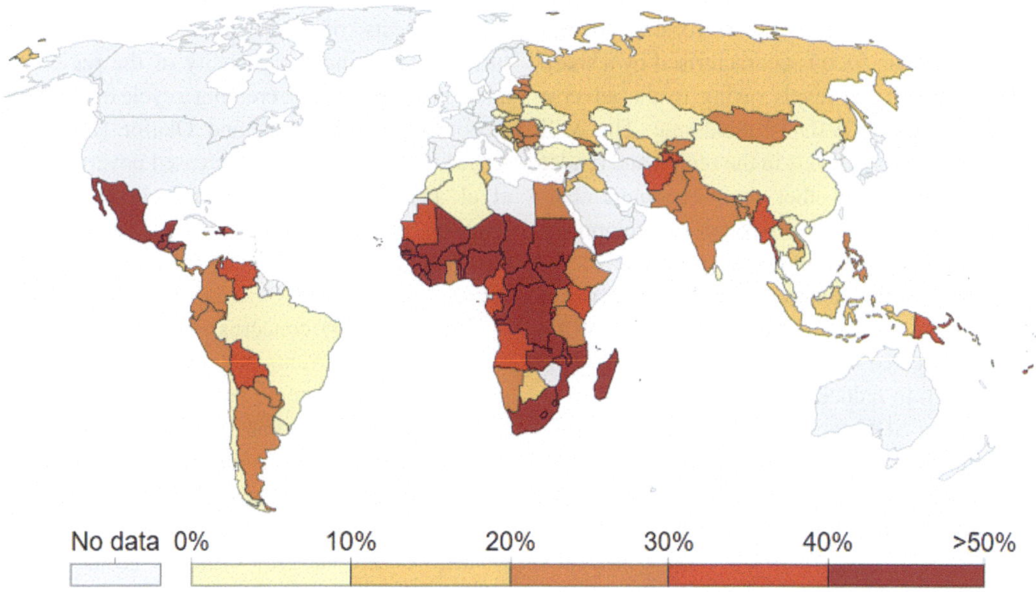

Fig. 19.3 Share of population living in poverty by national poverty lines, 2017. Source: Our World in Data (2019)

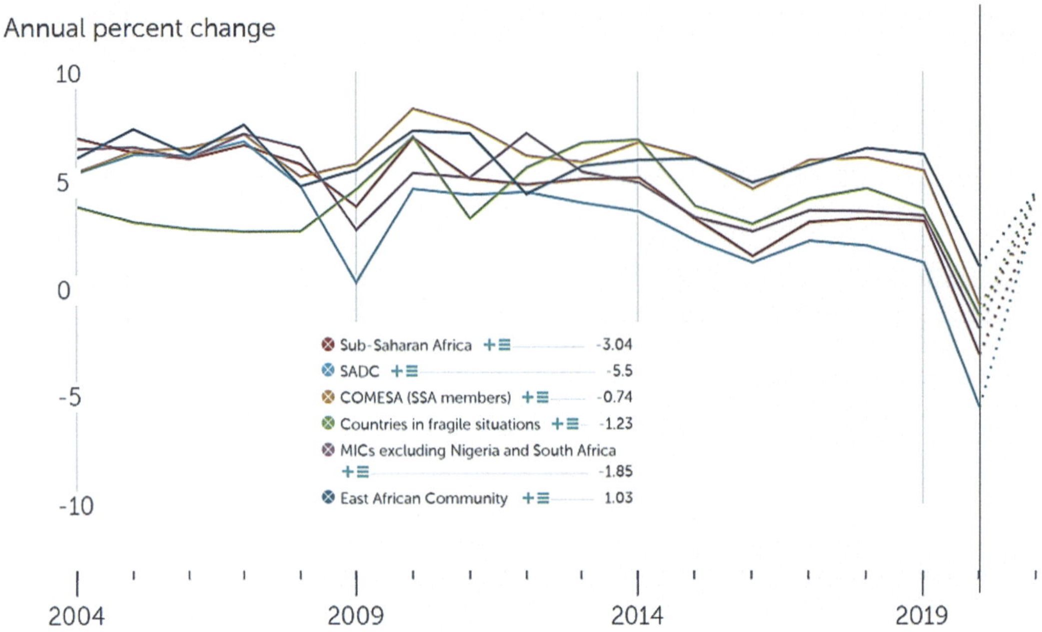

Fig. 19.4 African region real GDP trends 2004–2021 annual percentage change. Source: Authors, adapted from International Monetary Fund (2020)

Evidence from the International Monetary Fund (2020) shows that the gross domestic product (GDP) in SSA has also been in general decline over the years (Fig. 19.4), which could be the attributing factor behind the increasing poverty levels. The year 2020 is characterised by a sharp increase in poverty levels owing to the adverse impacts of the COVID-19 pandemic. This will worsen the poverty levels in the region and render many vulnerable to other socio-economic challenges. The Southern African Development Community (SADC) is going to be one of the worst affected regions, as the GDP is expected to shrink by as much as −5.26%, while the SSA's GDP is expected to shrink by −3.21%. This will likely result in reduced social spending aimed at meeting the SDG targets and goals with worse effects on the poor.

It is clear that over the past 20 years or so there has been a concentration of poverty in Africa, which should be a concern for the continent's leadership, civil society and the world at large. Poverty in Africa will inhibit the continent from effectively responding to natural disasters such as climate change, to build the much-needed resilience against their impacts that have accelerated in recent past. This can have a spillover effect on other world regions. The increasing poverty in the region can also amplify the prospects of conflict/war and instability, further worsening the capacity of the region to address its challenges, creating a cycle of poverty. According to Okunlola and Okafor (2020) in Africa, the conflict led to increased poverty. Ilyas et al. (2017) concurred with these assertions, pointing out that there are parallels between terrorism, poverty and economic performance in Africa.

There are also concerns over the impact of extreme weather events that are undermining the global economy in some years, as can be seen in Fig. 19.5. The disasters are particularly disturbing for developing countries and fragile economies, as resources meant for development often end up being directed at reconstruction efforts. Extreme weather events such as droughts (Dube et al. 2020), floods (Winsemius et al. 2016) and fires (Carmenta et al. 2017) have been observed to have a particularly devastating impact on economies and marginalised communities.

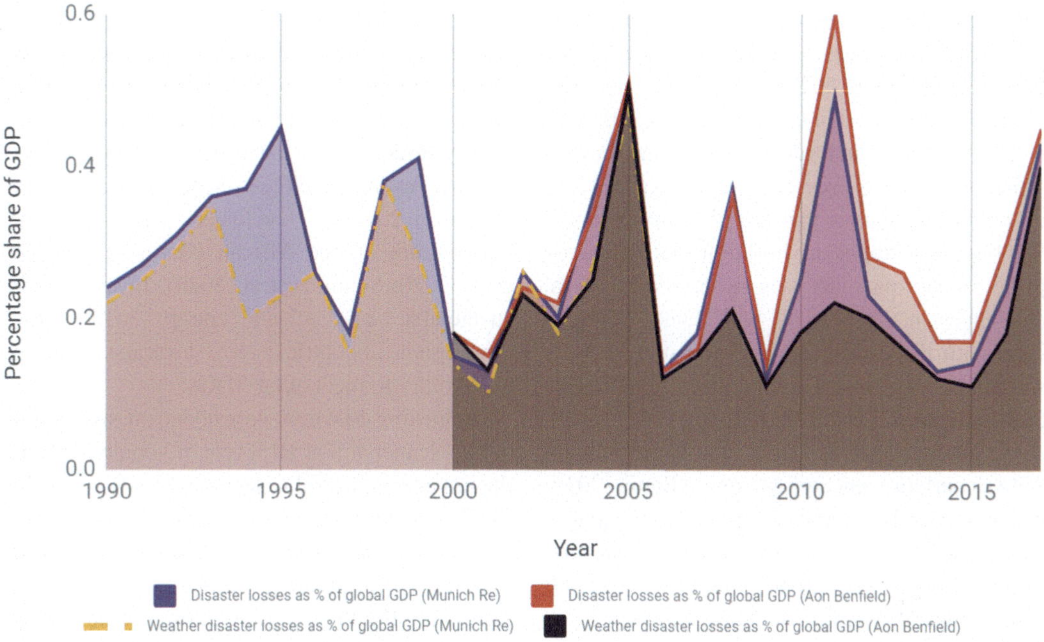

Fig. 19.5 Global economic losses from disasters as a share of GDP 1990–2017. Source: Authors, Data from Our World in Data (2020)

Therefore, this calls for the full implementation of the Sendai Framework for Disaster Risk Reduction 2015–2030 to reduce the shocks posed by natural disasters on global economies and assist in building resilient communities. Resilient communities are better prepared to deal with disasters, which ensures prosperous communities. Given the challenges faced by the world, there is a need for innovative and robust leadership capable of driving and steering the environmental protection and developmental agenda at international, national and local levels.

With a large proportion of people now living in urban areas, there are concerted efforts across the world to ensure that cities and other human settlements are inclusive, safe, resilient and sustainable (SDG 11). Achievement of SDG 11 is crucial, given the contribution of urban communities to global economies. Failure to ensure urban resilience, for example, could deepen inequalities and affect rural communities given the interlinkages between rural and urban communities. Poor governance and leadership have a detrimental effect on the management of the ever-increasing disasters. Vaidya and Chatterji (2020) argued that with over 55% of the global population living in urban areas and the same areas producing 85% of global GDP, achieving carbon emission reductions, reducing poverty and inequality and achieving economic growth, amongst other things, rely squarely on achieving urban sustainability.

Urban areas have been flagged as the key drivers of some of the global challenges, such as greenhouse gas (GHG) emissions that are key drivers of climate change. According to Satterthwaite (2008), there is a general belief that cities contribute between 75% and 80% of GHG emissions. Various activities in the urban space are responsible for some of the GHG emissions, such as energy demands for heating and cooling (Hussain et al. 2019), transport systems (Kou et al. 2020) and other service industries such as urban tourism (Dube and Nhamo, 2020). Cities are also blamed for a host of other types of air pollution (Shehzad et al. 2020; Bauwens et al. 2020). Further, cities are responsible for the bulk of ocean pollution (Lu et al. 2020), land pollution (Song et al. 2019) and water pollution (Zhang et al. 2019). The problem of water pollution has heightened challenges of water and sanitation supply in urban areas, particularly in the Global South (Adams et al. 2020).

Apart from a host of environmental challenges faced by urban and rural settlements, human settlements have several other social and economic challenges that are of concern to the SDG agenda. Rapid urbanisation that has taken place in the past decades ushered in several challenges in urban areas, with some failing to cope with the ever-increasing demands such as shortage of water, sanitation challenges (Satterthwaite 2016; Moretto et al. 2018), development of slums (Turok and Borel-Saladin 2018), urban unemployment (Batu 2016; Nakamura 2016), increased poverty, inequality and hunger and food insecurity (Crush and Frayne 2011; Tacoli 2019), amongst others. Of course, these challenges are more pronounced in the developing world and emerging economies. The fear is that the COVID-19 pandemic will likely worsen this situation.

One of the critical challenges that are concerning to urban communities, particularly in Africa, is the challenge of energy insecurity. With the African population expected to surpass that of China by 2025 (Fig. 19.6), the increase in population will need strategic leadership to achieve the SDGs, as this will present a new set of unique challenges. There is an anticipation that energy supply challenges will worsen in Africa if urgent steps are not taken to address the energy supply and demand gap. In recent years, energy insecurity has been triggered by extreme weather events such as droughts (Dube and Nhamo 2020b), cyclones and extreme temperature on the African continent (Kayaga et al. 2020; Schmitz et al. 2020). Investment in sustainable and clean energy will allow developing countries to leverage energy availability to meet other SDGs.

Achieving SDGs is dependent on strong and ethical leadership at all levels of society, both in the private and public sectors. Nowhere in the SDGs are these expressed as clearly as in SDG 16, which articulates the critical leadership imperatives required to achieve what SDGs aspire to. Figure 19.7 highlights some of the critical issues that need to be factored in to ensure successful implementation and governance of the various SDGs. One of the aspects that stand out from the analysis of SDG 16 is the need for transparent and accountable leadership. Accountability and transparency are critical in the fight against corruption and demand that those who are in leadership take responsibility for actions taken and/or lack thereof. Lack of transparency in other settings has often led to

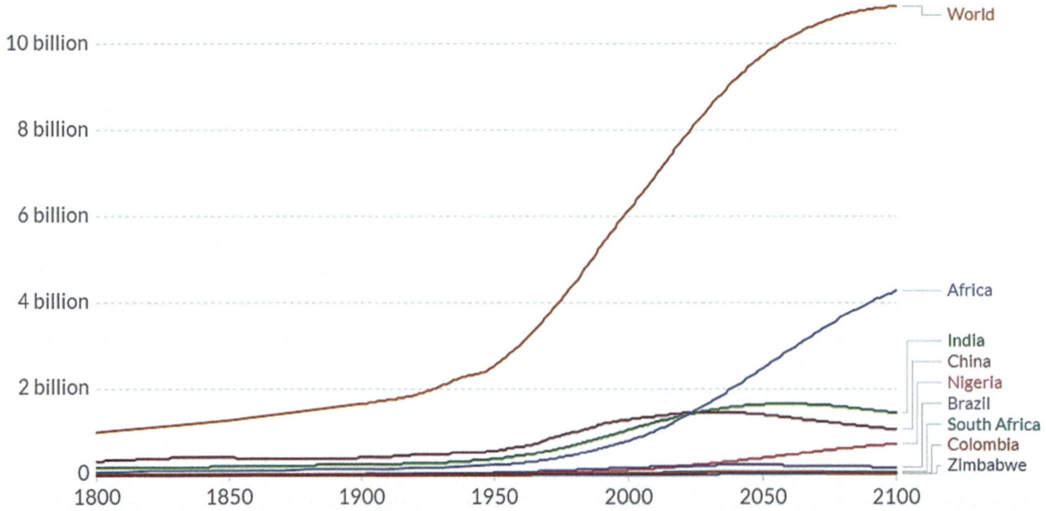

Fig. 19.6 Historical estimates of population, combined with the projected population in 2100. Source: Gapminder (2019) based on UN's medium variant scenario

Fig. 19.7 Sustainable Development Governance. Source: Authors

high levels of corruption and hindered progress towards the achievement of the SDGs.

In general, countries that are bedevilled by lack of transparency are ranked high on corruption and poverty, with citizens often complaining about the lack of or poor service delivery (Walton 2020). It can therefore be argued that accountability and transparency are critical drivers for ensuring good service delivery in key areas of water, sanitation, waste management, public health, education and other such crucial matters that are the focus areas of Agenda 2030.

Any institution that seeks to achieve the ideals sought by Agenda 2030 must ensure that it is inclusive in policy planning and implementation in a manner that ensures inclusivity and participation of citizenry and other stakeholders. Such a model allows for better planning and scrutiny of projects to avoid failure and resource wastage. The approach has to ensure that the interests of the marginalised and previously disadvantaged are taken into consideration. The inclusion of women (SDG 5), people with disabilities and indigenous people have been a critical debate, with well-documented benefits of such approaches. Dalu et al. (2020) bemoan practices that leave out women in developmental issues in South Africa. On the other hand, Luckhurst (2019) documents how inclusivity allowed the shaping of the G20 agenda, which allowed the world to better respond to the 2008 financial meltdown through policy contestation, advocacy and effective coalition-building amongst other responses. Similar approaches can be utilised by both private and public entities to address global challenges. Access to accurate and reliable information becomes a critical factor in allowing inclusivity and citizen participation in various projects.

Justice and respect for the rule of law is another key aspect that needs to be considered, as it feeds into the puzzle of accountability and transparency. Lack of the rule of law can result in a chaotic society. Lack of adherence to the rule of law can perpetuate inequality, drive away investments and inhibit socio-economic development. Chen (2017) argues that there is reversible causality between the rule of law, development and state environmental laws. On the other hand, lack of the rule of law that manifests itself as *coup d'état* adversely affects development due to the consequent chaos.

Lwabukuna (2016) argues that one of the major impediments that are negatively affecting the achievement of SDGs is the lack of the rule of law, which in turn negatively affects the development agenda. A just society is a peaceful society. Peace is a critical ingredient in driving development. Fragile states often battle to attract investment and to ensure a smooth flow of economic activities, which hampers progress and service delivery.

It is with that understanding that Meuleman and Niestroy (2015) called for a meta-governance guided by Common but Differentiated Governance (CBDG). The approach was aimed at ensuring that different governance styles are combined into successful governance frameworks to achieve the aspirations of the SDGs. An approach that is underpinned by the respect of the rule of law establishes governance support arrangements that assist the processes of developmental design, review, monitoring and evaluation of developmental projects, which in turn assist in decision-making for both government and private entities.

From the above deliberations, it is clear that society is incredibly invested in ensuring the success of SDGs that are society oriented to ensure a prosperous future. In this book, several cases have been profiled, and the next section deals with the key findings that emanated from this book.

3 Emerging Key Findings

As articulated in this book, there is a tendency (rightfully so) by progressive governments to ensure that the SDGs are streamlined into national policies. Chapter 2 unveiled how the progressive efforts by the Brazilian Government ensured that national policies are linked to SDG 16 and—most importantly—to another critical SDG, which speaks to aspects of inequality (SDG 10). The chapter highlights how the Brazilian model has ensured that no one is left behind in the SDG agenda (United Nations, 2015). This is crucial as sustainability can only be achieved when everyone is on board. Chapter 3 tackled another critical aspect that is closely linked to the SDG agenda. It dealt with how the Ugandan community is borrowing concepts and innovations from other parts of the world to ensure the localisation of the SDGs. In the chapter, the authors show how a Korean model known as Saemaul Undong (SMU) has activated community members to actively participate in SDG localisation projects in a manner that ensures inclusivity.

The 2030 Agenda for Sustainable Development requires the participation of multiple players, both private and public actors. Crucial to the achievement of several of the SDGs is the participation of civil society and non-governmental organisations (NGOs). Scheyvens et al. (2016) underscored the role of civil society, pointing out that they possess innovative skills, technology and efficiencies that can be used in leveraging the achievement of SDGs. Chapter 4 showed how Tsungirirai Welfare Organisation (TWO) was assisting rural communities in Masvingo, Zimbabwe, to tackle the issues of rural household poverty and rural hunger and ensure livelihood security. The chapter offered a classic model that can be used by countries in a similar context to Zimbabwe to harness efforts by NGOs to ensure household livelihood security. Through the assistance of the civil society, the Masvingo rural community realised skill transfer, which ensured self-sustenance and increased economic activity. This benefitted the community and the individuals involved in these projects.

In the same breath, Chapter 5 brought us to the critical aspects of innovative social programmes being conducted in Colombia. Crucial to the social innovation and transformation agenda is the issue of financial inclusion. Chapter 6 made a case for the need to ensure financial inclusion in Colombia so as to achieve sustainable development. In as much as financial inclusion has not been assigned as a stand-alone goal, it is given pre-eminence across several goals, which warrant the issue of critical attention. According to Fu et al. (2017), about 2 billion adults worldwide are economically excluded and lack access to formal savings, credit and insurance services. This becomes problematic in reducing inequality and poverty. For example, insurance is a critical imperative in responding to the threat posed by disasters and extreme weather events. It is in that light, therefore, that this chapter sheds some brilliant insights.

Indeed the transformative agenda will not be complete without a discussion about women empowerment. Dhar (2018) noted that the SDGs offered an opportunity for the continuation and consolidation of the progress made by the women's UN Beijing Platform for Action (1995) and the Millennium Development Goals to ensure gender equality. Regardless of the gender fight by civil society, in many instances women remain marginalised, with some spaces being exclusively male dominated. Razavi (2016) noted that in as much as SDG 5 offers some comfort in championing women's basic human rights, there are challenges along the way to get to where women want to be. Chapter 7, therefore, highlights some of the progress that has been made in ensuring gender transformation in the mining sector in South Africa. While mining is a crucial sector, the level of women participation remains very low (Monteiro et al., 2019). The chapter highlights the role of mentoring in ensuring gender transformation within the mining sector.

The efficient transport network is critical in driving the SDG agenda. With infrastructure being a challenge and a limiting factor in economic development across Africa, there is a need for a multi-stakeholder approach to ensure the availability of an efficient road network system to facilitate socio-economic development. Chapter 8 showcased the role of complexity leadership theory in solving problems that can emanate between parties in the construction and maintenance of roads. The chapter also outlined how partnerships can assist in achieving the SDGs such as SDG 17, partnership for global goals envisaged. Chapter 9 explored leadership schools of thought that are imperative for a functioning national statistical office capable of delivering on the mandate of providing timely accurate data in South Africa. The study demonstrated how trait leadership was used in the implementation of SDGs. The second phase of the SDG transition project was characterised by the neoclassical school. The chapter demonstrated the need for varying kinds of leadership in the localisation of SDGs by organisations. To date, Statistics South Africa is one of the national statistical offices to have a dedicated SDG tracker, which allows for real-time monitoring of SDG implementation progress.

There is an increasing demand for SDG-related information across the world. Access to reliable information and data is a challenge for many communities. According to Bawack (2018), the socio-economic success of communities is predicated on the value and collection of their libraries. Libraries play an important role as places for discovery, innovation and skill repository; as such, their significance in fighting poverty and other societal challenges can never be overemphasised. In Bradley's (2016) view, libraries are essential as they provide a platform for public literacy, information and technology, which are issues of considerable value to the SDG agenda. Chapter 10 highlighted how the University of South Africa (UNISA) created a one-stop-shop SDG portal on their library website to the benefit of the public. Material from a variety of disciplines has been put together, which assists students and researchers to better articulate and contribute to the SDG debate.

Chapter 11 highlighted how a multi-stakeholder approach could be used to achieve the SDGs. The chapter detailed how various leadership skills were employed in an Eswatini project to ensure the success of SDG-targeted projects. The study found that the key attributes required for successful leadership in achieving the SDGs include collaboration, pragmatic optimism, unassuming influence, single-minded conviction, consistent trust-building and patience. Such attributes are critical in enabling a conducive environment that ensures projects' success, transparency and accountability as accounted in earlier discussions.

The last part of the book deals with municipal issues as they relate to various SDGs. Chapter 12 looked at the transboundary implementation of integrated water resources management (IWRM)

(SDG target 6.5) in South America. The study highlights the criticality of water resource management from a transboundary and transdisciplinary approach and makes a case for IWRM. The case study provides real-life scenarios and can be applied in countries of similar context.

Reporting on SDG progress is one of the critical aspects that need to happen across society. Reports allow stakeholders to understand the progress that has been made for monitoring and evaluation purposes. Universities play a fundamental role in society, and their operations have a significant impact on society and the environment. There has been growing demand for universities to be more accountable with regard to sustainability. Organisations are encouraged under the United Nations Global Compact, a non-binding United Nations pact, to adopt sustainable and socially responsible policies, and to report on their implementation. One such organisation that has gone this route is one of the largest institutions of higher learning in Africa, the University of South Africa (UNISA). Chapter 13 documented the challenges that UNISA was facing in that regard.

As articulated earlier on, urban areas have a plethora of SDGs that they need to respond to as part of the SDG agenda process. Chapter 14 demonstrated how the model of SDG leadership could be utilised in Finnish municipalities to pursue Agenda 2030. The chapter highlighted the importance of coordination between different sectors and networks in line with the dictates that are critical in pursuing the fulfilment of the SDGs. Next, Chapter 15 focused on the role and challenges that can be realised in localising the SDGs in local governments and in traditional leadership setups. The chapter highlighted that conflict resolution capabilities, community mobilisation and good ecological resource management are critical in SDG implementation at a local level.

Chapter 16 articulated the critical challenges faced by municipalities in achieving SDGs and provided critical recommendations and options that can be taken to address some of the service delivery challenges faced by developing countries. Chief amongst the problems highlighted in this chapter is inadequate financing of health provisions, water and sanitation and energy challenges which have been worsened by climate change. The need for financial assistance for developing countries is a critical challenge that global leaders have to deal with for a smooth transition and to build resilient urban communities going forward.

Chapter 17 took the readers on a sustainability journey by documenting a bibliometric analysis of studies conducted thus far to provide an understanding of the key research areas being given attention by academics. The study found that a significant number of studies profiled are informed by political debates that occur at the United Nations Conferences. Key focus issues that have been given considerable attention by academics include those that are central to the SDGs, such as gender equity, climate action and global health. The chapter underscored the need for collaborations and synergies to address issues of sustainable development.

One of the critical challenges faced by the world in addressing the SDGs is the issue of youth unemployment. According to Hällsten et al. (2017), youth unemployment is a critical problem that needs urgent attention in many developing societies, especially in developing countries where general unemployment levels are high. Chapter 18 discussed how innovation and technology in rural Masvingo's Bikita district are being used to boost youth skills in agriculture, ensuring food security and livelihood security. The chapter provides some insights on how rural empowerment can be achieved through ICT adoption and usage by marginalised communities.

4 Policy Recommendations

The book mirrors several case studies that are of global and regional importance. One thing that became clear is that the quest to achieve sustainability for society through the pursuance of SDGs is central to global debates. With a number of countries facing various challenges that threaten society's well-being, there is a need for robust global leadership that is dedicated to pursuing a just, equitable and safe society for all. The new global order that is pressured by diseases, climate change, pollution and increased population demands a paced-up approach to pursue the achievement of the global goals. The attainment of sustainability requires differentiated management and governance thinking and a people-centred approach, something that is often lacking. Challenges such as conflict require urgent attention, as it diverts attention from critical social developmental issues and worsens problems of unemployment and service delivery. The problem of corruption aggravates the challenges of service delivery and takes away resources from critical sectors such as health, education, housing, water and sanitation, to mention a few.

Pursuing sustainability therefore requires a new kind of leadership, which is value oriented with foundations premised on ethical leadership that respect the rule of law. New business culture is needed to drive development where the focus is on efficiency so that society derives value from investments. The ethos of democracy is fundamental in reducing conflict and ensuring accountability and transparency amongst public and private officials. It reduces conflict and leakages that derail focus from critical developmental issues. Accountability, empathy and responsiveness to society's needs and aspirations are critical in achieving societal goals. Societal goals cannot be achieved in isolation; there is a need for creating synergies across various role players to achieve development and growth with equity that is inclusive. Collaborations and synergies are critical in pulling together resources and strengths to address common developmental issues. New governance and management models need to be developed and marketed to all role players to pace up the drive to ensure sustainability across the global village for the common good of the connected society. With a few years left to achieve the global goals, continuous monitoring and evaluation are critical. Concrete measures have to be put in place to ensure that everyone has access to reliable information to assist in decision-making processes going forward. With 2030 only 10 years away as this book is published, world governments must scale up implementation so that societies can benefit.

References

Adams, E. A., Stoler, J., & Adams, Y. (2020). Water insecurity and urban poverty in the Global South: Implications for health and human biology. *American Journal of Human Biology, 32*(1), e23368.

Batu, M. M. (2016). Determinants of youth unemployment in urban areas of Ethiopia. *International Journal of Scientific and Research Publications, 6*(5), 343-350.

Bauwens, M., Compernolle, S., Stavrakou, T., Müller, J. F., Van Gent, J., Eskes, H., & Yu, H. (2020). Impact of coronavirus outbreak on NO2 pollution assessed using TROPOMI and OMI observations. *Geophysical Research Letters, 47*(11), e2020GL087978.

Bawack, R. (2018). Academic Libraries in Cameroon: Achieving Agenda 2030 Goals. *International Information & Library Review, 50*(1), 63-66.

Bradley, F. (2016). A world with universal literacy: The role of libraries and access to information in the UN 2030 Agenda. *IFLA journal, 42*(2), 118-125.

Carmenta, R., Zabala, A., Daeli, W., & Phelps, J. (2017). Perceptions across scales of governance and the Indonesian peatland fires. *Global Environmental Change, 46*, 50-59.

Chen, M. J. (2017). Environmental governance: disentangling the relationship between economic growth and rule of law on environmental policy stringency. *Letters in Spatial and Resource Sciences, 10*(2), 253-275.

Crush, J. S., & Frayne, G. B. (2011). Urban food insecurity and the new international food security agenda. *Development Southern Africa, 28*(4), 527-544.

Dalu, M., Manyani, A., & Masunungure, C. (2020). Gender Inclusivity and Development in South African Public Urban Spaces. In R. Massey, & G. A (Eds.), *Urban Geography in South Africa* (pp. 239-250). GeoJournal Library. Springer, Cham. https://doi.org/10.1007/978-3-030-25

Dhar, S. (2018). Gender and sustainable development goals (SDGs). *Indian Journal of Gender Studies, 25*(1), 47-78.

Dube, K., & Nhamo, G. (2020). Greenhouse Gas Emissions and Sustainability in Victoria Falls: Focus on Hotels, Tour Operators and Related Attractions. *African Geographical Review*, 1-16. https://doi.org/10.1080/19376812.2020.1777437

Dube, K., & Nhamo, G. (2020b). Vulnerability of nature-based tourism to climate variability and change: Case of Kariba resort town, Zimbabwe. *Journal of Outdoor Recreation and Tourism, 29*, 100281.

Dube, K., Nhamo, G., & Chikodzi, D. (2020). Climate change-induced droughts and tourism: Impacts and responses of Western Cape province, South Africa. *Journal of Outdoor Recreation and Tourism*, 100319. https://doi.org/10.1016/j.jort.2020.100319

Fu, J., Queralt, J., & Romano, M. (2017). Financial inclusion and the 2030 agenda for sustainable development: a missed opportunity. *Enterprise Development and Microfinance, 28*(3), 200-211.

Gapminder. (2019). *Data Population*. Retrieved October 21, 2020, from https://www.gapminder.org/data/

Hällsten, M., Edling, C., & Rydgren, J. (2017). Social capital, friendship networks, and youth unemployment. *Social Science Research, 61*, 234-250.

Hussain, M., Butt, A. R., Uzma, F., Ahmed, R., Islam, T., & Yousaf, B. (2019). A comprehensive review of sectorial contribution towards greenhouse gas emissions and progress in carbon capture and storage in Pakistan. *Greenhouse Gases: Science and Technolo*. https://doi.org/10.1002/ghg.1890

Ilyas, S., Mehmood, B., & Aslam, R. (2017). Terrorism and poverty: double trouble for macroeconomic performance in African countries. *African Journal of Economic Review, 5*(1), 1-13.

Intergovernmental Panel on Climate Change (IPCC). (2018). *The Special Report on Global Warming of 1.5 °C (SR15)*. Retrieved August 22, 2020, from https://www.ipcc.ch/sr15/

International Monetary Fund. (2020). *Real GDP Growth* . Retrieved October 10, 2020, from https://www.imf.org/external/datamapper/NGDP_R_PCH@AFRREO/SSA/SADC/COMESA/FRC/MIC854/EAC-5

IPCC. (2019). *Special Report on the Ocean and Cryosphere in a Changing Climate*. IPCC. Retrieved August 8, 2020, from https://www.ipcc.ch/srocc/

Kayaga, S. M., Amankwaa, E. F., Gough, K. V., Wilby, R. L., Abarike, M. A., Codjoe, S. N., & Abdullah, K. (2020). Cities and extreme weather events: impacts of flooding and extreme heat on water and electricity services in Ghana. *Environment and Urban*, 0956247820952030. https://doi.org/10.1177/0956247820952030

Kou, Z., Wang, X., Chiu, S. F., & Cai, H. (2020). Quantifying greenhouse gas emissions reduction from bike share systems: a model considering real-world trips and transportation mode choice patterns. *Resources, Conservation and Recycling, 153*, 104534.

Lu, X. M., Peng, X., Qin, S., Xue, F., & Wu, J. N. (2020). Microbial community successional patterns in offshore sediments impacted by chemical pollution from Taizhou and Xiamen Cities in China. *Marine Pollution Bulletin, 150*, 110600.

Luckhurst, J. (2019). Governance networks shaping the G20 through inclusivity practices. *South African Journal of International Affairs , 26*(4), 521-547.

Lwabukuna, O. K. (2016). International Rule of Law and Development: Underpinnings of the MDGs and the Post-2015 SDGs Agenda. *Journal of Peacebuilding & Development, 11*(2), 89-94.

Meuleman, L., & Niestroy, I. (2015). Common but differentiated governance: A metagovernance approach to make the SDGs work. *Sustainability, 7*(9), 12295-12321.

Monteiro, N. B., da Silva, E. A., & Neto, J. M. (2019). Sustainable development goals in mining. *Journal of Cleaner Production, 228*, 509-520.

Moretto, L., Faldi, G., Ranzato, M., Rosati, F. N., Ilito Boozi, J. P., & Teller, J. (2018). Challenges of water and sanitation service co-production in the global South. *Environment and Urbanisation, 30*(2), 425-443.

Nakamura, A. (2016). Migration between Different Types of Urban Areas and Urban Unemployment. 大阪市立大学経済学会經濟學雜誌= *Journal of Economics, 117*(2), 93-104.

Nhamo, G., Dube, K., & Chikodzi, D. (2020). *Counting the Cost of COVID-19 on the Global Tourism Industry* (1 ed.). Nature Switzerland: Springer International Publishing. https://doi.org/10.1007/978-3-030-56231-1

Okunlola, O. C., & Okafor, I. G. (2020). Conflict–Poverty Relationship in Africa: A Disaggregated Approach. *Journal of Interdisciplinary Economics*,. https://doi.org/10.1177/0260107920935726

Our World in Data. (2019). *Global Extreme Poverty*. Retrieved October 10, 2020, from https://ourworldindata.org/extreme-poverty#citation

Our World in Data. (2020). Global economic losses from disasters as a share of GDP. Retrieved July 11, 2020, from https://ourworldindata.org/grapher/economic-losses-from-disasters-share-gdp?country=~OWID_WRL

Razavi, S. (2016). The 2030 Agenda: challenges of implementation to attain gender equality and women's rights. *Gender & Development, 24*(1), 25-41.

Rogers, C. D., & Hunt, D. V. (2019). Realising visions for future cities: an aspirational futures methodology. *Proceedings of the Institution of Civil Engineers-Urban Design and Planning, 172*(4), 125-140.

Satterthwaite, D. (2008). Cities' contribution to global warming: notes on the allocation of greenhouse gas emissions. *Environment and urbanisation, 20*(2), 539-549.

Satterthwaite, D. (2016). Missing the Millennium Development Goal targets for water and sanitation in urban areas. *Environment and Urbanisation, 28*(1), 99-118.

Scheyvens, R., Banks, G., & Hughes, E. (2016). The private sector and the SDGs: The need to move beyond 'business as usual. *Sustainable Development, 24*(6), 371-382.

Schmitz, W. I., Schmitz, M., Canha, L. N., & Garcia, V. J. (2020). Proactive home energy storage management system to severe weather scenarios. *Applied Energy, 279*, 115797.

Shehzad, K., Sarfraz, M., & Shah, S. G. (2020). The impact of COVID-19 as a necessary evil on air pollution in India during the lockdown. *Environmental Pollution, 266*, 115080.

Song, Y., Kirkwood, N., Maksimović, Č., Zheng, X., O'Connor, D., Jin, Y., & Hou, D. (2019). Nature based solutions for contaminated land remediation and brownfield redevelopment in cities: a review. *Science of the Total Environment, 663*, 568-579.

Tacoli, C. (2019). The urbanisation of food insecurity and malnutrition. *Environment and Urbanisation*. https://doi.org/10.1177/0956247819867255

Turok, I., & Borel-Saladin, J. (2018). The theory and reality of urban slums: Pathways-out-of-poverty or cul-de-sacs? *Urban Studies, 55*(4), 767-789.

United Nations . (2020). *Decade of Action*. Retrieved October 9, 2020, from https://www.un.org/sustainabledevelopment/decade-of-action/

United Nations. (2015). *Agenda 2030 on Sustainable Development*. Retrieved July 11, 2020, from https://sustainabledevelopment.un.org/content/documents/21252030%20Agenda%20for%20Sustainable%20Development%20web.pdf

Vaidya, H., & Chatterji, T. (2020). SDG 11 Sustainable Cities and Communities. In I. Franco, T. D. Chatterji, & J. Tracey (Eds.), *Actioning the Global Goals for Local Impact. Science for Sustainable Societies* (pp. 173-185). Singapore: Springer.

Walton, G. W. (2020). Good Governance, Corruption, and Papua New Guinea's Public Service. *The Palgrave Handbook of the Public Servant*, 1-19. https://doi.org/10.1007/978-3-030-03008-7_65-1

Winsemius, H. C., Aerts, J. C., Van Beek, L. P., Bierkens, M. F., Bouwman, A., Jongman, B., & Ward, P. J. (2016). Global drivers of future river flood risk. *Nature Climate Change, 6*(4), 381-385.

Zander, K. K., Cadag, J. R., Escarcha, J., & Garnett, S. T. (2018). Perceived heat stress increases with population density in urban Philippines. *Environmental Research Letters, 13*(8), 084009.

Zhang, R., Lu, C. C., Lee, J. H., Feng, Y., & Chiu, Y. H. (2019). Dynamic environmental efficiency assessment of industrial water pollution. *Sustainability, 11*(11), 3053.

Index

G. Nhamo et al. (eds.), *Sustainable Development Goals for Society Vol. 1*, Sustainable Development Goals Series, https://doi.org/10.1007/978-3-030-70948-8

C

D

The manufacturer's authorised representative in the EU is Springer Nature Customer Service Centre GmbH, Europaplatz 3, 69115 Heidelberg, Germany. If you have any concerns regarding our products, please contact ProductSafety@springernature.com

Printed and bound by CPI Group (UK) Ltd, Croydon, CR0 4YY

30/09/2026

02225692-0019